LIGHT LOCALISATION AND LASING

The properties of quasi-random and random photonic systems have been extensively studied over the last two decades, but recent technological advances have opened new horizons in the field, providing better samples and devices. New optical characterization techniques have enhanced understanding of the novel and fundamental properties of these systems.

This book examines the full hierarchy of these systems, from 1D to 2D and 3D, from photonic crystals and random microresonator chains to quasicrystals. It treats photon transport as well as photon generation and random lasing, and deals with semiconductors, organics, and glass materials.

Presenting basic and state-of-the-art research on this fascinating field, this collection of self-contained chapters is an ideal introductory text for graduate students entering this field, as well as a useful reference for researchers in optics, photonics, and optical engineering.

MHER GHULINYAN is a Scientist at the Center for Materials and Microsystems, Fondazione Bruno Kessler, Italy. His main research interests are in the field of complex dielectric systems and resonator optics.

LORENZO PAVESI is Professor of Experimental Physics, Chairman of the Department of Physics, and Head of the Nanoscience Laboratory at the University of Trento, Italy. His research activity is concerned with the optical properties of semiconductors and with silicon photonics.

LIGHT LOCALISATION AND LASING

Random and Quasi-Random Photonic Structures

Edited by

M. GHULINYAN

Fondazione Bruno Kessler, Italy

and

L. PAVESI

University of Trento, Italy

CAMBRIDGE
UNIVERSITY PRESS

University Printing House, Cambridge CB2 8BS, United Kingdom

Cambridge University Press is part of the University of Cambridge.

It furthers the University's mission by disseminating knowledge in the pursuit of
education, learning, and research at the highest international levels of excellence.

www.cambridge.org
Information on this title: www.cambridge.org/9781107038776

First published 2015

Printed in the United Kingdom by T. J. International Ltd, Padstow

A catalogue record for this publication is available from the British Library

Library of Congress Cataloging-in-Publication Data
Light localisation and lasing in random and quasi-random photonic
structures / edited by Mher Ghulinyan and Lorenzo Pavesi.
pages   cm
Includes bibliographical references and index.
ISBN 978-1-107-03877-6
1. Photonics – Materials.   2. Optoelectronic devices.   I. Ghulinyan, Mher,
editor of compilation.   II. Pavesi, Lorenzo, editor of compilation.
TA1522.L53 2014
621.36′5–dc23
2014018695

ISBN 978-1-107-03877-6 Hardback

Contents

4 Ordered and disordered light transport in coupled microring resonators

SHAYAN MOOKHERJEA

5 One-dimensional photonic quasicrystals

MHER GHULINYAN

6 2D pseudo-random and deterministic aperiodic lasers

HUI CAO, HEESO NOH, AND LUCA DAL NEGRO

Contributors

Hui Cao *Department of Applied Physics, Yale University, 15 Prospect Street, New Haven, CT 06520, USA*

Luca Dal Negro *Division of Materials Science and Engineering, Department of Electrical and Computer Engineering & Photonics Center, Boston University, Brookline, MA 02446, USA*

Azriel Genack *Department of Physics, Queens College and Graduate Center of the City University of New York, 65–30 Kissena Blvd Flushing, NY 11367*

Mher Ghulinyan *Center for Materials & Microsystems, Fondazione Bruno Kessler, I-38123 Povo, Italy*

Ad Lagendijk *Complex Photonic Systems (COPS), MESA+ Institute for Nanotechnology, University of Twente, 7500 AE Enschede, the Netherlands*

Alexandra Ledermann *Optics Expert, Visteon Electronics Germany-GmbH, An der Raumfabrik 33b, D-76227 Karlsruhe, Germany*

Marco Leonetti *ISC-CNR, UOS Sapienza, P. A. Moro 2, 00185 - Rome, Italy and Instituto de Ciencia de Materiales de Madrid (CSIC) Cantoblanco 28049 Madrid, Spain*

Cefe López *Instituto de Ciencia de Materiales de Madrid (CSIC) Sor Juana Inés de la Cruz 3, 28049 Madrid, Spain*

Shayan Mookherjea *University of California, San Diego, Electrical and Computer Engineering, 9500 Gilman Dr. MC 0407, La Jolla, CA 92093, USA*

Allard P. Mosk *Complex Photonic Systems (COPS), MESA+ Institute for Nanotechnology, University of Twente, P.O. Box 217, 7500 AE Enschede, the Netherlands*

Heeso Noh *Department of Nano and Electronic Physics, Kookmin University, Seoul, Korea*

Michael Renner *Department of Physics and Research Center OPTIMAS, University of Kaiserslautern, Erwin-Schrödinger-Str. 56, 67663 Kaiserslautern, Germany*

Zhou Shi *Department of Physics, Queens College and Graduate Center of the City University of New York, 65–30 Kissena Blvd Flushing, NY 11367*

Georg von Freymann *Department of Physics and Research Center OPTIMAS, University of Kaiserslautern, Erwin-Schrödinger-Str. 56, 67663 Kaiserslautern, Germany; Fraunhofer – Institute for Physical Measurement Techniques (IPM), Erwin-Schrödinger – Str. 56, 67663 Kaiserslautern, Germany*

Willem L. Vos *Complex Photonic Systems (COPS), MESA+ Institute for Nanotechnology, University of Twente, P.O. Box 217, 7500 AE Enschede, the Netherlands*

Léon A. Woldering *Transducer Science and Technology (TST), MESA+ Institute for Nanotechnology, University of Twente, 7500 AE Enschede, the Netherlands*

Preface

This book is the result of our interest in understanding, mastering, and engineering randomness in photonic systems. It is a natural consequence of what we did in the past. In the late 1980s, while Lorenzo Pavesi was working on semiconductor superlattices he noticed that for some energies the vertical transport through the superlattice minibands was inhibited due to disorder (L. Pavesi *et al.* 1989. *Phys. Rev. B*, **39**, 7788). Then, working on the recombination dynamics of excitons in porous silicon, he further noticed that the random arrangement of silicon quantum dots has a strong influence on the recombination dynamics of excitons (L. Pavesi *et al.* 1993. *Phys. Rev. B*, **48**, 17625). After Mher Ghulinyan came to Trento in 2002, we developed the techniques to fabricate free-standing porous silicon dielectric multilayers of any stacking sequence (M. Ghulinyan *et al.* 2003. *J. Appl. Phys.*, **93**, 9724). This was the first time that we had the chance to design at will one-dimensional periodic, aperiodic, or random photonic systems. A fascinating new physics opened up for us: that of the analogy of photon propagation in complex dielectric systems with carrier transport in random electronic systems. Our latest results in the field are associated with sequences of ring resonators where randomness causes the formation of resonant coupling between different rings with the possibility of yielding the optical analog of the electromagnetic induced transparency (M. Mancinelli *et al.* 2011. *Opt. Express*, **19**, 13664), or chaotic photon propagation.

Over all these years, we have had the chance to interact with many researchers active in the field of periodic, quasiperiodic, and random photonic systems. From these interactions the idea of this book was born. We have therefore collected together a series of self-contained chapters to cover the whole field with the specific aim of introducing the different aspects, showing the current status of the research, and envisaging future directions. All invited authors have responded to this challenge with great enthusiasm and professionalism.

The book opens with Chapter 1 by W. L. Vos, A. Lagendijk, and A. P. Mosk, which introduces the field and covers the timeline between the early studies on these systems and the very latest achievements in the field. An extended historical overview details the early stages of research, its progress throughout past decades, and finally focuses the reader's attention on recent advances, which are detailed in the subsequent chapters.

This is followed by Chapter 2 by A. Genack and Z. Shi, where the transport of light (classical photons) through a random medium is described. In particular, it is shown that the modes in a random medium are in complex correlation with each other and the overall transmission of the system critically depends on how much the different modes overlap. The role of mode statistics is crucial for both linear and nonlinear optical phenomena and this statistics determines, for example, the threshold of random lasing in an amplifying medium.

Chapter 3 by M. Leonetti and C. López deals with the phenomenon of random lasing in disordered highly scattering materials. Like conventional resonators, random lasers can display spectrally narrow emission lines and a threshold-like onset. However, these devices possess several interesting properties which make them different from conventional lasers: such as the fact that they provide a poly-directional output. Specifically, a random laser can be continuously driven from a configuration exhibiting weakly interacting electromagnetic resonances to a regime of collectively oscillating, strongly interacting modes.

Quasi-one-dimensional sequences of coupled resonators are the subject of Chapter 4 by S. Mookherjea. In this chapter, the focus is on the fundamental aspects of light propagation, including a study of non-idealities (e.g. disorder-induced deviations from ballistic transport) in chains of silicon microring resonators. The slowing down of light propagation is achieved through the phenomenon of light interference in a sequence of coupled resonators. Meanwhile, in disordered structures, the same light interference is also responsible for the localization of electromagnetic waves.

Chapter 5 by M. Ghulinyan enters into the topic of quasi-random optical systems (quasicrystals). Quasicrystals are aperiodic structures that are constructed following simple deterministic generation rules, and if made from dielectric material, can show fascinating properties which govern light transport through them. Quasicrystals exhibit an energy spectrum that consists of a self-similar set of eigenstates and their transmission spectrum contains forbidden frequency regions called "pseudo band gaps," similar to the band gaps of a photonic crystal. This chapter details the peculiarities of ultrashort light pulse propagation through pseudo band-edge states of Fibonacci-type photonic quasicrystals, showing interesting optical phenomena such as mode beating, strong pulse stretching and suppressed group velocities originating from the excitation of "critically localized" band-edge states.

Two-dimensional aperiodic systems, past, present, and future, are discussed in Chapter 6 by H. Cao, H. Noh, and L. Dal Negro. The recent developments in advanced nanolithographic techniques have allowed the realization of bi-dimensional arrays of metal nanoparticles, arranged in an aperiodic order in both dimensions in a plane. These plasmonic nanostructures can alter out-of-plane white light scattering, following the strict rules of underlying quasiperiodic order. New and fascinating features, such as colored photonic–plasmonic scattering, plasmon-enhanced structural coloration of metal films, mode patterns, dipole radiation, and lasing from optimized aperiodic structures are discussed.

Chapter 7 by A. Ledermann, M. Renner, and G. von Freymann focuses on three-dimensional (3D) quasicrystals and deterministic aperiodic structures. Both fabrication and characterization are presented. The quality of quasicrystalline order is investigated through the comparison of obtained diffraction patterns of structures with a local five-fold real-space symmetry axis, revealing a ten-fold symmetry, as required by theory for 3D structures. Importantly, this chapter reports on the realization of a high-quality silicon inverse quasicrystal operating at near-infrared frequencies.

The book closes with Chapter 8 by W. L. Vos and L. A. Woldering on 3D photonic band gap crystals. Three-dimensional photonic crystals with a 3D photonic band gap play a fundamental role in cavity quantum electrodynamics (QED), especially in phenomena where the local density of optical states is essential: spontaneous emission inhibition or enhancement of emitters embedded in a 3D band gap crystal, thresholdless laser action in a miniature photonic crystal cavity, and breaking of the weak-coupling limit of cavity QED. Finally, several exciting applications of 3D photonic band gap crystals are discussed, namely the shielding of decoherence for quantum information science, the manipulation of multiple coupled emitters including resonant energy transfer, lighting, and a possible spin-off to 3D nanofabrication for future high-end computing.

We are grateful to our past and present colleagues, students, and friends at the Nanoscience Laboratory of the Department of Physics of the University of Trento and at the Center for Materials and Microsystems of the Bruno Kessler Foundation in Trento, for maintaining an environment of scientific excellence and friendship over the years. We owe special thanks to the authors of the various chapters for their excellent work. In addition to thanking the authors, we would like to thank L. Barnes and N. Gibbons, the editorial assistants, for their help, assistance and patience.

LP. I dedicate this book to my children – Maria Chiara, Matteo, Michele, and Tommaso – who have taught me how the disorder of things does not necessarily lead to the clutter of minds, but rather that from chaos can arise creativity (any allusion to their bedrooms is intentional).

MG. I dedicate this book to my sons Davit and Michael, and to my wife Lilit, with whom I have taught our sons for years that order is good and disorder is bad (but I have it clear in my mind that certain disorder on certain length-scales is so rich and fascinating).

Mher Ghulinyan and Lorenzo Pavesi

1

Light propagation and emission in complex photonic media

WILLEM L. VOS, AD LAGENDIJK, AND ALLARD P. MOSK

1.1 General overview

In many areas in the physical sciences, the propagation of waves in complex media
plays a central role. Acoustics, applied mathematics, elastics, environmental sciences, mechanics, marine sciences, medical sciences, microwaves, and seismology
are just a few examples, and of course nanophotonics [6, 429, 448, 449]. Here,
complex media are understood to exhibit a strongly inhomogeneous spatial structure, which determines to a large extent their linear and nonlinear properties.
Examples of complex media are random, heterogeneous, porous, and fractal media.
The challenges that researchers in these areas must surmount to describe, understand, and ultimately predict wave propagation are formidable. Right from the start
of this field – in the early 1900s – it was clear that new concepts and major approximations had to be introduced. Famous examples of such concepts are the effective
medium theory [55, 60, 321] and the radiative transport theory [81, 491]. These
concepts are very much alive, even today. The fundamental challenge with these
approximate concepts is that often the length scales of the inhomogeneities in the
complex medium are comparable with the wavelength, whereas the range of valid-
ity of these approximations is restricted to situations where these length scales
are much larger than the wavelength. Consequently, many relevant situations arise
where either the effective medium theory, or radiative transport theory, or both,
fail dramatically. Examples of such situations are given in this introduction and
throughout this book.

The complexity of the medium that supports wave propagation can be classified in a number of ways. Many different types of spatial inhomogeneities in a
host matrix can be envisioned, varying from completely random, via aperiodic and

Light Localisation and Lasing, ed. M. Ghulinyan and L. Pavesi. Published by Cambridge University Press.
© Cambridge University Press 2015.

waveguide structures, to quasi-crystalline and even structures with long-range periodic order. The inhomogeneities could be of a self-organized form, or the fruit of precise engineering. An additional classification is whether or not the complexity is confined to the surface (in two dimensions (2D)) or is present throughout the volume of the medium (in three dimensions (3D)), as in porous media. A useful criterion for the classication of inhomogeneous media is whether the inhomogeneity is of a continuous nature, or stems from discrete scatterers. One can thus classify the topology of the inhomogeneities relevant to classical waves: if the inhomogeneities are connected from one side to the other, the medium has a network topology; if the inhomogeneities are completely surrounded by host material, the complex medium is said to have a Cermet topology. It appears that topology plays an important role in the scattering of various types of waves – such as scalar, elastic, electromagnetic – in complex media [120].

Ever since the breakthrough achievement of renormalization group theory [539] the dimensionality of a complex system has been known to be vital [80, 429]. Hence, the dimensionality of the problem is crucial to the study of waves in complex media. 1D and 2D systems have the intriguing property that waves are always localized, that is, a wave that starts at a certain spatial position always returns. As stated by the Mermin–Wagner–Hohenberg theorem, 2D systems have the lower critical dimension for most field theories [195, 330], allowing for the study and use of a wealth of scaling phenomena that are highly challenging to uncover. In a 3D world there is a striking phase transition between localized and extended states. Famous examples are Anderson localization, or the photonic band gap in 3D.

From scaling theory [2, 429], it is known that the extent of the complex system is crucial. The finite size determines the transport of the waves – specifically, the conductance. For extended states, it appears that the conductance scales with dimensionality minus two, times the logarithm of the system size. For localized states, the conductance decreases exponentially with system size. Therefore, the study of system size dependence of complex media provides an important key in distinguishing extended from localized states, and in characterizing the formation of gaps.

Advances made in the understanding of waves in complex media have led to a number of practical applications, and are generating new ones at an ever increasing pace [530]. Examples are found in remote optical sensing ("looking through a cloud"), inverse optical scattering, noninvasive medical imaging ("find the tumour in tissue"), applied optics (quality control of optical systems by controlling surface roughness), optical devices and (random) lasers, furthermore, in oil and mineral prospecting by seismic methods, in ultrasonic imaging and non-destructive testing ("find hairline cracks in an airplane wing"), material characterization, all the way to microwave propagation and detection in antennas, mobile phones, and radar.

Researchers working with waves in complex media can thus be found in many areas of the engineering sciences, physics, and chemistry. Concepts developed in one of the individual disciplines have found or are bound to find uses in other wave disciplines. It has thus become increasingly clear that multidisciplinary contacts are very fruitful. Important experimental discoveries and theoretical breakthroughs made in one of these individual wave disciplines frequently appear to have important consequences for all the other wave disciplines.

These rapid developments and cross-fertilization have led to an enormous increase in the understanding of wave propagation that has meanwhile advanced to a very high level [6, 38]. As a result, we are reaching the point where we can incorporate complex wave behavior in applied techniques such as lidar, sonar, and radar. The fast rise of the field of medical imaging with diffuse waves is also stimulated by this tremendous progress. In all cases, the advances pertain to new concepts, to new experimental discoveries, and to new techniques.

1.2 Light in complex photonic media

The exciting subject of light in complex photonic media has become a very active area of research and has grown to become a field in its own right. The field has its roots in the physics of electrons and spins in condensed matter [6, 11, 429, 501]. The essential optical properties of complex photonic media are determined by the spatially varying refractive index, analogous to the periodic potential for an electron in condensed matter. Large variations of the refractive index cause a strong interaction between light and the photonic medium. As a result, multiple scattering dominates the linear and nonlinear optical properties. As complex photonic media, we consider in this introductory chapter random media with short-range order, photonic crystals with long-range periodic order, arrays of cavities or waveguides, quasicrystals with orientational order yet without long-range order, or even aperiodic media. In this section we give a brief overview of these classes of complex media in the sequence of their historical appearance, and the main research directions are illustrated with a sample of key references.

1.2.1 Random media

The study of light propagation in random media is a field of research with a rich history starting in the 1980s [210, 491]. In daily life, we encounter light in disordered dielectric media, such as paint, milk, fog, clouds, or biological tissue. Light performs a random walk through the complex medium, as illustrated in Fig. 1.1(C). The average step size in the random walk is equal to the scattering mean free path ℓ_s. In the above mentioned media, the mean free path is (much) larger than the

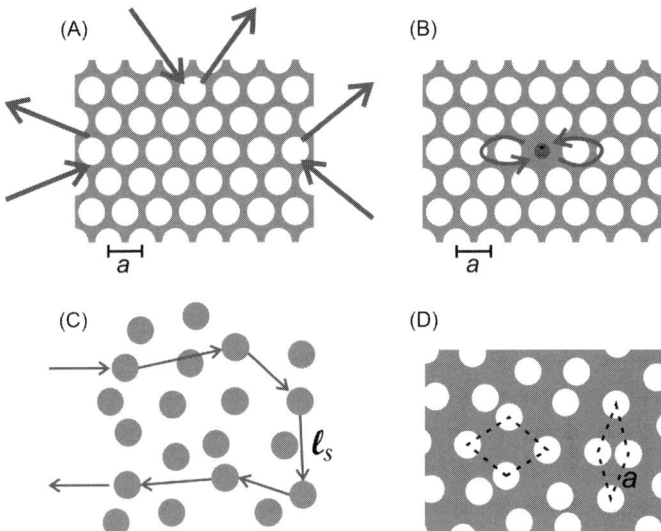

Figure 1.1 Schematic cross-sections of the main nanophotonic complex media discussed in this chapter. (A) A photonic band gap crystal: incident light within a particular frequency band – shown as arrows – is multiply diffracted by all crystal planes, irrespective of the direction of incidence. Hence, it is dark inside the crystal in the band gap as a result of interference. (B) A cavity in a photonic band gap crystal. (C) A random medium with mean free path ℓ_s step size between scattering events. (D) A quasicrystal, Penrose tiling. The two kinds of parallelograms ("fat" and "skinny") are indicated by dashed lines. The lattice parameter a is equal to the edge length of the constituent parallelograms.

wavelength, $\ell_s > \lambda$, and the transport of light is well described by diffusion, as though light scatters like particles [6, 501].

When the average distance between scattering events is reduced to become comparable to the wavelength of light $\ell_s \simeq \lambda$, interference cannot be neglected anymore. This realization came as a shock, since it was originally thought that random multiple scattering scrambles all phases. Experimentally, the signature of interference in complex media is observed as an enhanced backscatter cone in the intensity – even for relatively weakly interacting samples. The cone is the result of constructive interference between light that propagates along any random path in the complex medium, and light that propagates along the time-reversed path [488, 543]. The width of the cone is inversely proportional to the mean free path, and thus serves as a convenient probe thereof. The mean free path decreases when the interaction between light and the complex medium increases.

Ultimately, interference causes transport to grind to a complete halt, a phenomenon known as Anderson localization of light [11, 221, 263]. The diffusion of light tends to zero at the Anderson localization transition, as light inside the medium has a very high probability of returning to its original position. While many efforts have been devoted to the observation of Anderson localization of light in 3D, there have been only a few reports to date, using strongly scattering powders of GaAs [531] and titania [453, 465]. Currently, the consensus seems to be that the observations are compounded by optical absorption, nonlinear effects, or spurious fluorescence [493, 530]. Therefore, the quest for Anderson localization of light at optical frequencies in 3D is still open. Experimental techniques that are likely to settle debates on localization versus spurious effects use the probing of optical correlations, as demonstrated for microwaves [78], see Chapter 2. As we have seen above, Anderson localization is easier to observe in lower dimensions. This has led to a renewed interest in this striking wave phenomenon with the advent of quasicrystals and aperiodic media, as well as in waveguides and coupled cavity arrays; see the following subsections.

Optical waves that are transmitted through or reflected from a complex medium also reveal another interference phenomenon, namely speckle. By eye, speckle is observed as the grainy pattern in a light spot on a wall that is illuminated with a laser pointer. Speckle is a random interference pattern that depends very sensitively on the detailed arrangement of all scatterers in the complex medium. Interestingly, there has been a recent realization that the speckle pattern can be drastically modified by playing on the spatial wavefronts of the incident waves; see Section 1.2. The angular distribution of light in a speckle pattern is impossible to predict, a property that can be used in applying random media as physically unclonable objects for encryption [370]. By averaging over the positions of the scatterers, or by averaging over the optical frequencies, the interference washes out and a speckle pattern reverts to a diffuse distribution. Nevertheless, speckle has well-defined statistical properties (intensity distribution). Moreover speckle contains three types of correlation that depend sensitively on the light propagation inside the complex medium [501]. Firstly, if the frequency of the incident light is changed, the speckle pattern deforms as described by the short-ranged C_1 correlation function. Likewise, if the direction of the incident beam is changed, the speckle pattern deforms, again as described by the C_1 correlation function. Simultaneously, the speckle pattern follows the incoming beam, which is known as the memory effect. Secondly, long-ranged correlations occur even between different speckle spots, as described by the C_2 correlation function. Thirdly, infinite-range correlations occur when one averages over all incident and outgoing directions. These so-called C_3 correlations are also known as universal conductance fluctuations, and have been observed in beautiful experiments by Scheffold and Maret [419].

Separate from propagation and transport effects as described above, random complex media are also pursued for their effects on the local density of optical states (LDOS) which notably controls the emission of an embedded light source.

Inspired by work on photonic crystals – themselves, offspring of random media – it has been realized that random media may also open a 3D photonic band gap. Calculations supported by microwave experiments predict that a 3D photonic band gap will occur in amorphous diamond [206, 301]. This pursuit is also inspired by condensed matter, where it is well known that long-range periodic order is not crucial for sustaining a band gap, as both crystalline and amorphous silicon possess such a gap.

In random media, it has been predicted that the LDOS displays intricate spatial fluctuations that are described by the C_0 correlation function [427]. As opposed to the correlation functions C_1, C_2, C_3 above, that have analogies in transport of electrons, C_0 is peculiar to photonic media, as spontaneous emission is peculiar to photonic media. The first intriguing effects have recently been observed by studying light emitters inside random photonic media [46, 415], on the surface of random media [411], and even inside random plasmonic media [249].

Whereas in a conventional laser, mirrors provide feedback, a random laser uses multiple scattering of light as a feedback mechanism. In a random laser, a spontaneously emitted photon is amplified by stimulated emission. Simultaneously its path length in the gain medium is increased. Consequently, the emission spectrum narrows for increasing pump powers and the output power for the peak of the spectrum shows typical threshold behavior. In contrast with a conventional laser the emitted light from a random laser is omnidirectional. The idea of generating light inside a scattering medium by stimulated emission was proposed in 1968 by Letokhov [292]. After initial work, the field took off in the mid 1990s, fueled by a debate on whether laser action in scattering samples was truly due to diffusion or rather to single scattering [269, 534]. The *Nature* editor involved coined the term "random laser" which has remained in vogue ever since. A random laser can be conceived of as a multiple scattering medium with gain or as a laser system with a complex cavity configuration. Well-known laser physics effects are observed with random lasers, such as relaxation oscillations, intensity fluctuations, and mode coupling, as well as prominent multiple scattering phenomena such as enhanced backscattering and speckle; see the reviews [68, 529].

In pioneering studies Cao and co-workers detected narrow features in the output spectrum of random lasers, called "spikes" [71]. Interestingly, the community has not yet converged on an interpretation of these intriguing narrow spectral features. While impressive theoretical efforts have focused on 2D random laser systems [486], it remains to be seen how the theoretical concepts translate to 3D systems, and how these theories can be connected with experiments. Anderson localization, Fabry–Perot resonances, absorption-induced confinement, and photons traveling exceptionally far through the gain medium, are all examples of explanations put forward that illustrate the many proposed interpretations;

see [68, 529]. We recommend Chapter 3 for a further review on this exciting subject.

1.2.2 Photonic band gap crystals

Photonic crystals became a subject of intense research in the late 1990s and early 2000s [448, 449]. We will defer discussion of 3D band gap-induced cavity QED phenomena such as spontaneous emission inhibition, Purcell-enhanced emission, or thresholdless laser action, to Chapter 8 in this book.

The field of photonic band gap crystals is intimately linked to that of random media by the role of interference in modifying the transport of light. Indeed, one of the original drives for photonic band gaps was the prediction that Anderson localization is more easily reached in photonic band gap crystals with controlled disorder [222]. Moreover, photonic crystals are being pursued for flexible possibilities for controlling the propagation of light, ranging from simple filter action in reflectivity or transmission, to superprisms and negative refraction, to ultimately confined yet broadband waveguiding in 3D; see [306].

1.2.3 Waveguides and coupled cavities

One-dimensional (1D) waveguides in photonic crystals – mostly in 2D slab crystals – and arrays of cavities have attracted great interest since the mid 1990s, motivated by applications of slow light, and fueled by on-chip optical communication [27, 251]. For a long time, the effects of unavoidable variations in size and position of the building blocks on the propagation of light were neglected [240]. More recently, localization received revived interest in the second half of the 2000s. The interest arose in the context of randomness in coupled cavities, and randomness in photonic crystal waveguides [338, 482]. Recently, it has even been realized that in the presence of 1D Anderson localization in waveguides, spontaneous emission of embedded quantum dots is considerably enhanced [414]. An even stronger analogy between the behavior of light and that of electrons in condensed matter physics is the recent observation of Lifshitz tails in photonic crystal waveguides [201]. For a review of these exciting phenomena, see Chapter 4 in this book.

1.2.4 Quasicrystals and aperiodic media

Quasicrystals and aperiodic photonic media have attracted attention since the 2000s. The pursuit of quasicrystals has been motivated by the notion of opening photonic gaps at low index contrast, thereby allowing a wide choice of materials, including biomaterials. This motivation was fueled by reports of a 2D gap found to open at a low index contrast $m \geq 1.45$ [568]. Subsequent work revealed that a gap opens for $m \geq 2.6$ [563]. Nevertheless, 2D quasicrystals are generating exciting results and are being pursued for their strong light scattering properties and intricate random laser action. In addition, deterministic and multifrequency laser emission is being reported from aperiodic media [98]; see Chapter 6.

In the case of 1D quasicrystalline structures, pioneering work was reported by Ref. [102], where intricate pulse stretching was observed, as well as a strongly suppressed group velocity for frequencies close to a Fibonacci band gap. For a further review on this exciting subject, see Chapter 5.

In the case of 3D quasicrystals, initial studies were reported in the microwave regime, where crystals with different structures can conveniently be machined [316]. Work at optical frequencies was reported by Ref. [273], where the required nanostructures were made with advanced laser writing techniques. By studying the transport of light with ultrashort pulses, a very short mean free path was observed which points to incipient localization effects in these intriguing structures [273]. For an in-depth review see Chapter 7.

1.3 Shaping wavefronts in complex media

In complex media such as photonic crystals, light is strongly scattered both at the interfaces and in the bulk of the material. Because of unavoidable disorder in the structure, random scattering adds to the desired diffraction [240]. As a result, it is impossible to directly image or focus light deep inside strongly photonic nanostructures, and excitation and light collection from sources inside photonic crystals is typically inefficient.

Recently developed wavefront shaping methods, which are leading to a boost in activity of the optics of complex media, offer a method to counteract this unavoidable disorder, and even exploit it to obtain a high focusing resolution [345]. These methods make use of the fact that scattered light does not lose its coherence, so that the incident waves can be forced to interfere constructively at a chosen point, giving rise to a focus of high intensity. With the help of megapixel spatial light modulators (see Fig. 1.2), optical wavefronts are spatially shaped and adapted with sufficient precision to counteract and even exploit scattering. Even in the very first wavefront shaping experiments, light was focused through opaque media that were more than 15 mean free paths thick and it was shown that a focus can be created that is 1000 times more intense than the diffusive speckle background [509]. Using fluorescent or nonlinear probes, light can also be focused inside a scattering medium [510]. A particularly beneficial feature in emission experiments is that, due to reciprocity, a phase screen that optimizes

Figure 1.2 Schematic image of incident plane waves from the left. In a spatial light modulator (SLM) the plane wavefronts are modified into complex phase patterns (right) that propagate to a complex medium.

excitation will also optimize light collection, provided that the processes take place at sufficiently close wavelengths. By scanning the optimized spot, imaging with limited range is possible in several configurations [197, 507]. An important step towards even more flexible focusing and imaging was taken by Popoff and co-workers, who used wavefront shaping methods to measure a transmission matrix of a disordered nanophotonic system [393]. With the help of such a transmission matrix one can image through scattering media without scanning, and focus light at a chosen point in the transmission plane without further optimization.

Random nanophotonic materials exert a strong control over light, e.g. scattering it to high transversal wavevectors that are not accessible by normal refraction. Using wavefront shaping, this scattered light can be focused to a small spot and used for imaging at a resolution better than 100 nm [499]. The combination of wavefront shaping with specially designed plasmonic materials is expected to give rise to an even higher imaging resolution [166, 229, 283]. Developments in imaging using shaped wavefronts are progressing quickly. Recent work has enabled see-through imaging based on speckle correlations [44]. In this imaging modality, no calibration or probing in the object space is necessary, and the resolution of the image is essentially set by the diffraction limit.

While most of the work so far has involved narrowband light, recent progress in pulse shaping in disordered materials has shown that sufficient bandwidth is available to focus, delay, and compress femtosecond pulses through optically thick media [21, 230, 323]. This tailoring of broadband pulses is of special importance in photonic media in the cases of nonlinear excitation and ultrafast optical switching.

Contrary to the situation in biomedical settings, in nanophotonics one often deals with time-invariant structures and photostable emitters. This makes it possible to perform accurate optimizations of even small signals, as one can extend the integration time to obtain good signal to noise. This should enable wavefront-based optimization of entirely new signals such as the weak spontaneous emission from emitters deep inside a photonic band gap crystal.

1.4 Unified view of complex media: photonic interaction strength

While the various classes of complex media reviewed above have very different order, we propose here a unified view to characterize light propagation in all of these different kinds of complex media. To this end we define a figure of merit that gauges the interaction with light for all possible complex media. Such a figure of merit allows us to compare examples from the different classes of complex media when they are considered for a similar application. The photonic interaction strength S is defined as the dimensionless ratio of the polarizability α of an average scatterer in a complex medium to the average volume per scatterer V [237, 515, 517]:

$$S = \frac{4\pi\alpha}{V}, \tag{1.1}$$

where the prefactor depends on the chosen electrodynamic units. The photonic interaction strength can be interpreted as the ratio of the "optical volume," i.e. the volume that light "experiences" at each scatterer, to the physical volume of each scatterer. That these two volumes are not the same can be readily appreciated from the well-known fact that an atom on resonance – such as alkali atom near the D-lines – has a polarizability $\alpha \simeq \lambda^3 \simeq 1\,\mu m^3$, which is much greater than the physical volume of an atom of about $(0.3\,nm)^3 = 3 \times 10^{-11}\,\mu m^3$. Based on considerations for periodic systems [237], one can rewrite the photonic strength as

$$S = \frac{|\Delta\epsilon|}{\bar{\epsilon}}|f(\Delta\mathbf{k})|, \tag{1.2}$$

where $\Delta\epsilon$ is the spatial modulation of the dielectric function in the complex medium, $\bar{\epsilon}$ is the volume-averaged dielectric function, and $f(\Delta\mathbf{k})$ is the medium's structure factor evaluated at a dominant scattering vector $\Delta\mathbf{k}$ [80]. For dielectric media, Eq. (1.2) points to three main ways to increase the photonic strength. Firstly, increasing the dielectric contrast $\Delta\epsilon$, which rationalizes the pursuit of materials with a high refractive index contrast, such as semiconductors Si, GaP, and GaAs. Secondly, decreasing the average dielectric constant $\bar{\epsilon}$, which explains the prevalence of a minority of high-index material in strongly interacting photonic media. Thirdly, optimizing the geometrical factor $f(\Delta\mathbf{k})$, which directs choices to specific structural geometries such as the diamond structure for 3D photonic crystals [192], or which motivates the pursuit of quasicrystals [316].

Let us provide typical magnitudes of S for well-known examples of complex media. For Anderson localization and photonic gap behavior at optical frequencies the strength must be in the range $S > 0.15$ to 0.20. Atoms in a crystal lattice scatter X-rays only very weakly since the refractive index for X-rays hardly differs from 1; hence the photonic interaction strength is very small, $S \approx 10^{-4}$. This result explains why the vast literature on X-ray scattering does not consider Anderson localization or photonic band gaps for X-rays. A widely studied optical material such as an opal made from silica or polystyrene nanospheres has a moderate photonic interaction strength near $S = 0.06$, on account of the moderate refractive index contrast. Complex photonic media made from high refractive index semiconductors have high photonic strengths of more than 0.2. Among the highest photonic strengths are those in a narrow frequency range about an atomic resonance: for Cs atoms near $\lambda = 850$ nm the strength is about $S = 1.0$.

Interestingly, the photonic interaction strength is related to the characteristic length optical scale relevant to the particular complex medium under study. For random photonic media, the characteristic length optical scale is the scattering mean free path ℓ_{sc}, the average step size in the random walk of light. In photonic crystals with long-range periodic order, the characteristic length optical scale is the Bragg attenuation length ℓ_{Bragg}, that is, the distance needed to build up Bragg interference such that transmission in a gap has decreased to $1/e$. Indeed, the characteristic length optical scale ℓ_X (X = sc, Bragg, etc.) is inversely related to the photonic interaction strength by

$$\ell_X = \frac{\lambda}{\pi S}. \tag{1.3}$$

This relation shows that a strongly interacting complex photonic medium has a short characteristic length scale ℓ_X. For instance, in a strongly interacting random medium it takes only a short distance to develop the optical interference required for Anderson localization. Since any real complex photonic medium is necessarily finite, it follows that for a complex medium to reveal the desired characteristics (e.g. localization, band gaps), its extent L should exceed the characteristic length scale: $L > \ell_X$.

Acknowledgments

We thank Léon Woldering and Florian Sterl for help with the illustrations. It is a great pleasure to thank all our colleagues at COPS for many years of pleasant and fruitful collaborations, including Cock Harteveld, Femius Koenderink, Peter Lodahl, Pepijn Pinkse, all of our postdocs, PhD students and undergraduate students, and many, many others. We also thank colleagues elsewhere such as Sergey Skipetrov, Bart van Tiggelen, Diederik Wiersma, and many others. This work is part of the research program of the "Stichting voor Technische Wetenschappen (STW)," and the "Stichting voor Fundamenteel Onderzoek der Materie (FOM)," which are supported financially by the "Nederlandse Organisatie voor Wetenschappelijk Onderzoek (NWO)," and is also supported by the ERC.

2

Transport of localized waves via modes and channels

ZHOU SHI AND AZRIEL GENACK

2.1 Introduction

Suppressed transport and enhanced fluctuations of conductance and transmission are prominent features of random mesoscopic systems in which the wave is temporally coherent within the sample [6, 9, 501, 525]. The associated breakdowns of particle diffusion and of self-averaging of flux were first considered in the context of electronic conduction and for many years thought to be exclusively quantum phenomena [1, 2, 9, 11, 12, 113, 114, 476, 477, 525]. Independently, however, wave localization was demonstrated theoretically for a statistically inhomogeneous waveguide [156]. Over time, it became increasingly apparent that localization and mesoscopic fluctuations reflected general wave properties and might therefore be observed for classical waves as well [7, 22, 54, 78, 106, 130, 144, 145, 154, 156, 175, 221, 223, 225, 241, 263, 326, 353, 419, 462, 466, 487, 488, 501, 531, 535, 543]. In particular, the level and transmission eigenchannel descriptions proposed, respectively, by Thouless [476, 477] and Dorokhov [113, 114] to describe the scaling of conductance in electronic wires at zero temperature, are essentially wave descriptions involving the character of quasi-normal modes of excitation within the sample and speckle patterns of the incident and transmitted field. Quasi-normal modes, which we will refer to as modes, are resonances of an open system. These modes decay at a constant rate due to the combined effects of leakage from the sample and dissipative processes. Eigenchannels of the transmission matrix, are obtained by finding the singular values of the field transmission matrix and they represent linked field speckle patterns at the input and output of the sample surfaces. Eigenchannels are linear combinations of phase coherent channels impinging upon and emerging from the sample. Examples of such channels may be propagating transverse modes of an empty waveguide, or transverse momentum

Light Localisation and Lasing, ed. M. Ghulinyan and L. Pavesi. Published by Cambridge University Press.
© Cambridge University Press 2015.

states in the leads attached to a resistor. In measurements, input and output channels are often combinations of source and detectors at positions on the incident and output planes, respectively. Whereas modes are biorthogonal field speckle patterns over the volume of the sample [87, 293], eigenchannels are biorthogonal field speckle patterns at the input and output planes of the sample. When there is no risk of confusion we will refer to eigenchannels as channels. Although levels and channels have not been observed directly in electronic systems, these approaches have served as powerful conceptual guides for calculating the statistics and scaling of conductance [9].

Recent measurements of spectra of transmitted field patterns and of the transmission matrix of microwave radiation propagating through random multimode waveguides have made it possible to determine the eigenvalues of modes and channels as well as their speckle patterns in transmission in mesoscopic samples [432, 520]. These experiments were carried out in a multimode copper tube filled with randomly positioned dielectric elements, which is directly analogous to a resistive wire in the zero-temperature limit in which dephasing vanishes. The study of modes and channels promises to provide a comprehensive description of transport and to clarify long-standing puzzles regarding steady state and pulsed propagation.

In this chapter, we will discuss studies of wave localization and strong fluctuations of the electromagnetic (EM) field, intensity, total transmission and transmittance, also known as the "optical" conductance from the perspectives of modes and channels. These approaches are useful in numerous applications. The mode picture is of particular use in considering emission, random lasing, and absorption, while the channel framework is indispensable in optical focusing, imaging, and transmission fluctuations. We will describe lasing in disordered liquid crystals [246] and in random stacks of glass cover slips [333] in which the mode width falls below the typical spacing between modes. The lasing threshold is then suppressed by the enhancement of the pump intensity and by the lengthening of the dwell time of emitted light within the sample. Measuring the transmission matrix allows us to study the fluctuations of transmittance over a random ensemble. The statistics of transmittance can be described using an intuitive "Coulomb charge" model [464]. Measurements of the transmission matrix make it possible to obtain the statistics of transmission in single samples at a particular incident frequency [110], which are crucial for focusing and imaging applications [109, 110, 345, 393]. These statistics, as well as the contrast in focusing, are given in terms of the participation number of transmission eigenvalues and the size of the measured transmission matrix [109, 110].

In the next section (Section 2.2), we discuss the analogies between the transport of electrons and classical waves. Spectra of intensity and total transmission and transmittance are presented based on the measurements of the field transmission coefficient. A modal analysis of spectra of field speckle patterns on the output surface of a multimode waveguide and along the length of a single mode waveguide is described in Section 2.3. Pulse propagation in mesoscopic samples is discussed in terms of the distribution of mode decay rates and the correlation between modal speckle patterns in transmission. The role of modes in lasing in nearly periodic liquid crystals and in random slabs is described in Section 2.4. In Section 2.5, we describe the statistics of transmission eigenvalues and their impact on the statistics of transmission for ensembles of random samples and in single instances of the transmission matrix. The manipulation of transmission eigenchannels to focus radiation is described in Section 2.6. We conclude in Section 2.7 with a discussion of the prospects of a complete description of transport in terms of modes and channels.

2.2 Analogies between transport of electrons and classical waves

Anderson [11] showed over 50 years ago that the electron wavefunction would not spread throughout a disordered three-dimensional crystal once the ratio of the width of distribution of the random potential at different sites, relative to the coupling between sites, passes a threshold value. At lower levels of disorder, electrons diffuse in the band center but are localized in the tail of the band. Ioffe and Regel [209] pointed out shortly thereafter that an electron wavefunction could not be considered to be properly propagating if it were scattered after traveling less than a fraction of a wavelength, so that for traveling waves, $\ell > \lambda/2\pi$, where ℓ is the mean free path. This gives the criterion for localization in three dimensions, $k\ell < 1$, where k is the wavenumber. Although not explicitly noted at the time, this criterion for localization applies equally to classical and quantum waves.

In subsequent work, Thouless [476] considered the electronic state in the system as a whole rather than the strength of scattering within the medium. He argued that in bounded samples, the weight of electron states at the boundaries of the sample relative to points in the interior would be a useful measure of the extension of electron states within the sample. Since localized states would be peaked within the sample remote from the boundaries, their energies could be expected to be insensitive, even to substantial changes at the boundary such as those engendered in a periodic system when the boundary conditions for repeated sections of a random system are changed from periodic to antiperiodic [476]. When the energy shift is less than the typical spacing between states, the state is localized within the sample. An associated measure of electron localization, which relates to the properties of the states and not to the impact of some hypothetical manipulation of the sample, is the typical width of an electron level relative to the average spacing between levels. When the electron wavefunction is exponentially peaked within the sample, electrons are remote from the boundary and their escape from the sample is slow. The linewidth of the level is then smaller than the spacing between neighboring levels, which is the inverse of the density of states of the sample as a whole. This indicates that electron localization is achieved when the dimensionless ratio of the level width to level spacing, the Thouless number, $\delta = \delta E / \Delta E$, falls below unity. Pendry [379] has described the electron localization process as one in which "electrons can be forced to abandon their predilection for momentum" in favor of space as the defining characteristic. The inhibition in transport manifested in localization in space then leads to a lengthened escape from the sample and narrow linewidth, manifested in terms of sharp spikes in energy "one per electron" [379] as opposed to a continuous spectrum. For diffusing electrons, the wavefunction extends throughout the sample. Energy then readily leaks out of the sample and levels are consequently short-lived with line widths greater than the typical spacing between levels. Thus the electron localization threshold lies at $\delta = 1$. The Thouless number may equally well be defined for classical waves as the ratio of the typical frequency width to the spacing of quasi-normal modes, $\delta = \delta\omega / \Delta\omega$, where ω is the angular frequency. The level width is the inverse of the Thouless time τ_{Th} in which a mode leaks out of the sample. Wave localization is signaled by exponentially long dwell times for the wave. Such long decay times contribute little to the linewidth, which could be dominated by spectrally broad modes peaked near the sample boundary with short decay times. To most meaningfully capture the dynamics of a mode, it is therefore natural to identify $\delta\omega$ with the average of the inverse Thouless time, $\delta\omega \equiv \langle \tau_{Th}^{-1} \rangle \equiv \langle 1 / \Gamma_n^0 \rangle^{-1}$, where $\langle \ldots \rangle$ indicates an average taken over modes for an ensemble of samples and Γ_n^0 is the leakage rate of energy in the nth mode of the sample [476, 520]. $\delta < 1$ is a universal criterion for localization in any dimension for any type of wave.

Thouless [476, 477] was concerned with the coupling between adjacent regions in finite samples and so with the scaling of δ as an indicator of the changing character of the electron states with sample size. However δ cannot easily be measured in electronic systems and has been measured only recently for classical waves [520]. Using the Einstein relation, which gives the conductivity in terms of a product of the electron diffusion coefficient D and the density of states, which is $1/\Delta E$ divided by the sample volume, Thouless [477] showed that δ was equal to the conductance G in units of the quantum of conductance, $\delta = G/(e^2/h)$. Thouless [477] argued therefore that the dimensionless conductance would scale exponentially for localized waves as would be expected for δ. He showed that the resistance of a wire at $T = 0$ behaves ohmically [477], with the resistance increasing linearly with length L, and the dimensionless conductance varying as $g = N\ell/L$, only up to a length $\xi = N\ell$, at which $\delta = 1$. Here, N is the number of independent channels that couple to the resistor, $N \sim Ak_F^2/2\pi$, where A is the cross-sectional area of the sample, k_F is the electron wavenumber at the Fermi level, ℓ is the electron mean free path, and ξ is the localization length. For $L < \xi$, electrons diffuse with a residence time within the conductor of $\tau_{Th} \sim L^2/D$ [476, 477]. The level width would then be $\delta E \sim \hbar/\tau_{Th} \sim \hbar D/L^2$, while the level spacing ΔE would scale inversely with the volume of the wire as $1/L$. As a result, for $L < \xi$, δ scales as $1/L$. For $L > \xi$, electrons would be localized and δ and g would fall exponentially, while the resistance would increase exponentially with L. Abrahams *et al.* [2] showed that only above two dimensions is it possible for transport to be diffusive at all length scales. For lower dimensions, localization always sets in as the size of the sample increases, independent of the scattering strength, so that a transition between diffusive and localized transport can only occur above two dimensions [2].

The scaling of average conductance and fluctuation in conductance may also be calculated within the framework of random matrix theory [38, 113, 114, 327, 464]. The field in outgoing channel b is connected to the field in all possible incident channels a via the transmission matrix field t, $E_b = \sum_{a=1}^N t_{ba} E_a$. Taking the two independent polarization states into account, the number of propagating modes in the empty waveguide leading to the sample is $N = 2\pi A/\lambda^2$, where A is the illumination area and λ is the wavelength of the incident wave. Summing over all possible incoming and transmitted channels yields the transmittance $T = \sum_{a,b=1}^n |t_{ba}|^2 = \sum_{n=1}^N \tau_n$ [208], where the τ_n are the eigenvalues of the matrix product $tt^\dagger$. The transmission eigenvalue can also be found using the singular value decomposition of the transmission matrix $t = U\Lambda V^\dagger$. Here, U and V are unitary matrices and Λ is a diagonal matrix with elements $\lambda_n = \sqrt{\tau_n}$. The ensemble average of T is equal to the dimensionless conductance, $\langle T \rangle = g$ [268]. Random matrix theory predicts that, for diffusive waves, the transmission eigenvalue follow the bimodal distribution, $\rho = \frac{g}{2\tau\sqrt{1-\tau}}$ [114, 327, 464]. Most of the contribution to T comes from approximately g eigenvalues that are larger than $1/e$, while most of the eigenvalues are close to zero. The characteristics of these "open" [207] and "closed" channels were first discussed by Dorokhov [113, 114]. He considered the scaling of each of the transmission eigenvalues in terms of the associated localization lengths ξ_n. He found that the average spacing between inverse localization lengths of adjacent eigenchannels in a sample made up of N parallel chains with weak transverse coupling to neighboring chains was constant and equal to the inverse of the localization length $1/\xi$.

Localization of quantum and classical waves in quasi-one-dimensional (Q1D) samples with lengths much greater than the transverse dimensions occurs at a length at which even the highest transmission eigenvalue τ_1 falls below $1/e$. Thus localization will always be achieved as the sample length is increased in Q1D samples [477]. It is difficult, however, to localize EM waves in three-dimensional dielectric materials. As opposed to s-wave scattering prevalent in electronic systems, EM waves experience p-wave scattering and cannot be trapped by a confining potential. The scattering cross-section only becomes appreciable once the size of the scattering element becomes comparable to the wavelength. But once the scattering length is comparable to the wavelength, the mean free path cannot fall significantly below the scatterer size and it is hard to satisfy the Ioffe–Regel condition for localization in three dimensions, $k\ell < 1$. For smaller scattering elements such as spheres of radius a, the Rayleigh scattering cross-section is proportional to a^6, while the density of spheres is proportional to $1/a^3$. As a result, the inverse mean free path for fixed volume fraction of particles is proportional to a^3. For high particle density and $a \ll \lambda$, the sample acts as an effective medium with mean free path $\ell \sim 1/a^3$. It is therefore not possible to achieve strong scattering with $k\ell < 1$ by crowding together small scattering elements [223]. In ordered structures, however, EM bands appear and a photonic band gap (PBG) with vanishing density of states can be created in appropriate structures with sufficiently strong contrast in dielectric constant. John [222] has pointed out that disturbing the order in such structures would create localized states within the frequency range of the band gap, in analogy with the Urbach tail at the edge of the electronic band gap in semiconductors.

Although it has proven to be more difficult to localize EM radiation than electrons in three dimensions, transport of EM radiation can be probed in ways that are often closer to the theoretical paradigm of Anderson localization than is the case for electronics. The particles of classical waves do not interact as do electrons, dephasing is negligible even at room temperature and an ensemble of statistically equivalent random samples can be created. For classical waves, localization is most easily achieved in low-dimensional systems such as masses on a string [187], a single optical fiber [426], single- [424] and multi-mode waveguides [78], surfaces [106], layered structures [43, 561], and in highly anisotropic samples [265, 400, 422], particularly samples in which longitudinal structure along the direction of wave propagation is uniform. Anderson localization can be expected for EM radiation at the edge of the conduction or pass band in three-dimensional systems. Transport at the Anderson threshold has been observed for ultrasound in a slab of brazed aluminum beads [198].

An experimentally important difference between classical and quantum transport is that coherent propagation is the rule for classical waves such as sound, light, and microwave radiation in granular or imperfectly fabricated structures, whereas the electrons are only coherent at ultralow temperatures in micron-sized samples. For classical waves in static samples, the wave is not inelastically scattered by the sample so that the wave remains temporally coherent throughout the sample even as its phase is random in space. In contrast, mesoscopic features of transport arise in disordered electronic systems only in samples with dimensions of several microns at ultralow temperatures. Mesoscopic electronic samples are intermediate in size between the microscopic atomic scale and the macroscopic scale. In contrast, monochromatic classical waves are generally temporally coherent over the average dwell time of the wave within large samples. It is therefore possible to explore the statistics of mesoscopic phenomena with classical waves. Such studies may also be instructive regarding the statistics of transport in electronic mesoscopic samples. Measurements can be made in both the frequency and time domains. The impact of weak localization can be investigated in the time domain by measuring transmission following an excitation pulse or by Fourier transforming spectra of the field multiplied by the spectrum of the exciting pulse.

The connection between electronic and classical transport emerges as well from the equivalence proposed by Landauer [208, 268] of the dimensionless conductance g and the transmittance T, known as the "optical" conductance. The transmittance is the sum over all incident and outgoing channels of the transmission coefficient of flux. The phase of electrons arriving from a reservoir in different channels is randomized over the time of the measurement and the conductance is related to the incoherent sum of transmission coefficients over all channels, $g = \langle T \rangle = \langle \sum_{a,b=1}^{N} T_{ba} \rangle = \langle \sum_{a=1}^{N} T_a \rangle$. Measurements have been made of the statistics of transmission coefficients of the field, t_{ba}, intensity, $I_{ba} = |t_{ba}|^2$, and total transmission, $T_a = \sum_{b=1}^{N} T_{ba}$, for a single incident channel, a, and for the transmittance, T.

The experimental setup for measurements of microwave transmission in Q1D geometry described in this chapter is shown in Fig. 2.1. Measurements are carried out in ensembles of random Q1D samples contained in a copper tube. The samples are random mixtures of alumina spheres of diameter 0.95 cm and index of refraction $n = 3.14$ at a volume fraction 0.068. Source and detector antennas may be translated over a square grid of points covering the incident and output surfaces of the sample and rotated between two perpendicular orientations in the planes of the sample boundaries. Spectra of the field transmission coefficient polarized along the length of a short antenna are obtained from the measurement of the in- and

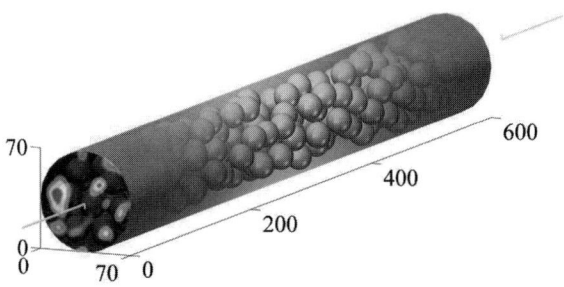

Figure 2.1 Copper sample tube containing random medium with microwave source and detector antennas. An intensity speckle pattern is produced with a single source location.

out-of-phase components of the field using a vector network analyzer. The intensity for a single polarization of the wave is the sum of the squares of the in- and out-of-phase components of the field. The sum of intensity across the output face for two perpendicular orientations of the detector antenna gives the total transmission. The field speckle pattern for each antenna position on the sample input is obtained by translating the detection antenna over the output surface. An example of an intensity speckle pattern formed in transmission is shown at the output of the sample tube in Fig. 2.1. The tube is rotated and vibrated momentarily after measurements are completed for each sample configuration to create a new and stable arrangement of scattering elements. In this way, measurements are made over a random ensemble of realizations of the sample. Field spectra can be Fourier transformed to yield the temporal response to pulsed excitation.

Spectra of intensity, total transmission, and transmittance normalized by the ensemble average values $s_{ba} = T_{ba}/\langle T_{ba}\rangle$, $s_a = T_a/\langle T_a\rangle$, and $s = T/\langle T\rangle$ in a single random configuration in two different frequency ranges are shown in Fig. 2.2. Fluctuations of relative intensity are noticeably suppressed in the higher frequency range as the degree of spatial averaging increases. For the ensemble represented in Fig. 2.2, $\text{var}(s_a) = 0.13$ in the high frequency range and 3.88 in the low frequency range. Since waves are localized for $\text{var}(s_a) > 2/3$ [78], this indicates that the wave is localized in the low frequency range.

It is instructive to consider the spectra in Fig. 2.2 from both the mode and channel perspectives. When the wave is localized, distinct peaks appear when the incident radiation is on resonance with a mode. The resonance condition holds for all source and detector positions and therefore sharp peaks remain even when transmission is integrated over space. When the wave is diffusive, many modes contribute to transmission at all frequencies and for all source and detector positions. The relative coupling strengths of a single polarization component of the intensity into and out of each of these modes have negative exponential distribution and phases of the field transmission coefficient are random so that relative fluctuations will be suppressed with increased spatial averaging. From the channel perspective, many orthogonal transmission channels contribute to transmission for diffusive waves, and the overlap with channels varies with source and detector positions. Fluctuations in the incoherent sum of this random jumble of orthogonal channels are therefore suppressed upon averaging over space. This suppresses the variance of transmission by a degree related to the number of channels that contribute substantially to transmission. This may be expressed quantitatively in terms of the participation number of eigenvalues of the transmission matrix, $M \equiv (\sum_{n=1}^{N} \tau_n)^2 / \sum_{n=1}^{N} \tau_n^2$. For diffusive waves, $\text{var}(s_a) \sim 1/M$ and relative fluctuations are enhanced, since the number of effective channels M is smaller than the number of independent channels N. We will see below that for diffusive

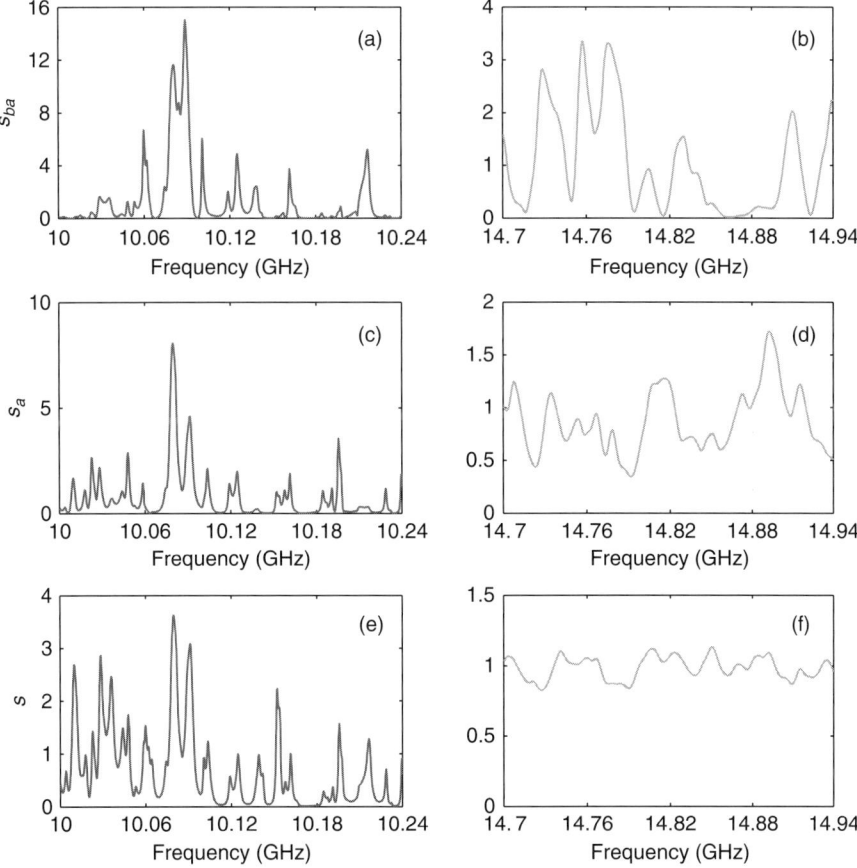

Figure 2.2 Spectra of transmitted microwave intensity, total transmission, and transmittance relative to the ensemble average value for each in a single random configuration. The wave is localized in (a), (c), and (e) and diffusive in (b), (d), and (f).

waves the spectrum of transmission eigenvalues is rigid, so that the number of transmission eigenvalues above $1/e$ fluctuates by approximately unity and fluctuations of conductance T are of order unity [8, 9, 73, 207, 278, 380, 525]. This results in universal conductance fluctuations which are independent of the sample size for Q1D samples [525].

The localization transition may be charted in terms of a variety of related localization parameters, all of which can be measured for classical waves. In addition to δ and the average over a random collection of samples of the dimensionless conductance, $g = \langle T \rangle$, measurements of fractional fluctuations of intensity or total transmission characterize the nature of the wave in random systems. In the diffusive limit, the variance of total transmission relative to the average value of total transmission over a random ensemble of statistically equivalent samples is inversely related to g, $\mathrm{var}(s_a) = 2/3g$ [78, 241, 353, 466, 501]. Since the wave is localized for $g < 1$, localization occurs when var$(s_a) > 2/3$. Perhaps the most easily experimentally accessible localization parameter is the variance of fractional intensity, which in the same limit is var$(s_{ab}) = 1 + 4/3\,g$. The localization threshold at $g = 1$ corresponds to var$(s_{ab}) = 7/3$. var(s_a) and var(s_{ab}) remain useful localization parameters even for localized waves. Fluctuations are relatively insensitive to absorption as compared with measurements of absolute transmission [78]. Mesoscopic fluctuations are directly tied to intensity correlation within the sample [130, 154, 326, 418, 423, 462]. The fractional correlation of intensity at two points on the output surface or between two transmission channels, b and b', is equal to the variance of relative total transmission, $\kappa = \langle \delta s_{ba} \delta s_{b'a} \rangle = \mathrm{var}(s_a)$. It is equal to $\langle M^{-1} \rangle$ in the diffusive limit, which is enhanced over the value of $1/N$ that would be expected if mesoscopic correlation were not present.

The relationships between key localization parameters mentioned above arise since the nature of propagation in disordered Q1D samples, in which the wave is thoroughly mixed in the transverse directions, depends only on a single dimensionless parameter [2]. For diffusive waves, $\delta = g = 2/3\,\mathrm{var}(s_a) = 2/3\,\kappa = 2/3\,\langle M^{-1} \rangle$. The relationships var$(s_a) = \kappa$ and var$(s_{ab}) = 2\mathrm{var}(s_a)+1$ hold through the localization transition, but the relationship between the other variables does not. However, we anticipate these relationships will change in a manner that can be described in terms of a single parameter. Other classical wave measurements that indicate the closeness to the localization threshold are coherent backscattering [7, 488, 535, 543] and the transverse spread of intensity in steady state or in the time domain [198, 265, 400, 422, 453]. The width of the coherent backscattering peak gives the transverse spread of the wave on the incident surface and hence the transport mean free path ℓ, from which the value of $k\ell$ can be found.

2.3 Modes

We find that the fields at any point in the sample may be expressed as a superposition of the field associated with the excitation of all of the modes in the sample. This superposition is a sum of products for each mode of the j polarization component of the spatial variation of the mode, $a_{n,j}(\mathbf{r})$, and the frequency variation of the mode, which depends only upon the central frequency of the mode ω_n and its linewidth, Γ_n,

$$E_j(\mathbf{r}, \omega) = \sum_n a_{n,j}(\mathbf{r}) \frac{\Gamma_n/2}{\Gamma_n/2 + i(\omega - \omega_n)} = \sum_n a_{n,j}(\mathbf{r})\varphi_n(\omega). \qquad (2.1)$$

The frequency variation of the nth mode $\varphi_n(\omega)$ is given by the Fourier transform of $\exp(-\Gamma_n t/2)\cos(\omega_n t)$ for $t > 0$. Equation (2.1) can be fit simultaneously to the field at a large number of points on the output speckle patterns in a single config-uration, since all spectra share a common set of ω_n and Γ_n. Armed with the values of ω_n and Γ_n, we find $a_{n,j}(\mathbf{r})$ and hence the speckle pattern for each of the modes.

The transmission spectrum is determined by the variation with position of the field amplitudes $|a_{n,j}(\mathbf{r})|$ and phases over the transmitted speckle patterns for the modes. The contribution of individual modes to transmission can be seen in the spectrum of total transmission near the single strong peak at 10.15 GHz, shown in Fig. 2.3(a) in a random sample of length $L = 61$ cm. The asymmetrical shape for the line in both intensity and total transmission indicates that more than a sin-gle mode contributes to the peak. The modal analysis of the field spectra shows that three modes contribute substantially to transmission over this frequency range. Spectra of the total transmission for the three modes closest to 10.15 GHz acting independently are plotted in Fig. 2.3(a). The integrated transmission for the 28th

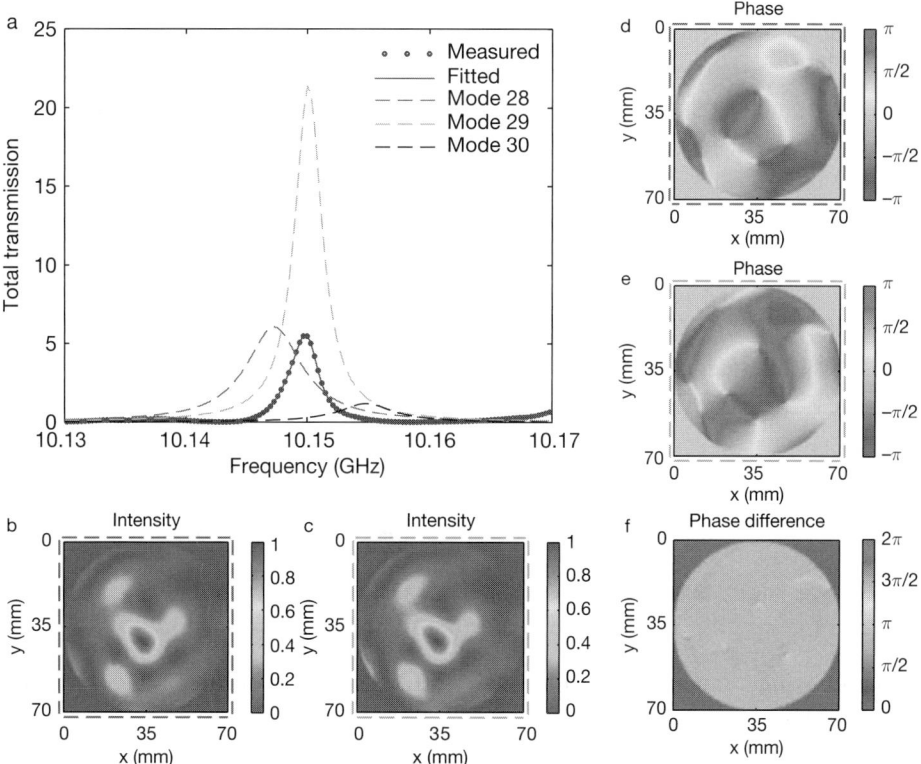

Figure 2.3 (a) Three modes contribute to the asymmetric peak in the total trans-mission spectrum. Modes 28, 29, and 30 are in order of increasing frequency. Intensity speckle patterns for modes 28 and 29 are shown in (b) and (c) and the corresponding phase patterns are shown in (d) and (e). (f) The phase in mode 29 is shifted by nearly a constant of 1.02π rad relative to mode 28 (Ref. [520]).

and 29th mode found in the spectrum starting at 10 GHz are each greater than for the measured peak, indicating that these modes interfere destructively. The intensity and phase patterns for these two modes are shown in Figs. 2.3(b–e). Aside from a difference in the average value of transmission, the intensity speckle patterns of the two modes are nearly the same. The distributions of phase shift at 10.15 GHz for the two modes are also similar, except for a constant phase difference between them of $\Delta \varphi = 1.02\pi$ rad. The similarity between the speckle patterns for these overlapping modes suggests that these modes are formed from coupled resonances within the sample which overlap spatially and spectrally. We expect that such overlapping resonances peaked at different locations will hybridize to form modes of the system. Such modes may be close to symmetric and antisymmetric combinations of the two local resonances. This would produce similar intensity speckle patterns at the output with a phase shift of $\sim \pi$ rad between the modes. The similarity in the intensity speckle patterns of these adjacent modes and the uniformity of the phase shift across the patterns of these modes allow for interference between modes across the entire speckle pattern. The similarity between modes is most evident in a sample configuration such that a particular pair of nearest neighbor modes are as close as possible in frequency. This is a point of anticrossing which arises because of level repulsion inside the sample [51, 259]. The magnitude of the field inside a 1D sample is seen to be the same throughout the sample. Modes are orthogonal by virtue of a change in phase close to π rad along the length of the sample. At the anticrossing, the fields at the sample output for the two modes are the same except for a change in phase of π rad. The Thouless number, which equals the dimensionless conductance for diffusive waves, provides a key measure of the dependence of transmission on the underlying characteristics of modes. The modal decomposition of transmission spectra for the ensemble from which the configuration analyzed in Fig. 2.3 is drawn gives $\delta = 0.17$.

The statistics of level spacing was first considered by Wigner [536] in the context of nuclear levels probed in neutron scattering. He conjectured the eigenvalues of the Hamiltonian matrix would have the same statistics as the spacing of eigenvalues of a large random matrix with Gaussian elements. Agreement was found between the spacing between peaks in the scattering cross-section and Wigner's surmise for the spacing of random Hamiltonian matrices. However, the analysis of spectra of nuclear scattering cross-sections was done in samples with relatively sharp spectral lines. We have seen above that even when $\delta < 1$, a number of lines may coalesce into a single peak. A comparison of level spacing statistics in samples with different values of modal overlap δ in which the phase of the scattered wave can be measured is therefore of interest in forming a picture of the statistics of transmission. Progressively stronger deviations from the Wigner surmise are found for decreasing values of δ.

It has not been possible to access the field distribution within the interior of multiply scattering three-dimensional samples, but spatial distributes can be examined in one- and two-dimensional samples [187, 259, 424]. The presence of both isolated and overlapping modes within the same frequency range has been observed in measurements of field spectra along the length of slotted single-mode random waveguides. The waveguides contained randomly positioned binary dielectric elements and a smaller number of low index Styrofoam elements. Measurements were carried out in the frequency range of a pseudogap associated with the first stop band of a periodic structure of consecutive binary elements. The density of states is particularly low in the frequency range of the band gap so that $\delta < 1$. When spectrally isolated lines are found, they are strongly peaked in space and their intensity spectrum at each point in the sample is Lorentzian with the same width at all points within the sample. When modes overlap spectrally, however, spectral lines have complex shapes which vary with position within the sample and the spatial intensity distribution is multiply peaked. Mott [347] argued that interactions between closely spaced levels in some range of energy in which $\delta < 1$ would be associated with two or more centers of localization within the sample. Pendry [378] showed that the occasional overlap of electronic states would dominate transport since regions in which the value of the electron wavefunction is high would not be far from both the input and output boundaries. Since the wave can then find a ready path through the sample, such modes are relatively short-lived and spectrally broad. This enhances the contribution of coupled resonances to transport. Such multiply peaked and spectrally overlapping excitations within the sample, termed "necklace states" by Pendry, are important in transmission since they arise in the localized regime and transport through isolated modes is typically small and over a narrow linewidth.

The variation in space and frequency of the amplitude of the waves within the pseudogap in a single random configuration is shown in Fig. 2.4. An additional ripple is observed in Fig. 2.4(b) in the intensity variation through the sample, corresponding to a phase shift of π rad, each time the frequency is tuned through a mode. The analysis of field spectra inside the waveguide within the pseudogap into the modes and a background which varies slowly in frequency is shown in Fig. 2.5. The slowly varying background shown in Fig. 2.5(a) is the fit of a polynomial in the difference in frequency from a point in the middle of the spectral range considered. This background is presumably related to on-resonance excitation of many modes on either side of the band gap. The mode structure within the single-mode waveguide sample changes when a spacing is introduced between two parts of the sample and is gradually changed. A succession of mode hybridizations is observed with increasing spacing as a single mode tends to shift in frequency until it encounters the next mode. As the spacing is increased, the mode that had been

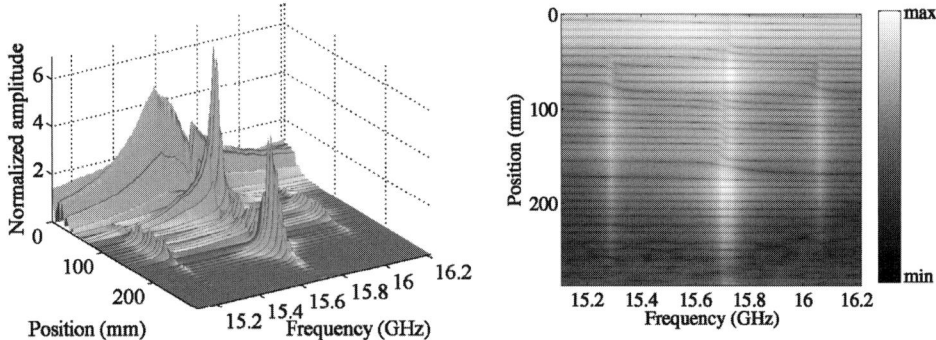

Figure 2.4 (a) Spectra of the field amplitude at each point along a random sample with spectrally overlapping peaks normalized to the amplitude of the incident field. (b) Top view of (a) in logarithmic presentation (Ref. [424]).

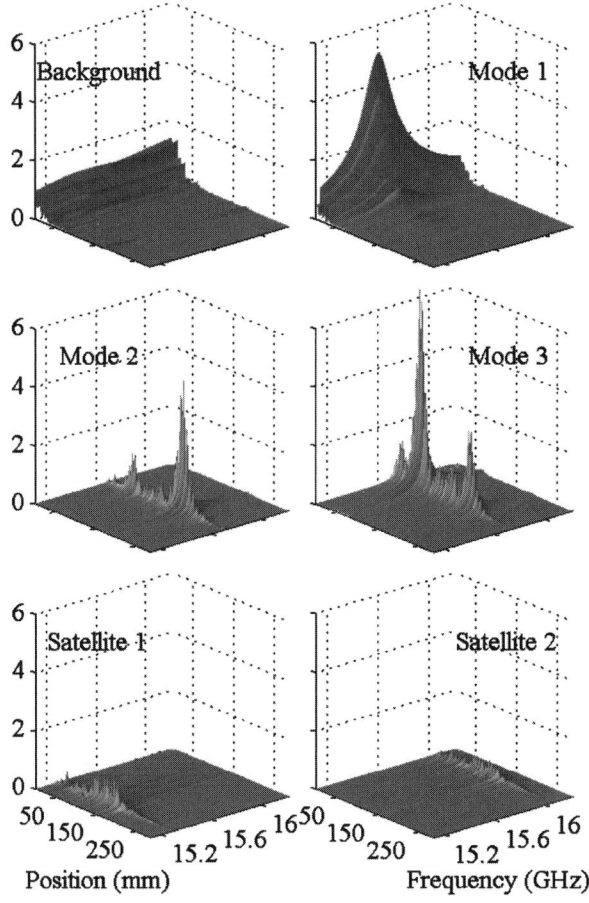

Figure 2.5 Decomposition of field pattern in Fig. 2.4 into a slowly varying polynomial term and five modes (Ref. [424]).

moving becomes stationary and the next mode begins to move [51]. Simulations have shown that changing the index of a single scatterer leads to mode hybridization in 2D random systems and may change the character of a mode from a single peak to multiple peaks [259].

One-dimensional localization has also been observed in optical measurements in single-mode optical fibers [426] and in single-mode channels that guide light within photonic crystals [305, 446, 482]. When the structure bracketing the channel is periodic, the velocity of the wave propagating down the channel experiences a periodic modulation so that a stop band is created. When disorder is introduced into the lattice, modes with spatially varying amplitude along the channel are created. Modes near the edge of the band gap are long lived and readily localized by disorder. An example of spectra of vertically scattered light versus frequency for light launched down a channel through a tapered optical fiber is shown in Fig. 2.6. The inset shows the disordered sample of holes with random departure from circularity in silicon-on-insulator substrates at a hole filling fraction of $f \sim 0.30$.

The modal decomposition method described above can be applied for localized waves for which modal overlap is relatively small. The impact of modes can be seen in the changing decay rate of transmission following pulsed excitation, even for diffusive waves. The slowing down of the decay rate with time will become more pronounced in samples in which the wave is more strongly localized [248, 335, 457]. In the diffusive limit, the transverse extent of the modes is large and the wave

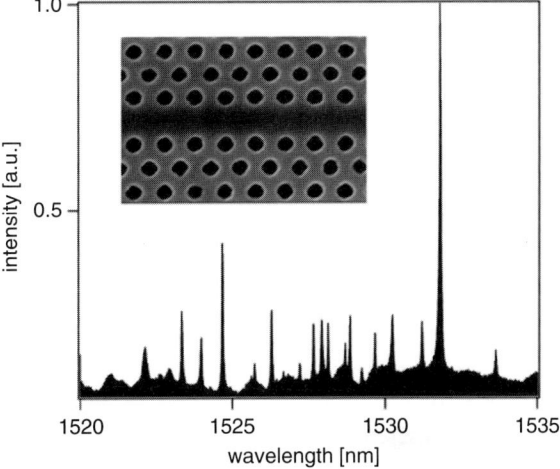

Figure 2.6 Spectrum of wave transmitted to a region within a single-mode photonic crystal waveguide near the short wavelength edge of the first stop band at $\sim$1520 nm. The channel surrounded by irregular holes is shown in the inset (Ref. [482]).

is coupled to its surroundings through a large number of speckle spots. One expects therefore that the decay rate of all modes will approach the diffusive limit, equal to the decay rate of the lowest diffusion mode [151, 523], $1/\tau_1 = \pi^2 D/(L + 2z_0)^2$, after a time τ_1 in which higher-order modes with decay rates $1/\tau_n = n^2 \pi^2 D/(L + 2z_0)^2$ have largely decayed. Here, n is the order of the diffusion mode and z_0 is the length beyond the boundary of the sample at which the intensity inside the sample extrapolates to zero. We find that pulsed transmission deviates increasingly from the diffusion model in nominally diffusive samples with $g > 1$ as the value of g decreases and the measured value of κ increases. Strong deviations from pulsed transmission in diffusing samples far from the localization threshold are observed for microwave radiation, light, and ultrasound.

Measurements of pulsed transmission through a random sample of alumina spheres at low density in samples of different length and absorption with values of $\kappa = 0.09, 0.13, 0.25$, and 0.125 are shown in Fig. 2.7 in samples A–D, respectively [79]. The decay rate of intensity is seen to deviate from the constant rate of the diffusive limit and is seen in Fig. 2.7(b) to decrease at a nearly constant rate. A linear falloff of the decay rate would be associated with a Gaussian distribution of decay rates for the modes of the medium [335]. A slightly more rapid decrease in the decay rate is associated with a slower than Gaussian falloff of the distribution of mode decay rates. The slowing down of the decay rate at long times reflects the survival of more slowly decaying modes [79, 335]. The distribution of modal decay rates is related to the Laplace transform of the transmitted pulse intensity.

Sample D is the same as sample B, except for the increased absorption due to a titanium foil inserted along the length of the sample tube. The variation with time of the decay rate in sample D is the same as that in sample B, except for an additional constant decay rate in sample D due to absorption. This shows that at

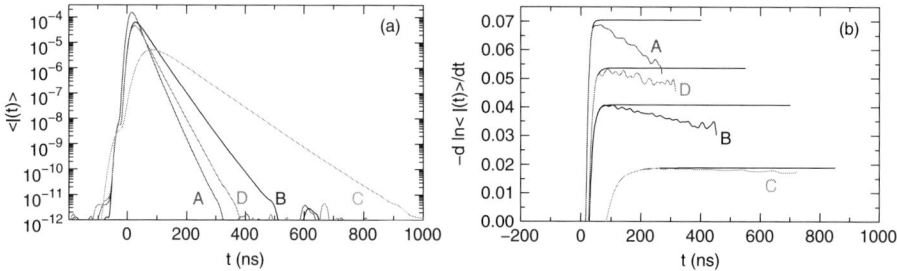

Figure 2.7 (a) Average pulsed transmitted intensity in samples of alumina spheres with lengths, $L = 61$ cm (A), 90 cm (B and D), and 183 cm (C). Sample D is the same as sample B except for the insertion of a titanium foil along the length of the sample tube D to increase absorption. (b) Temporal derivative of the intensity logarithm gives the rate γ of the intensity decay due to both leakage out of the sample and absorption (Ref. [79]).

the low level of absorption, scattering rates are not affected by absorption and the effect of absorption simply introduces a multiplicative exponential decay which is the same for all trajectories at a given time. Thus the degree of renormalization of transport due to weak localization involving the interference of waves following time-reversed trajectories that return to a point in the medium is not affected by absorption. We note that the fractional reduction of the decay rate is greater at a given time delay in shorter samples with higher values of g. This is because the length of trajectories of partial waves within the medium is the same for all samples, but the number of crossings of trajectories is greater when the paths are confined within a smaller volume.

The temporal variation of transmission can also be described in terms of the growing impact of weak localization on the dynamic behavior of waves, which can be expressed via the renormalization of a time-dependent diffusion constant or mean free path [86]. The decreasing decay rate has also been explained using a self-consistent local diffusion theory for localization in open media [442, 443]. The theory uses a one loop self-consistent calculation of an effective diffusion coefficient that falls with increasing depth inside the sample. The spatial varia-tion of the local diffusion coefficient reflects the increasing fraction of returns of wave trajectories to a point with greater depth due to the lengthened dwell time in the sample. This theory gives excellent agreement with recent measurements of non-diffusive decay of a transmission pulse of ultrasound transmission through a sample of aluminum spheres, as seen in Fig. 2.8 [198]. The value of g is just beyond the localization threshold as determined from measurements of var(s_{ba}),

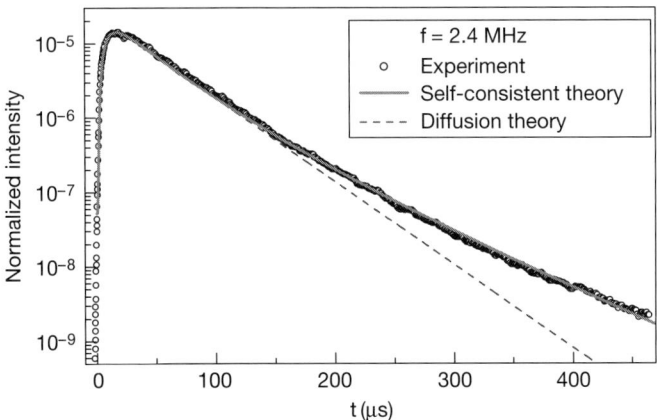

Figure 2.8 Averaged time-dependent transmitted intensity $I(t)$ normalized so that the peak of the input pulse is unity and centered on $t = 0$, at representative frequencies in the localized regimes. The data fit the self-consistent theory (solid curve). For comparison, the dashed line shows the long-time behavior predicted by diffusion theory (Ref. [198]).

which is slightly above the value of 7/3 predicted as the localization threshold. The intensity distribution on the output surface of the sample is found to be multifractal as predicted near the threshold for Anderson localization [128, 335]. However, the value of $k\ell$ in this sample is $\sim$1.5 which would indicate that the wave is diffusive. Measurements of the spread of intensity on the sample output with increasing time delay show a trend towards an exponential decay of intensity at later times, supporting the localization of the wave. A similar approach to an exponential decay of intensity with transverse displacement on the sample output from the point of injection of the pulse on the incident surface has been observed by Sperling *et al.* [453] in optical measurements through a slab of titania particles. When $k\ell < 4.5$, the variance of the spatial intensity distribution reaches a peak value and then actually falls. This is taken as support of localization of the wave at $k\ell > 1$. However, this criterion for localization is tied directly to the Thouless criterion for localization $\delta = 1$. It might also be that at later times longer-lived modes, which are more confined in space, are more heavily represented and dominate the spatial distribution. Although each of these modes will not spread in time, the modes that survive with increasing time delay would be the more strongly confined modes and would lead to a falling variance of the spatial intensity distribution in time. Such states may be prelocalized with a slower falloff in space than exponential, but still faster than for diffusive waves [17, 350].

The slowing of the spread of the transmitted wave in the transverse direction can also be seen in transverse localization in samples which are uniform in the longitudinal direction. This has been observed in a 2D periodic hexagonal lattice with superimposed random fluctuations [422]. The structured sample is created by first illuminating the photorefractive sample with a hexagonal optical pattern and then with a random speckle pattern of varying strength. A transition from a diffusive to a localized wave in the transverse plane is seen in the output plane, with the ensemble average of the spatial intensity distribution changing from a Gaussian to an exponential function centered on the input beam as the thickness of the sample increases. Since the wave incident upon the sample, which is uniform along its length, is paraxial, it is not scattered in the longitudinal direction and travel time through the sample is proportional to the sample thickness. Transverse localization is also observed in an array of disordered waveguide lattices [265].

Measurements of pulsed microwave transmission in more deeply localized waves transmitted through a Q1D sample of random alumina spheres of thickness approximately 2.5 times the localization length, are shown in Fig. 2.9 [564]. The impact of absorption was removed statistically by multiplying the average measured intensity distribution by $\exp(t/\tau_a)$ [78, 524]. For times near the peak of the transmitted pulse, diffusion theory corresponds well with the measurements of $\langle I(t) \rangle$. For times up to four times $\tau_D = L^2/\pi^2 D$, the decay rate of the lowest

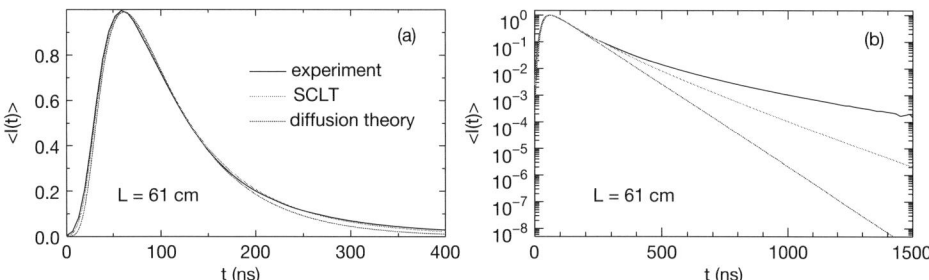

Figure 2.9 (a) Fit of self-consistent localization theory (SCLT) (dashed curve) at early times to the average intensity response (solid curve) to a Gaussian pulse with $\sigma_\nu = 15$ MHz in a sample of length $L = 61$ cm and the result of classical diffusion theory (dotted curve); (b) semilogarithmic plot of $\langle I(t) \rangle$ reaching to longer times. In (a) and (b), the curves are normalized to the peak value (Ref. [564]).

diffusion mode, pulsed transmission is in accord with self-consistent localization theory, but transmission decays more slowly for longer times. This indicates the inability of this modified diffusion theory to capture the decay of long-lived localized states. Such states are included in a position-dependent diffusion [480] that is in accord with simulations of the steady-state intensity distribution within random systems [375]. The difference between self-consistent localization theory and the theory for position-dependent diffusion is seen to be precisely in the ability of the latter to include the impact of long-lived resonant states [375, 480].

Destructive interference between neighboring modes, together with the distribution of mode transmission strengths and decay rates, can explain the dynamics of transmission. The average temporal variation of total transmission due to an incident Gaussian pulse is found by composing the response to the pulse from the Fourier transform of the product of the field spectrum and the Gaussian pulse. The progressive suppression of transmission in time by absorption may be removed by multiplying $\langle T_a(t) \rangle$ by $\exp(t/\tau_a)$ to give, $\langle T_a^0(t) \rangle = \langle T_a(t) \rangle \exp(t/\tau_a)$ [78, 524]. With the influence of absorption removed, decay is due solely to leakage from the sample. The measured pulsed transmission corrected for absorption is shown as the solid curve in Fig. 2.10 and is compared with the incoherent sum of transmission for all modes in the random ensemble corrected for absorption, $\sum_n T_{an}^0(t)$, shown as the dashed curve in Fig. 2.10. $\sum_n T_{an}^0(t)$ is substantially larger than $\langle T_a^0(t) \rangle$ at early times, but converges to $\langle T_a^0(t) \rangle$ soon after the peak. Although transmission associated with individual modes rises with the incident pulse, transmission at early times is strongly suppressed by destructive interference of modes with strongly correlated field speckle patterns, such as those shown in Fig. 2.3. At later times, random frequency differences between modes leads to additional random phasing between modes, and averaged pulsed transmission approaches the incoherent sum of decaying modes. The decay of $\langle T_a^0(t) \rangle$, shown as the solid curve in Fig. 2.10,

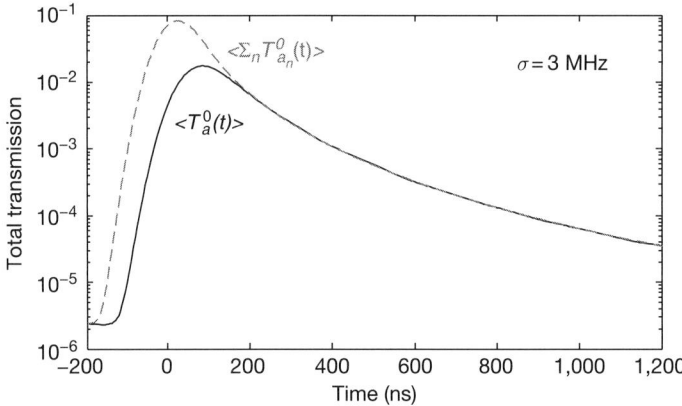

Figure 2.10 Semilogarithmic plot of the ensemble average of pulsed transmission and the incoherent sum of transmission due to all modes in the random ensemble (Ref. [520]).

is seen to slow considerably with time delay, reflecting a broad range of modal decay rates.

Measurements by Bertolotti *et al.* [43] of pulsed infrared transmission through random layers of porous silicon, with different porosity produced by controlled electrochemical etching of silicon, show that the pulse profiles depend on the degree of spectral overlap of excited modes. As the number of layers increases, the spectra become sharper since propagation of a paraxial beam in the structure is essentially one-dimensional and δ falls with sample thickness. When the pulse excites an isolated resonance peak, the decay rate of the falling edge of the transmitted pulse is seen in Figs. 2.11(a) and (b) to be larger for the narrower spectral peak. The delayed rise in transmission seen in Fig. 2.11(a) suggests, however, that more than a single mode is involved, since the interaction of a pulse with a single mode would be expected to lead to a prompt rise in transmission as the sample is being excited, after which intensity decays at a constant rate. A symmetrical profile for the transmitted pulse is seen in Fig. 2.11(c), when the spectrum of the exciting pulse overlaps several spectral lines. These spectrally overlapping states form necklace states with a series of intensity peaks along the sample which provide a path for the wave through the medium and would be expected to be short-lived.

The relationship of pulsed transmission of microwave radiation through a sample of random dielectric spheres with $\delta = 0.43$ to the transmission spectrum is shown in Fig. 2.12 [564]. The decay is slow when the spectrum of the pulse overlaps a single narrow mode, and fast when two peaks fall within the spectrum of the incident pulse. In this case, the transmitted intensity is significantly modulated at the frequency difference between the modes.

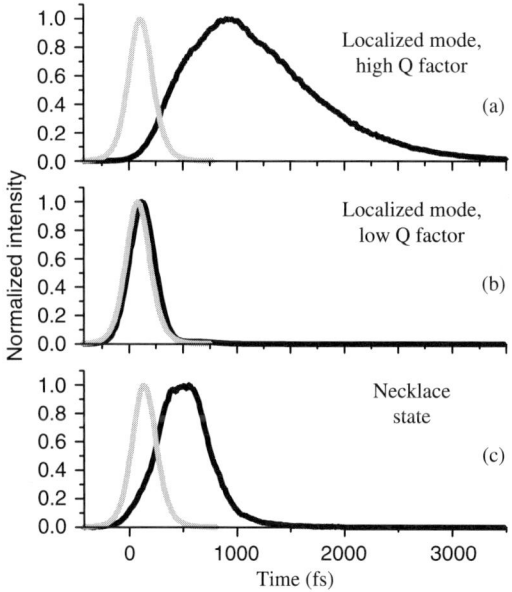

Figure 2.11 Time-resolved transmission data. In (a) and (b) the sample is excited on resonance with a sharp transmission peak, with small and large linewidths. In (c) a nearly symmetric pulse shape is observed that exhibits a fast decay time and a relatively large delay, typical for multiple-resonance necklace states. Sample thickness: 250 layers. Gray curves: instrumental response (cross-correlation between probe and gate), corrected for the delay introduced by the effective refractive index of the sample (Ref. [43]).

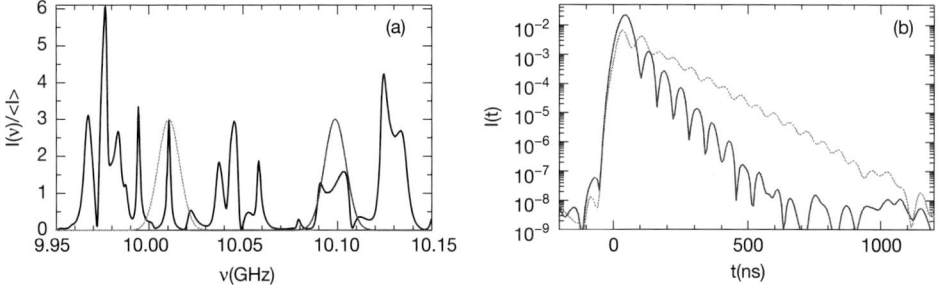

Figure 2.12 (a) Transmitted intensity spectrum in a random sample of $L = 40$ cm and Gaussian spectra of incident pulses peaked at the center of the isolated line and overlapping lines. (b) Intensity responses to the Gaussian incident pulses with spectral functions shown in (a) (Ref. [564]).

In addition to the increasing suppression of the leakage rate observed in diffusive samples with increasing delay, the variance of relative intensity fluctuations and the degree of intensity correlation also increase with time delay [77, 85, 519]. The field correlation function with displacement and polarization rotation in pulsed transmission is the same as in steady state [77]. This reflects the Gaussian statistics within the speckle pattern of a given sample configuration and time delay. The intensity correlation function at a given time delay depends on the square of the field correlation function and the degree of intensity correlation in the same way as in steady state, but the degree of extended range correlation $\kappa_\sigma(t)$ depends on delay time and on the spectral bandwidth of the pulse σ. The probability distribution functions of intensity at various delay times are also given using the formulas derived for steady-state propagation and depend upon a single parameter, which is the variance of the ratio of total transmission to its average over the ensemble, which equals the degree of intensity correlation at that time, $\mathrm{var}(s_a(t)) = \kappa_\sigma(t)$. The time variation of $\kappa_\sigma(t)$ reflects the number of modes and the degree of correlation in the speckle pattern of the modes. Since strong correlation in the speckle patterns of a number of modes tends to produce a single transmission channel formed from these modes and the correlation at any time is directly related to the number of channels contributing to transmission at that time, modal speckle correlation tends to increase the degree of intensity correlation.

For narrow band excitation, $\kappa_\sigma(t)$ falls before increasing at later times, since transmission at early times is dominated by a subset of short-lived modes among all of the modes overlapping the spectrum of the pulse which promptly convey energy to the output [519]. At later times, only the long-lived modes still contribute to transmission and so the degree of correlation increases with time at late times. But at intermediate times, when both short- and long-lived modes contribute to transmission, the number of modes and hence the number of channels contributing is relatively high. Hence the number of channels, which is directly related to κ, is also high so that $\kappa_\sigma(t)$ reaches a minimum.

The changing distribution of modes contributing to transmission is seen in the time–frequency spectrogram for a sample with $L = 61$ cm and $\delta = 0.17$ in Fig. 2.13. The spectrogram is formed from measurements of transmission at the output of a sample for an incident Gaussian pulse with width σ of its Gaussian spectrum as the central frequency of the pulse is tuned. The decay rate of the peak intensity of the mode at long times when isolated modes emerge in the time–frequency spectrogram is equal to the linewidth of the mode Γ_n to within an experimental error of 10%.

2.4 Lasing in localized modes

The nature of propagation varies according to the character of the modes of the medium. For $\delta > 1$, transport can be described in terms of diffusing particles of the wave with transmission falling inversely with sample thickness, while for $\delta < 1$, transport is via tunneling through localized or multipeaked modes with average transmission falling exponentially. Since the location and intensity of pump

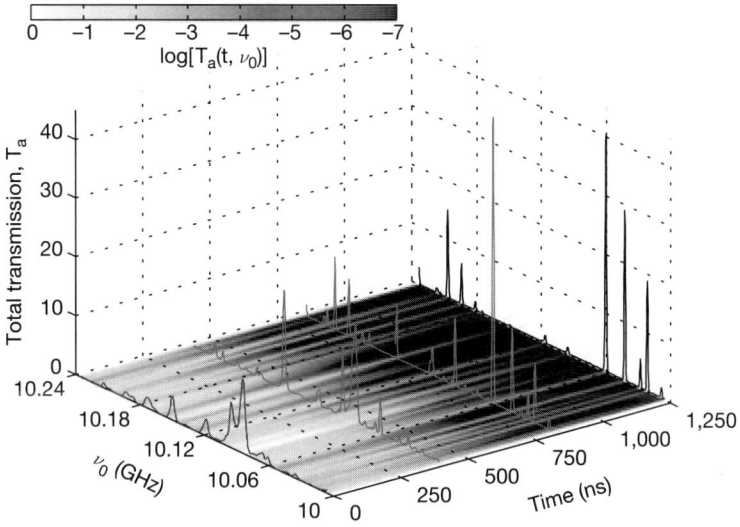

Figure 2.13 Logarithm of time–frequency spectrogram of total transmission plotted in the xy plane following the color bar. The central frequency ν_0 of the incident Gaussian pulse of linewidth $\sigma = 50.85$ MHz is scanned. Each of the four spectra of total transmission at different delay times is normalized to the total transmission at that time (Ref. [520]).

excitation within the sample and the lifetime of emitted photons within the gain region in which stimulated emission occurs depend upon the character of modes, lasing characteristics depend crucially upon the value of δ.

For $\delta \gg 1$ in random amplifying samples, nonresonant random lasing occurs. This can be described in terms of the densities of diffusing pump and emission photons and their coupling to energy levels whose occupation is described in terms of rate equations [41, 232, 533]. Very different behavior manifests in the regime $\delta \sim 1$, which can arise in strongly scattering but still diffusive samples no more than a few wavelengths thick [71, 139], and in 2D samples [504] in which the laser beam is tightly focused to create a small excitation volume [13, 552]. A small number of spectral peaks may then be observed in emission. These peaks sharpen up in the presence of gain due to enhanced stimulated emission in longer-lived, spectrally narrow modes [71].

Letokhov [292] considered the lasing threshold in a spherical sample with uniform gain, which is directly analogous to the critical condition for a nuclear chain reaction. Lasing occurs when on average more than one new photon is created for each photon that escapes the medium. Lasing was subsequently considered in granular media and in colloidal samples composed of dielectric particles in dye solution. Lasing in amplifying colloids reported by Lawandy *et al.* [269] is of particular interest since the strength of scattering and amplification could be

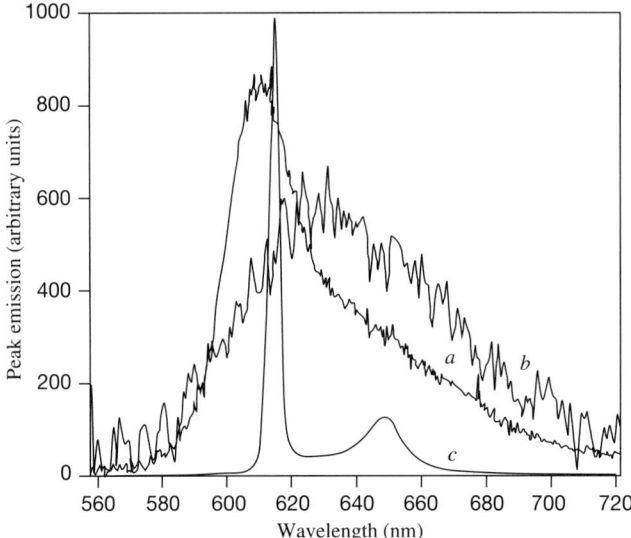

Figure 2.14 (a) Emission spectrum of a 2.5×10^{-3} M solution of R640 perchlo-rate in methanol pumped by 3 mJ (7 ns) pulses at 532 nm. (b) and (c) Emission spectra of the TiO$_2$ particles (2.8×10^{10} cm^{-3} colloidal dye solution pumped by 2.2 µJ and 3.3 mJ pulses, respectively. Emission: (b) scaled up 10 times, (c) scaled down 20 times (Ref. [269]).

controlled independently. A narrowing of emission and a shortening of the emit-ted pulse were observed above a certain threshold in pump power. A comparison of the emission spectrum in the neat dye solution and in the colloidal solutions is shown in Fig. 2.14. The original studies were carried out in weakly scattering samples excited over transverse dimensions much greater than the sample thick-ness, which was itself not much thicker than a mean free path. Wiersma *et al.* [534] suggested that the observations reported could be due to scattering of light in transverse directions and its subsequent redirection out of the sample by another scattering event.

The lasing threshold typically is not suppressed substantially below the threshold for amplified stimulated emission in a neat solution of dye. Light penetrates to a depth within the sample equal to the absorption length of the pump radiation $L_a = \sqrt{D\tau_a} = \sqrt{\ell\ell_a/3}$, which is the same as the exponential falloff with thickness of transmission in diffusing systems with loss [150]. Here $1/\tau_a$ is the absorption rate, ℓ the transport mean free path, $\ell_a = v\tau_a$ the length of the trajectory in which the intensity falls to $1/e$ due to absorption, and v is the transport velocity [489]. The excitation region illustrated in Fig. 2.15 is near the boundary so that the typical length of the paths of emitted photons would be comparable with that of the pump photons of $\sim\ell_a$ [152]. But this is the length over which stimulated emission occurs in a neat solution. So the lasing threshold may not be lowered below the value at

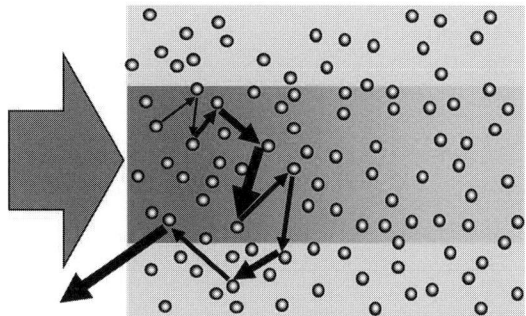

Figure 2.15 Possible path for a photon emitted and amplified within a dye medium containing random scatterers. The lighter region indicates the volume pumped by the laser (Ref. [152]).

which appreciable amplified spontaneous emission occurs in a neat dye solution. The lasing threshold could be lowered by increasing the residence time inside the medium in samples with a shorter mean free path at the emission frequency than at the pump frequency or by internal reflection at the boundary. Above the threshold, the transition pumped may be saturated so that absorption is suppressed and the wave can penetrate deeper into the sample.

The lasing threshold can be dramatically suppressed, however, for localized waves. When $\delta < 1$, the intensity within the sample may grow exponentially when on resonance with a localized state far from the boundary. The excited region will then be in the middle of the sample and so emission will be into modes that overlap the excitation mode and are similarly peaked in the middle of the sample, and so are long-lived. This was demonstrated in low-threshold lasing excited by a beam incident normally upon stacks of glass cover slips of approximate thickness $100\,\mu m$ and intervening air layers of random thickness and with Rhodamine 6G dye solution between some of the slides [333]. A plane wave incident upon parallel layers of random thickness is a one-dimensional medium and will be localized in the medium. In the present circumstance the layers are not perfectly parallel and so light is scattered off the normal. This leads to a delocalization transition with a crossover at a thickness at which the transverse spread of an incident ray is equal to the size of the speckle spots formed [561]. Beyond this thickness, the sample becomes three-dimensional with regard to propagation of the initially normally incident beam. The spread of the beam is abetted by the thick layers used and could be reduced dramatically if layers of thickness $\sim\lambda/4$ were used.

Emission spectra excited by a pulsed Nd:YAG laser at 532 nm in a stack of cover slides with intervening dye solution recorded with a 0.07 nm-resolution grating spectrometer are shown in Fig. 2.16. The broad emission spectrum of the neat dye solution in Fig. 2.16(a) is compared with the emission spectrum from the

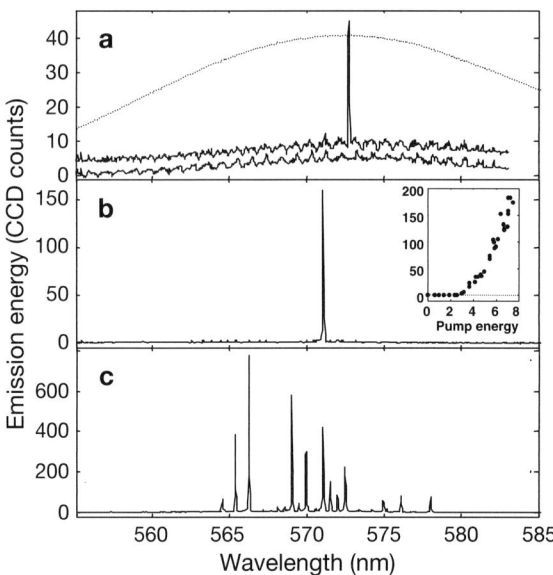

Figure 2.16 (a) Spectra of spontaneous emission in neat solution and of spontaneous emission and near-threshold lasing from Rhodamine 6G placed between layers of a glass slide stack at different laser pump energies. (b) Lasing in a single line above threshold and (c) in multiple lines well above threshold. The inset in (b) shows the sharp onset of lasing above threshold for excitation at a particular portion of the glass stack (Ref. [333]).

random stack with interspersed dye layers slightly below and above the lasing threshold. Below threshold, the spectrum shows resolution-limited peaks of the electromagnetic modes of the system. Above threshold, a collimated emission beam perpendicular to the sample layers was observed. The lasing spectrum with baseline shifted up for clarity is shown in Fig. 2.16(b). An abrupt change in the output power with increasing pump energy occurred at the lasing transition seen in the inset in Fig. 2.16(b). At low pump energies lasing occurred in a single narrow line (Fig. 2.16(b)), while at higher energies multimode lasing was observed (Fig. 2.16(c)) with wavelength and intensity varying randomly with the position of the pump beam on the sample surface. The lasing threshold was low enough for lasing to be observed with a chopped continuous wave Argon-ion laser beam at 3 W at 514.5 nm.

The role of resonance with localized modes at both pump and emission wavelengths is seen in the strong correlation of pump transmission and output laser power. Such strong correlation is the opposite to what would be expected for a nonresonant random laser in which peak emission would correspond with maximal absorption and so with reduced transmission.

Low-threshold lasing via emission into long-lived modes excited by a pump laser which penetrates deeply into a sample can be realized in periodic and nearly periodic structures. For 1D samples or for layered structures in which the dielectric function is modulated only along a single direction, gaps emerge in the frequency spectrum along the modulated direction. This is the case even when the layers are anisotropic with an orientation that varies with depth. The states at the edge of the band are long-lived and can be excited via emission from excited states of dopants in the periodic structure or of the structure itself when pumped by an external beam falling within the frequency range of the pass band. A coherent beam perpendicular to the layers then emerges without special alignment.

In an infinite structure, the group velocity vanishes as the band-edge is approached. This leads to the expectation of a lowered lasing threshold at the edge of a photonic band gap [117]. But in periodic structures of finite thickness, states at the band-edge are standing Bloch waves with a low number n of anti-nodes in the medium rather than traveling waves [247]. The intensity of the wave in each of these states is modulated by an envelope function, $\sin^2 n\pi x/L$, where x is the depth into the sample of thickness L. Lasing properties are determined by the modes with increasingly narrow linewidths and intensity at anti-nodes at resonance as the band-edge is approached. The width of modes increases as n^2 away from the band gap.

Band-edge lasing was demonstrated in dye-doped cholesteric liquid crystals (CLCs) [246]. Lasing from dye-doped CLCs was observed earlier and attributed to lasing at defect sites in the liquid crystal [205]. Roughly parallel rod-shaped molecules in CLCs with average local orientation of the long molecular axis, in a direction called the director, rotate with increasing depth into the sample. This periodic helical structure can be either right- or left-handed. The indices for light polarized parallel and perpendicular to the director are the extraordinary and ordinary refractive indices, n_e and n_o, respectively. For sufficiently thick films, the reflectance of normally incident, circularly polarized light with the same sign of rotation as the CLC structure is nearly complete within a band centered at vacuum wavelength $\lambda_c = nP$, where $n = (n_e + n_o)/2$ and P is the pitch of the helix, equal to twice the structure period. The reflected light has the same sign of rotation as the incident beam. The bandwidth is $\Delta\lambda = \lambda_c \Delta n/n$, where $\Delta n = n_e - n_o$. For circularly polarized light of opposite circular polarization, the wave is freely transmitted. In measurements on dye-doped cholesteric liquid crystal (CLC) films, the rate of spontaneous emission is inhibited within the band and enhanced at its edge for light polarized with the same handedness as the chiral structure. Light of opposite chirality is unaffected by the periodic structure. This makes it possible to make a direct measurement of the density of photon states by comparing the emission spectra of oppositely polarized radiation. The observed suppression of the density of states within the band and the sharp rise at the bandedge are shown

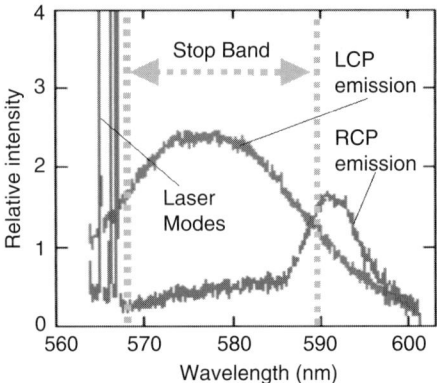

Figure 2.17 Left and right circularly polarized emission spectra from a right-handed dye-doped CLC sample as well as lasing emission at the short-wavelength edge of the reflection band. The height of the lasing lines is ~50 units (Ref. [246]).

in Fig. 2.17 and seen to be in good agreement with the calculated density of states in a 1D structure. The left circularly polarized (LCP) emission spectrum in this right-handed structure is due to the spontaneous emission of the PM-597 dye. RCP emission is suppressed in the stop band and peaked at the bandedges. The RCP light seen within the reflection band does not vanish because the emitted LCP light is converted to RCP light in Fresnel reflection from the surfaces of the glass sample holder. Multiple lasing lines are seen at the short-wavelength band-edge.

The lasing peaks in Fig. 2.17 do not correspond precisely with the modes of a perfectly periodic CLC structure. These modes are seen in transmission spectra in Fig. 2.18 in a dye-doped CLC sample which was carefully prepared and allowed to equilibrate. A comparison between transmissions measured with a tunable narrowband dye laser in a $37\,\mu$m thickness CLC sample with moderate absorption and simulations for a periodic system is shown in Fig. 2.18. The simulated spectrum is displaced vertically for visibility. In a nondissipative sample, the resonance transmission of all modes reaches unity. In Fig. 2.18, transmission through modes closest to the band-edge is most suppressed by absorption since these modes are longest-lived. Since the modes closest to the band-edge are longest-lived in nearly periodic systems, these states are most susceptible to being localized by disorder. Such localized states are often longer-lived than the corresponding states of a periodic system and so disorder can help as well as hinder lasing.

Simulations in random amplifying systems show that it is possible to maximize the lasing intensity at a particular frequency in the spectrum of a random laser by iteratively feeding back the intensity at a selected frequency to vary the intensity distribution of the pump beam [29]. The modes of the sample are not substantially modified in the lasing transition, but the spectral properties of the modes excited by

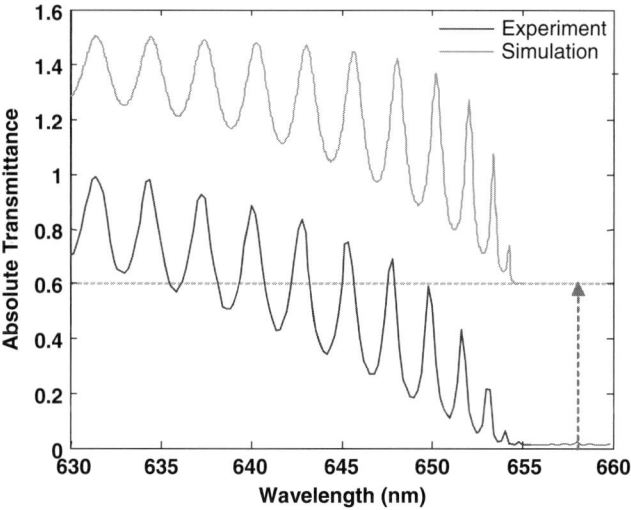

Figure 2.18 Comparison of measurements of transmission spectra in dye-doped CLC taken by Valery Milner in the lower curve, with simulations in the upper curve. The linewidth narrows as the index n of the mode away from the band-edge decreases. Differences between the frequencies of laser lines seen in Fig. 2.17 and frequencies of lines in high quality CLC samples in this figure are due to disorder in the sample of Fig. 2.17.

the pump beam are selected by the spatial profile of the pump beam. Türeci *et al.* have shown that modes of passive diffusive systems interact via the gain medium to create a uniform spacing between laser spectra [486]. In contrast, isolated modes of localized lasers interact weakly and emit at frequencies pegged to the modes of the passive systems [456].

2.5 Channels

Transmission through a disordered medium is fully determined by the transmission matrix t [38, 114, 327, 464]. The optical transmission matrix was measured by Popoff *et al.* [393] with the use of a spatial light modulator (SLM) and an interference technique to find the amplitude and phase of the optical field. Measuring the transmission matrix allows one to focus the transmitted light at a desired channel at the output surface by phase conjugating the transmission matrix [109, 393]. In this way, the transmitted field from different input channels arrives in phase at the focal spot and interferes constructively. The presence of a random medium can increase the number of independent channels that illuminate a point so that the focused intensity and resolution are enhanced [88, 499, 508]. Because of the enormous number of channels in optical experiments typically, only a small portion of the

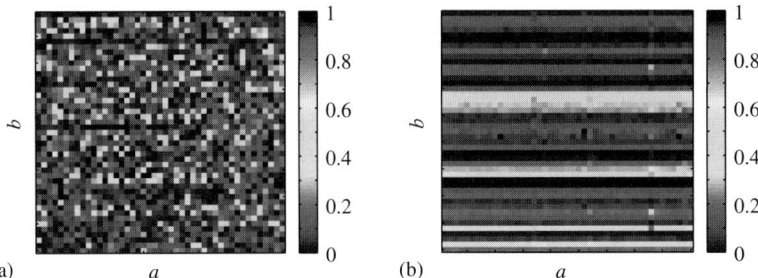

Figure 2.19 Intensity normalized to the peak value in each speckle pattern generated by sources at positions a are represented in the columns with index of detector position and polarization b for (a) diffusive and (b) localized waves (Ref. [110]).

transmission matrix is measured. In this case, the distribution of the singular values of the transmission matrix follows the quarter circle law, which is characteristic for uncorrelated Gaussian fluctuations of the elements of the transmission matrix [324, 500].

Measurements of microwave radiation propagation through random media confined in a waveguide allow us to measure the field on a grid of points for the source and detector [432]. The closest spacing between points is approximately the distance at which the field correlation function vanishes so that the fields at different points on the gird are only weakly correlated. The number of independent channels N supported in the empty waveguide is ~66 in the frequency range 14.7–14.94 GHz in which the wave is diffusive, and ~30 from 10–10.24 GHz in which the wave is localized within the sample. To construct the transmission matrix, $N/2$ points are selected from each of two orthogonal polarizations. A representation of intensity patterns in typical transmission matrices for both diffusive and localized waves at a given frequency is presented in Fig. 2.19. Each column presents the variation of intensity across the output surface at points b for a source at points a with two orthogonal polarizations. The intensity in each column shown in Fig. 2.19 is normalized by its maximum value. For localized waves, intensity patterns in each column are similar, indicating that transmission is dominated by a single channel. In contrast, no clear pattern is seen for diffusive waves since many channels contribute to the intensity at each point.

In Fig. 2.20, we show a spectrum of the optical transmittance and the underlying transmission eigenvalues from a single random realization for both localized and diffusive waves. This confirms that the highest transmission channel dominates the transmittance for localized waves while several channels contribute to transmission for diffusive waves. Thus for localized waves, the incident wave from different channels couples to the same eigenchannel and excites the same pattern

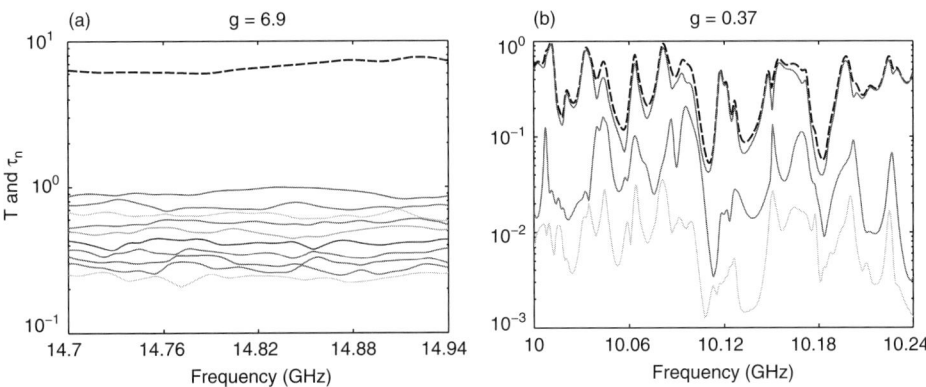

Figure 2.20 Spectra of the transmittance T and transmission eigenvalues τ_n for (a) a diffusive sample of $L = 23$ cm with $g = 6.9$ and (b) a localized sample of $L = 40$ cm with $g = 0.37$. The black dashed line gives T and the solid lines are spectra of τ_n (Ref. [432]).

in transmission as seen in Fig. 2.19(b). In contrast, the transmission patterns for incident waves for different incident channels are the sum of many orthogonal eigenchannels, so that the transmitted patterns are weakly correlated.

Dorokhov [113, 114] showed that, the spacing between the inverse of the localization length for adjacent eigenchannels is equal to the inverse of the localization length of the sample, $1/\xi_{n+1} - 1/\xi_n = 1/\xi$. For localized waves, this is equivalent to $\langle \ln \tau_n \rangle - \langle \ln \tau_{n+1} \rangle = L/\xi = 1/g_0$, where g_0 is the bare conductance that one would obtain in the absence of wave interference and the transport can be described in terms of diffusion of particles. In Fig. 2.21, we show that $\langle \ln \tau_n \rangle$ falls linearly with respect to the channel index n for both diffusive and localized waves. We denote the constant spacing between adjacent values of $\langle \ln \tau_n \rangle$ as $1/g''$, $\langle \ln \tau_n \rangle - \langle \ln \tau_{n+1} \rangle = 1/g''$. This supports the conjecture that g'' is the same as the bare conductance.

We expect that the bare conductance should be influenced by the wave interaction at the sample interface [264, 297, 565]. The wave interaction at the sample boundary can be described by a diffusion model [297] in which the incident wave is replaced by an isotropic source at a distance z_p from the interface in which the wave direction is randomized, and with a length z_0, which is the length beyond the sample boundary at which the intensity inside the sample extrapolates to zero. z_0 was found by fitting the time of flight distribution of a wave through random media [79, 146]. Once the surface effect is taken into account, the bare conductance is given as: $g = \eta \xi / L_{eff}$, where η is a constant of order of unity and L_{eff} is the effective sample length. The constant value of $g'' L_{eff}$, seen in Fig. 2.22 is consistent with g'' being the bare conductance and gives the localization length

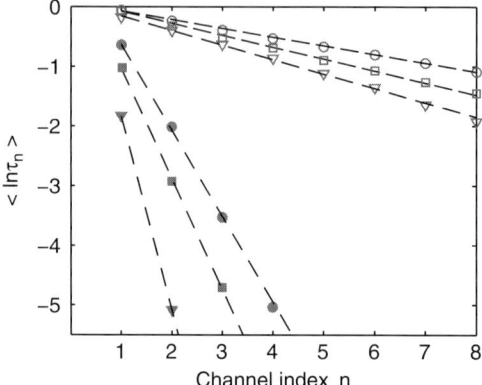

Figure 2.21 Variation of $\langle \ln \tau_n \rangle$ with channel index n for sample lengths $L = 23$ (circle), 40 (square), and 61 (triangle) cm for both diffusive (open symbols) and localized (solid symbols) waves fitted, respectively, with dashed lines (Ref. [432]).

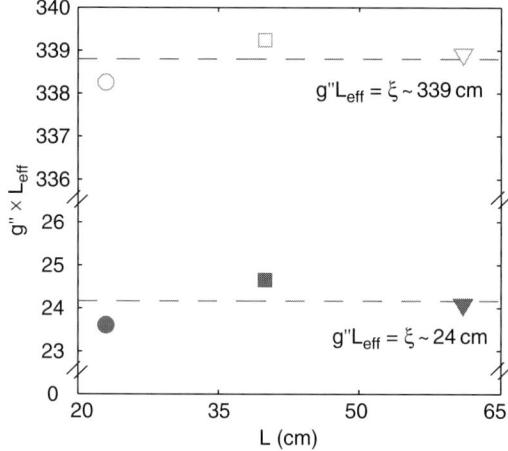

Figure 2.22 The constant products of $g''L_{eff}$ for three different lengths, for both diffusive and localized samples, give the localization length ξ in the two frequency ranges (Ref. [432]).

for the samples at two frequency lengths. The absolute values of the transmittance T and of the underlying transmission eigenvalues τ_n are obtained by equating $\langle T \rangle = Cg''$ for the most diffusive sample of length $L = 23$ cm, at which the renormalization of dimensionless conductance due to wave localization is negligible. The normalization factor C is used to determine the values of g for other samples.

The probability density of individual $\ln \tau_n$ of the first few eigenchannels and their contribution to the overall density $\ln \tau$ is given in Fig. 2.23 for the most diffusive sample with $g = 6.9$. Aside from the fall of the probability $P(\ln \tau)$ near

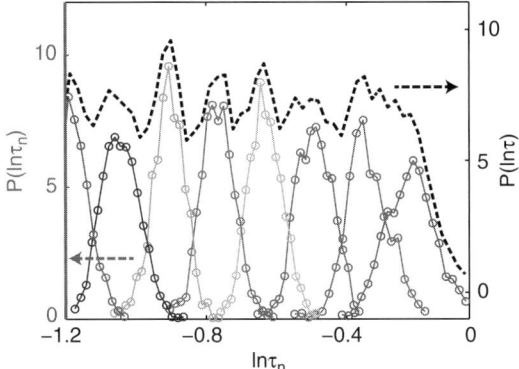

Figure 2.23 Probability density of $\ln \tau_n$ (lower curves) and the density of $\ln \tau$ (top dashed curve), $P(\ln \tau) = \sum_n P(\ln \tau_n)$ for the diffusive sample with $g = 6.9$ (Ref. [432]).

$\ln \tau \sim 0$, which reflects the restriction $\tau_1 \leq 1$, $P(\ln \tau)$ is nearly constant with ripples spaced by $1/g''$. The nearly uniform density of $P(\ln \tau)$ corresponds with a probability density $P(\tau) = P(\ln \tau)\frac{d\ln \tau}{d\tau} = g/\tau$. This distribution has a single peak at low values in contrast with the predicted bimodal distribution, which has a second peak at nearly unity [114, 327, 351, 464]. This may reflect the fundamental difference in measuring the transmission matrix based on scattering between independent discrete points instead of waveguide modes. In theoretical calculations in which scattering between waveguide modes is used, all of the transmitted energy can be captured. However, only a fraction of energy transmitted through the disordered medium is captured when the TM is measured on a grid of points. As a result, full information is not available and the measured distribution of transmission eigenvalues does not accurately represent the actual distribution in the medium. In particular the bimodal distribution of transmission eigenvalues is not observed. This has been suggested in recent simulation of a scalar wave propagation in Q1D samples, based on a recursive Green's function method. Goetschy and Stone [167] have recently calculated the impact of degree of control of the transmission channels on the density of transmission eigenvalues. The matrix t is mapped to $t' = P_2 t P_1$, where P_1 and P_2 are $N \times M_1$ and $M_2 \times N$ matrices which eliminate $N - M_1$ columns and $N - M_2$ rows, respectively, of the original random matrix t. Therefore, only $M_1(M_2)$ channels are under control on the input (output) surface, respectively, and the degree of control on the input and output surfaces is measured by M_1/N (M_2/N). As a result, the density of transmission eigenvalues for diffusive samples changes from a bimodal distribution to the distribution characteristic of uncorrelated Gaussian random matrices, when the degree of control is reduced [234, 393, 432]. Nevertheless, key aspects of the statistics of wave propagation and the limits of control of the transmitted wave can be explored using measurements of the transmission matrix.

Measurement of the transmission matrix allows us to explore the statistics of transmittance, the most spatially averaged mesoscopic quantity. The importance of sample-to-sample fluctuation of conductance in disordered conductors was first recognized in conduction mediated by localized states, but was soon observed in universal conduction fluctuations which give a constant value for the variance of the conductance in diffusive samples [3, 8, 12, 207, 278]. For diffusive waves, for which a number of transmission eigenchannels contribute substantially to the transmittance, the probability distribution of T is Gaussian, with variance independent of the mean value of T and of sample dimensions. In the localization limit, $L/\xi \gg 1$, in which transmittance T is dominated by the largest transmission eigenvalues, $T \sim \tau_1$, the single parameter scaling (SPS) theory of localization predicts that the probability distribution of the logarithm of transmittance in 1D samples is a Gaussian function with a variance equal to the average of its magnitude, $\text{var}(\ln T) = -\langle \ln T \rangle$. Therefore, the scaling of the average conductance and the entire distribution of conductance are determined by the single parameter $|\langle \ln T \rangle|/L = 1/\xi$. In recent work, the ratio, $\mathcal{R} \equiv -\text{var}(\ln T)/\langle \ln T \rangle$, is found to approach unity in Q1D samples, showing that propagation in Q1D in this limit is one-dimensional [434].

In the Q1D geometry, there is no phase transition between localization and diffusion since the wave always becomes localized when the length is increased. Instead, there exists a crossover from diffusive to localized regime. For samples just beyond the localization threshold, where only a few transmission eigenchannels contribute appreciably to the transmittance, numerical simulation [148, 317, 385, 444, 451] and random matrix theory calculations [349] by Muttalib and Wölfle found a one-sided log-normal distribution for the transmittance. The source of this unusual probability distribution of conductance can be understood with the aid of the charge model proposed by Stone *et al.* [464]. The charge model was first introduced by Dyson [118] to visualize the repulsion between eigenvalues of the large random Hamiltonian. In this model, transmission eigenvalues τ_n are associated with positions of parallel line charges at x_n and their images at $-x_n$ embedded in a compensating continuous charge distribution. The transmission eigenvalues are related to the x_n via the relation $\tau_n = 1/\cosh^2 x_n$. The repulsion between two parallel lines of charge of the same sign with potential $\ln|x_i - x_j|$ mimics the interaction between eigenvalues of the random matrix. The oppositely charged jellium background provides an overall attractive potential that holds the structure together. The repulsion between charges for diffusive waves is the origin of universal conductance fluctuation. For localized waves, the charges are separated by a distance greater than the screening length due to the background charges, so the "Coulomb" interaction is screened. The repulsion between the first charge at x_1 is associated with the highest transmission eigenvalues τ_1 and its image placed at $-x_1$ provides a ceiling of unity.

We have recently reported microwave measurements of the probability distribution of the "optical" transmittance T in the crossover from diffusive to localized waves [434]. A Gaussian distribution is found for the diffusive waves and nearly a log-normal distribution for deeply localized waves. Just beyond the localization threshold, a one-sided log-normal distribution is observed for an ensemble with $g = 0.37$. In this ensemble, an exponential decay of $P(T)$ is found for high values of transmittance, as was found in simulations and calculations [169]. The rapid falloff of $P(T)$ for $T > 1$ is due to the requirement that two eigenvalues need to be high. This requires that two charges as well as their images be close to the origin. The probability for high values of T is therefore greatly suppressed due to the repulsion between these charges.

Measurements of the transmission matrix provide the opportunity to investigate the statistics in single disordered samples as opposed to the statistics of ensembles of random samples. Such statistics are essential in applications such as imaging and focusing through a random medium. In the Q1D geometry, in which the wave is completely mixed within the sample, the statistics of the intensity relative to the average over the transmitted speckle pattern, $T_{ba}/(\sum_{b=1}^{N} T_{ba}/N) = NT_{ba}/T_a$, are independent of source or detector positions [76, 560]. Because of the Gaussian distribution of the field in any single speckle pattern, the probability distribution of relative intensity is $P(NT_{ba}/T_a) = \exp(-NT_{ba}/T_a)$. Thus the statistics of relative intensity are universal, the statistics of transmission in a sample with transmittance T would be completely specified by the statistics of total transmission T_a relative to its average (T/N) within the sample.

We find in random matrix calculations that the variance of normalized total transmission within a large single instance of transmission matrix is equal to the eigenchannel participation number [110],

$$var(NT_a/T) = M^{-1}. \tag{2.2}$$

These results can be compared with measurements in samples of small N by grouping together measurements in collections of samples with similar values of M. We show in Fig. 2.24 that the average of $var(NT_a/T)$ in subsets of samples with given M^{-1} is in excellent agreement with Eq. (2.2). $var[var(NT_a/T)/M^{-1}]$ is seen in the insert of Fig. 2.24 to be proportional to $1/N$, indicating that fluctuations in the variance over different subsets are Gaussian with a variance that vanishes as N increases.

The central role played by M can be appreciated from the plots shown in Fig. 2.25 of the statistics for subsets of samples with identical values of M, but drawn from ensembles with different values of g. The distributions $P(NT_a/T)$ obtained for samples with M^{-1} in the range 0.17 ± 0.01 selected from ensembles with $g = 3.9$ and 0.17 are seen to coincide in Fig. 2.25(a) and thus to

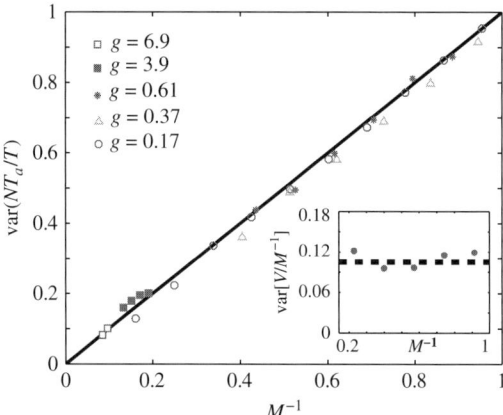

Figure 2.24 Plot of $\mathrm{var}(NT_a/T)$ computed within transmission matrices over a subset of transmission matrices with a specified value of M^{-1} drawn from random ensembles with different values of g. The straight line is a plot of $\mathrm{var}(NT_a/T) = M^{-1}$. In the inset, the variance of V/M^{-1} is plotted versus M^{-1}, where $V = var(NT_a/T)$ (Ref. [110]).

depend only on M^{-1}. The curve in Fig. 2.25(a) is obtained from an expression for $P(s_a)$ for diffusive waves given in Refs. [241, 353], in terms of a single parameter $g = 2/3\mathrm{var}(s_a)$, but with substitution of $2/3M^{-1}$ for g. The dependence of $P(NT_a/T)$ on M^{-1} alone and its independence of T is also demonstrated in Fig. 2.25(b) for M^{-1} over the range 0.995 ± 0.005 from measurements in samples of different length with $g = 0.37$ and 0.17. Since a single channel dominates transmission in the limit, $M^{-1} \to 1$, we have $NT_a/T = |v_{1a}|$, where v_{1a} is the element of the unitary matrix V which couples the incident channel a to the highest transmission channel. The Gaussian distribution of the elements of V leads to a negative exponential distribution for the square amplitude of these elements, and similarly to $P(NT_a/T) = \exp(-NT_a/T)$, which is the curve plotted in Fig. 2.25. In Fig. 2.25, we plot the relative intensity distributions $P(N^2T_{ba}/T)$ corresponding with the same collection of samples as in Fig. 2.25. The curves plotted are the intensity distributions obtained by mixing the distributions for $P(NT_a/T)$ shown in Fig. 2.25 with the universal negative exponential function for the intensity of a single component of polarization.

In addition, we find in microwave measurements in Q1D samples that the SPS ratio $\mathcal{R}$ is equal to the average of M weighted by T, $\langle MT \rangle / \langle T \rangle$, which approaches unity for $L \gg \xi$ [434]. The statistics of relative transmission within a single transmission matrix depends only upon the single parameter M, while the transmittance T serves as an overall normalization factor. Therefore, the statistics of intensity and total transmission over random ensemble is given by the joint probability distribution of T and M.

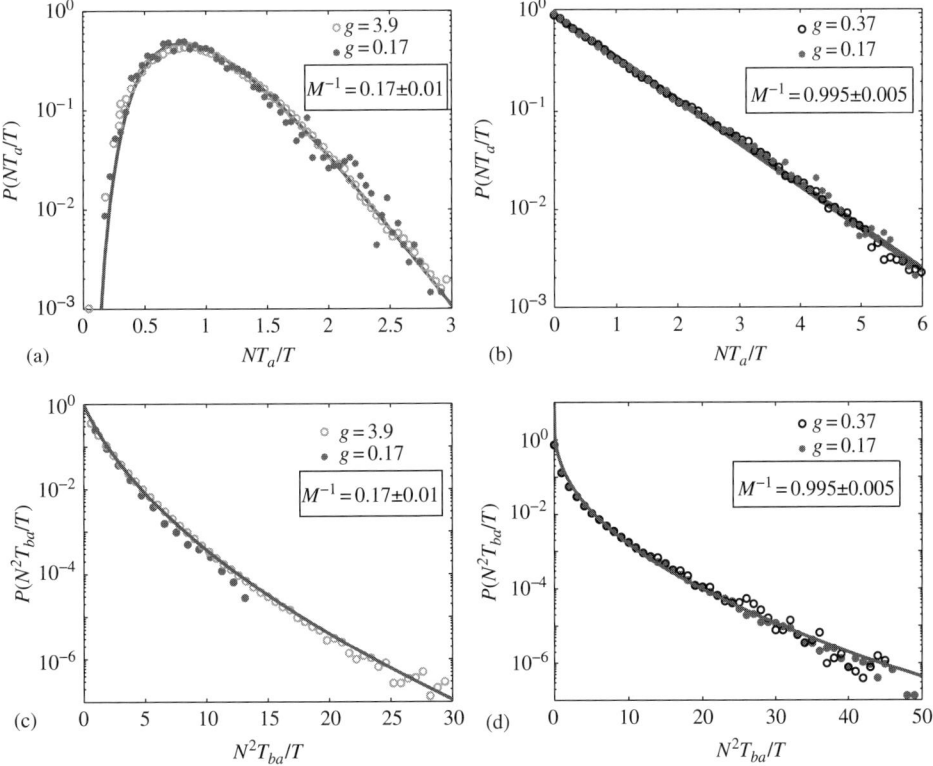

Figure 2.25 (a) $P(NT_a/T)$ for subsets of transmission matrices with $M^{-1} = 0.17 \pm 0.01$ drawn from ensembles of samples with $L = 61$ cm in two frequency ranges in which the wave is diffusive (circles) and localized (filled circles). The curve is the theoretical probability distribution of $P(s_a)$ in which var(s_a) is replaced by M^{-1} in the expression for $P(NT_a/T)$ in Refs. [353 and 241]. (b) $P(NT_a/T)$ for M^{-1} in the range 0.995 ± 0.005 computed for localized waves in samples of two lengths: $L = 40$ cm (circles) and $L = 61$ cm (filled circles). The straight line represents the exponential distribution, $\exp(-NT_a/T)$. (c, d) The intensity distributions $P(N^2 T_{ba}/T)$ are plotted under the corresponding distributions of total transmission in (a) and (b) (Ref. [110]).

2.6 Focusing

Focusing waves through random media was first demonstrated in acoustics by means of time reversal [133]. The amplitude and phase of the transmitted signal in time for an incident pulse from a source are picked up by arrays of transducers. The recorded signal is then played back in time and a pulse emerges at the location of the source. Recently, Vellekoop and Mosk [509] focused monochromatic light through opaque media by shaping the incident wavefront. Employing a genetic algorithm with a feedback from the intensity at the target point to adjust the phase of the incident wavefront, the intensity at the focus was enhanced by three orders of magnitude. The wavefront shaping method has been extended to focus optical pulses through random media at a spatial target at a selected time delay [21, 230].

In order to focus a wave at a target channel β, once the field transmission matrix has been measured, one simply conjugates the phase of the incident field relative to the transmitted field at β, yielding $t_{\beta a}^*/\sqrt{T_\beta}$ for the normalized incident field. Here, the incident field is normalized by $T_\beta = \sum_{a=1}^{N} |t_{\beta a}|^2$ to set incident power to unity. In this way, the fields from different incident channels a arrive at the target in phase and interfere constructively. Random matrix calculations confirmed by microwave measurements show that the contrast between the average intensity at the focal spot $\langle I_\beta \rangle$ and the background intensity $\langle I_{b \neq \beta} \rangle$, $\mu = \langle I_\beta \rangle / \langle I_{b \neq \beta} \rangle$, depends upon the eigenchannel participation number M and the size of the measured transmission matrix N [110],

$$\mu = \frac{1}{1/M - 1/N}. \tag{2.3}$$

This expression for the contrast is confirmed in measurements shown in Fig. 2.26. This expression is still valid when the size of measured transmission

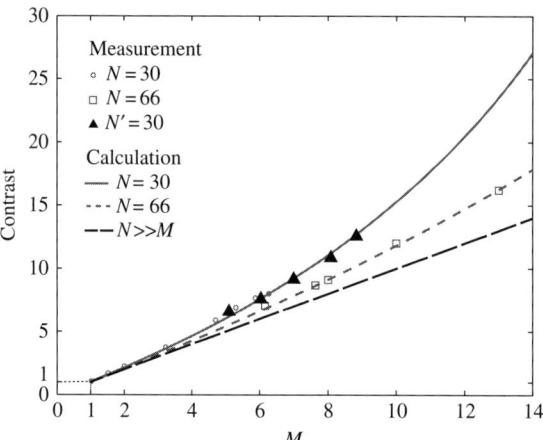

Figure 2.26 Contrast in maximal focusing versus eigenchannel participation number M. The open circles and squares represent measurements from transmission matrices $N = 30$ and 66 channels, respectively. The filled triangles give results for $N' \times N'$ matrices with $N' = 30$ for points selected from a larger matrix with size $N = 66$. Phase conjugation is applied within the reduced matrix to achieve maximal focusing. Equation (2.3) is represented by the solid and dashed curves for $N = 30$ and 66, respectively. In the limit $N \gg M$, the contrast given by Eq. (2.3) is equal to M, which is shown as the long-dashed line (Ref. [110]).

matrix N' is smaller than N and the corresponding M' is correspondingly smaller than M. This is demonstrated by constructing a matrix of size $N' = 30$ from the measured transmission matrix of size $N = 66$ and calculating the contrast by phase conjugating the transmission matrix of size N'. The contrast computed falls on the curve for $N = 30$ for different values of M'. These results may be applied to optical measurements of the transmission matrix in which the size of the measured matrix N' is generally much smaller than N. In the limit $N \gg M$, the contrast approaches M.

These results indicate that localized waves cannot be focused via phase conjugation because the value of M is close to one. This is shown in Fig. 2.27, in which phase conjugation has been applied to focus the transmitted wave at the center of the output surface for both diffusive and localized waves. Only for diffusive waves does a focal spot emerge from the background. We have recently demonstrated the use of phase conjugation to focus pulsed transmission through random media. By phase conjugating a time-dependent transmission matrix at a selected time delay, a pulse can be focused in space and time [431].

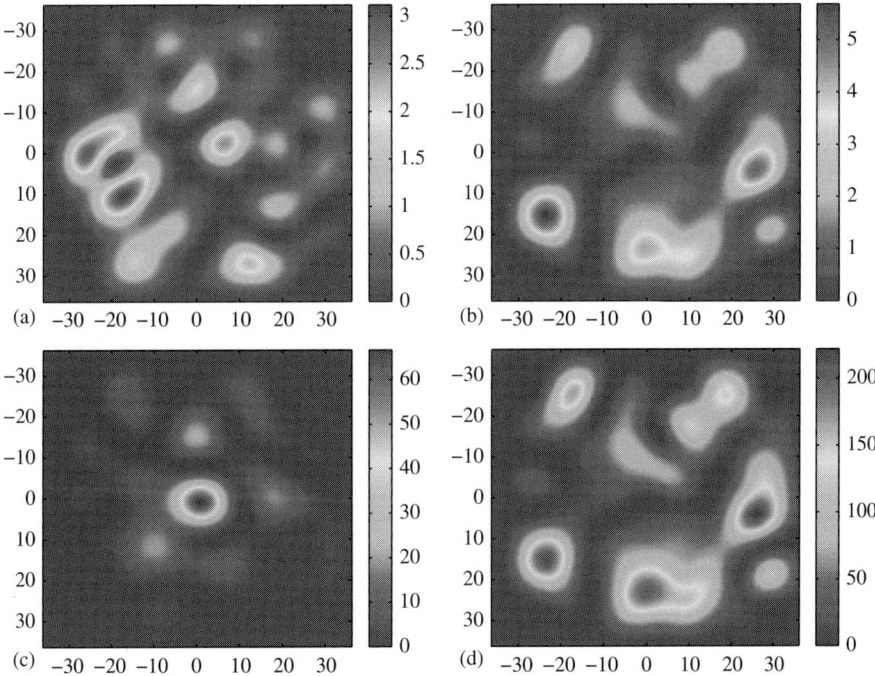

Figure 2.27 Intensity speckle pattern generated for $L = 23$ cm for diffusive waves (a) and for $L = 61$ cm for localized waves (b), normalized to the average intensity in the respective patterns. Focusing at the central point at the same frequency as in (a) and (b) via phase conjugation is displayed in (c) and (d) with 66 and 30 input points, respectively (Ref. [109]).

2.7 Conclusion

In this chapter, we have explored the mode and channel approaches to waves in random media. We believe that each of these approaches has the potential to provide a full description of transmission and its relation to the wave within the sample and that this will be of use in a wide variety of applications.

In recent work, we have considered four statistical characteristics of modes that have proven to be particularly promising for explaining steady-state and pulsed transmission and will be reported elsewhere. These characteristics are the statistics of the spacing and widths of modes, the degree of correlation in the speckle patterns of modes, and the mode transmittance. Correlation between speckle patterns includes correlation between the intensity patterns of modes, as well as the average phase difference and the standard deviation of phase shift between these patterns. The mode transmittance represents the transmittance integrated over frequency for a particular mode and is obtained from a modal decomposition of the transmission matrix based on measurements of field transmission spectra between sets of points on the incident and output surfaces. The analysis of waves into modes is of particular interest in emission and lasing since it gives the density of states, which is a key factor in the emission cross-section as well as the lifetimes of modes. The relationship between modes and transmission eigenchannels can be elucidated by expressing the transmission eigenchannels at each frequency as the sum of projections of the eigenchannels of the transmission matrix for the individual modes upon the transmission eigenchannel [433]. Precisely exciting a particular mode with a desired spatial distribution provides promise for control over energy deposition and collection within random media.

The relationship between modes and transmission eigenchannels can be seen from the equality of the density of states obtained from the sum of the contributions of modes and of eigenchannels. The density of quasi-normal modes or resonances of a region per unit angular frequency, is the sum over Lorentzian lines, $\rho(\omega) = 1/\pi \sum_n \frac{\Gamma_n/2}{(\Gamma_n/2)^2 + (\omega - \omega_n)^2}$. This is found from the central frequencies and linewidths determined from a modal decomposition of fields at any points in the medium. The density of states can also be obtained from the sum of the contributions of each transmission eigenchannel, which are the derivatives with angular frequency of the composite phase shift of the eigenchannel, $\rho(\omega) = 1/\pi \sum_{n=1}^{N} \frac{d\theta_n}{d\omega}$. The phase derivative is the intensity weighted phase derivative between all channels on the incident and output surfaces [111]. $\frac{d\theta_n}{d\omega}$ is the transmission delay time for the nth transmission eigenchannel. When a complete measurement of the transmission matrix is made, $\frac{d\theta_n}{d\omega}$ is the integral of intensity inside the sample for the corresponding eigenchannel. The eigenchannel delay time and the associated intensity integral inside the sample increases with the transmission eigenvalue τ_n. The density of states may be accurately measured from the transmission matrix as long as $N' > M$. We have also explored the distribution of transmission eigenvalues and seen that in a particular transmission matrix, the statistics of relative transmission depend only upon M. The absolute distribution within a single matrix then depends upon these two parameters M and T. Thus the distribution over a random ensemble of all transmission quantities depends only upon the joint distribution of M and T. This represents a considerable simplification from the joint distribution of the full set of transmission eigenvalues τ_n. Manipulation of the incident beam with knowledge of the transmission matrix makes it possible to achieve maximal focusing in a single transmission matrix, with the peak intensity depending only upon T and the contrast depending upon the value of M in the measured matrix and the dimension of this matrix. Knowledge of the spectra of both modes and channels may advance control over the wave projected within and through opaque samples for applications in imaging, and energy collection and delivery.

Acknowledgments

We would like to thank Jing Wang, Matthieu Davy, Patrick Sebbah, Valery Milner, Victor Kopp, Andrey Chabanov, Zhao-Qing Zhang, and Jerry Klosner for many stimulating discussions and for contributions to many of the results reviewed here. We thank the National Science Foundation for support under Grant Number DMR-1207446.

3

Modes structure and interaction in random lasers

MARCO LEONETTI AND CEFE LÓPEZ

3.1 From lasers to random lasers

Lasers are among the most useful and popular of all optical devices, with countless applications ranging from biology to astronomy. First predicted by Letokhov [292] and measured experimentally by Lawandy [269] and others [174, 319], random lasers [495, 529] connected for the first time the physics of ordinary lasers with that of disordered systems, boosting spectacularly in the early 1990s the interest of the scientific community in complex photonics. The possibility of using intrinsically disordered structures to create novel optical systems is attractive, not only from the applications point of view, but also fundamentally, allowing scientists to connect the theoretical paradigms of complexity, nonlinearity, disorder, and even glass physics with photonics.

   Laser stands for Light Amplification by Stimulated Emission of Radiation, thus the amplifier is its fundamental element. A coherent optical amplifier is capable of increasing the amplitude of an optical field while maintaining its phase. If a monochromatic beam is injected into such a device the output will have the same frequency, while the phase can be the same or shifted by a fixed amount. In contrast, an amplifier that increases the intensity of an optical wave without preserving the phase is an incoherent amplifier. An amplifier based on stimulated emission is coherent: the stimulation process allows a photon in a given mode to induce an atom lying in an excited state to undergo a transition to a lower energy level, emitting a photon that is identical to the exciting photon, thus preserving frequency, direction, polarization and phase. If stimulation happens in a material in which the population is inverted (i.e. the majority of the atoms lie at the excited level), then an avalanche process in which every photon creates a duplicate of itself, is ignited exponentially, increasing the amplitude in the mode. In a single-pass amplifier, that is, a system in which light makes a single run through

Light Localisation and Lasing, ed. M. Ghulinyan and L. Pavesi. Published by Cambridge University Press.
© Cambridge University Press 2015.

a medium previously prepared in the population inverted state, the total amplification $G(\omega)$ (the ratio between the input and the output intensities) depends on the input radiation frequency ω, increases exponentially with the optical path inside the gain material L, and depends also on the imaginary part of the susceptibility χ'', which usually has a Lorentzian lineshape centered around the resonance frequency ω_a [438]:

$$\chi''(\omega) = \frac{\chi_0''}{1 + [2(\omega - \omega_a)/\Delta\omega_a]^2}. \tag{3.1}$$

Because stimulated emission occurs only when the photon energy is equal to the atomic transition energy, the amplifier itself already works as a bandpass, selecting for amplification only those atoms that generate photons with an energy corresponding to the stimulating photon.

The single pass gain $G(\omega)$ is thus

$$G(\omega) = \exp\left[\frac{\omega L \chi_0''}{c} \times \frac{1}{1 + [2(\omega - \omega_a)/\Delta\omega_a]^2}\right]. \tag{3.2}$$

The second element of a standard laser is the resonator. The function of this element is two-fold: on the one hand it re-injects the light in the gain material, increasing the effective optical path length for amplification, and on the other it works as a filter when emission takes place in broad wavelength bands. For commercial lasers, in which a pair of mirrors is used to form the cavity, only the resonant wavelengths of the cavity modes, which give rise to standing waves between the mirrors, are selected for amplification. If the emission band is narrower than the mode spacing, only one mode will be amplified.

The greatest difference between a standard laser and a random laser resides in the resonator that is replaced by a multiple scattering medium. That is, feedback is implemented with a multiple scattering process that prevents the radiation from escaping the gain medium. Although this kind of resonator fulfills the first of the resonator's tasks, effectively enlarging the amplification path length, it does not carry out the second: in principle, all frequencies are returned back into the gain medium with and no resonant selection is performed. Indeed, this suppression of resonant effects was pursued in the initial stages that led to the concept of random lasers [10].

In practice, the disordered structure confines the light in the gain (population inverted) area forcing photons to flee by diffusing throughout very long random walks (see an artist's view of the random lasing phenomenon in Fig. 3.1). The simplest way to obtain such a diffusive laser in a laboratory is to mix a gain medium with a strongly scattering one. Earlier successful experiments involved titanium dioxide powder (one of the strongly scattering materials that possess a refractive

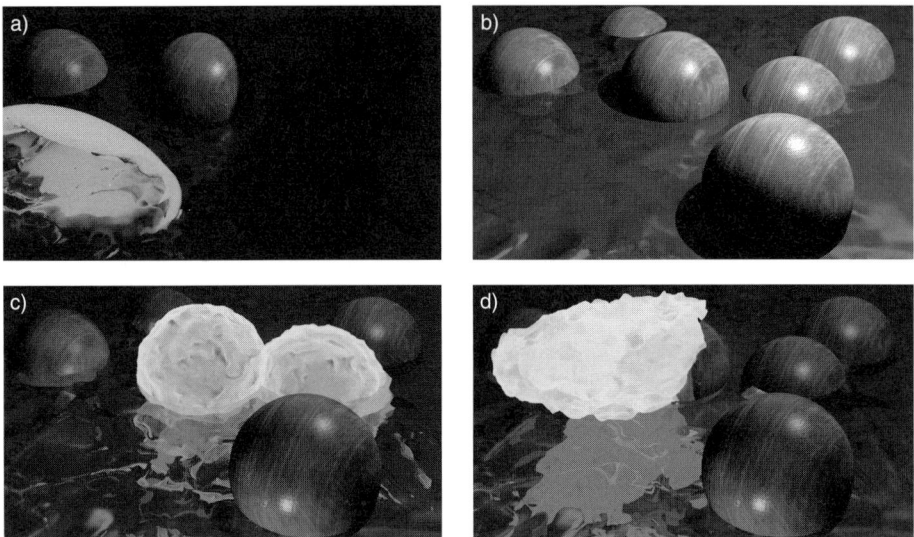

Figure 3.1 An artist's view of the random lasing process. In panel (a) an excitation
pulse impinges on a sample consisting of a set of dielectric particles submerged in
a pool of liquid dye. In (b) the active medium has absorbed the pulse and started
to glow due to fluorescence emission. In (c) light is trapped into a resonant cavity
consisting of two scatterers, while in (d) part of the resonant light is emitted in the
far field.

index up to 2.7) mixed with a rhodamine doped solution, or ground laser powder,
acting as both lasing and scattering material. By pumping such a system with a
pulsed laser the resulting emission is equal to the active media fluorescence (hundreds of nanometers in width), while a line-narrowed smooth, laser-like emission
spectrum is found at high pumping power.

Nearly two decades have passed since the first measurement of the random lasing
phenomenon and the simplest formulation of diffusive theoretical models. Many
novel experiments and theoretical formulations have been proposed since then, and
the physical understanding of this extremely complex system has grown with time.
In the following we will detail the various regimes of random lasing that have been
studied so far, each of which can disclose different aspects of this complex and
debated phenomenon.

3.2 Random lasing regimes

3.2.1 Diffusive random lasers

The first (and the simplest) approach to describing a random laser was that
proposed by Letokhov [292], based on light diffusion with gain. From this simple

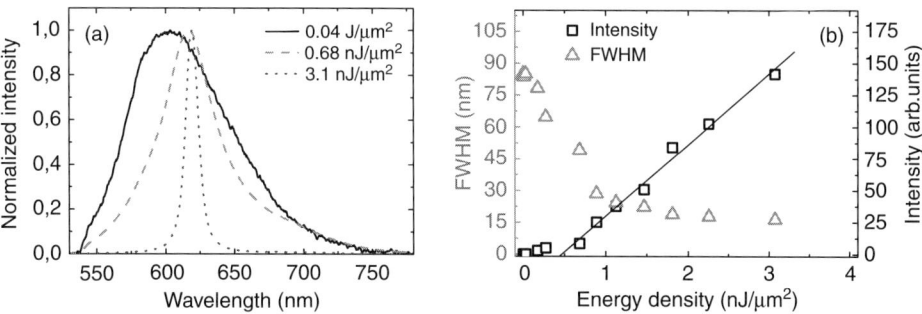

Figure 3.2 Typical energy behavior of random lasing emission. (a) Emission spectrum at different energies from a macroscopic random lasing. The fluorescence broadband emission is replaced by the line-narrowed peak at higher pumping energies. In (b): full width at half maximum and emission intensity as a function of the pump energy. The linear fit of the high energy points allow to individuate the kink corresponding to the lasing threshold. Further increase of pumping energy leads to saturation due to lack of enough excited molecules and departure from the linear behavior. Data from [290].

model it may be derived that the total gain is proportional to the number of active atoms and thus to the volume of the system; whereas the losses are determined by the diffusion across the boundary of the system and are thus proportional to the total surface. Since the volume grows as the cube of the size, whereas the surface grows as the square, their ratio is linear so that there will be a size below which surface (loss) dominates, while for a larger system volume (gain) will prevail. The critical volume above which gain overcomes losses is the threshold volume (see a typical random lasing spectrum and line narrowing behavior in Fig. 3.2).

In the two competing effects involved there are several factors to be taken into account: material gain determines how rapidly intensity is amplified, and scattering determines the propability that a photon escapes from the system, that is, how long the photon paths tend to be; whereas absorption determines the likelihood of a photon being absorbed. The relevant length scales that describe a diffusive random laser [533] are the scattering mean free path ℓ_s, defined as the average distance between two scattering events; the transport mean free path ℓ, defined as the average distance a wave travels before its direction of propagation is randomized; the gain length ℓ_g; and the amplification length ℓ_{amp}. The gain length is defined as the path length over which the intensity is amplified by a factor e. The amplification length is defined as the rms average distance between the beginning and ending points for paths of length ℓ_g :

$$\ell_{amp} = \sqrt{\frac{\ell\ell_g}{3}}.$$

(3.3)

The amplification length l_{amp} and gain length ℓ_g are the analogues of the absorption length ℓ_{abs} and the inelastic length ell_i. For an amplifying random medium one can define a critical volume above which the system becomes unstable. For example, in the case of a slab geometry the critical thickness L_{cr} is given by

$$L_{cr} = \pi \ell_{amp} = \pi \sqrt{\frac{\ell \ell_g}{3}}. \tag{3.4}$$

The relevant magnitude here is the light transport mean free path in relation to the sample size. When the mean free path is much larger than the sample size, ordinary lasing or amplified spontaneous emission is obtained, whereas if the mean free path is comparable with the sample size, light undergoes many scattering events prior to leaving the sample, which results in random lasing. An extraordinary regime is expected when the scattering mean free path is comparable to or shorter than the wavelength [531] in which exponentially localized modes are responsible for lasing. This limit has been observed in one- and two-dimensional lasers but has never been reached in 3D systems.

In many active media, gain is wavelength dependent (rhodamine, for example, one of the most efficient laser dyes, has an absorption band tens of nanometers broad, peaked at around 600 nm) so that diffusion theory predicts line-narrowing. This is a result of the fact that transitions with energy in the center of the gain peak are more efficient catching available photons faster with respect to less efficient transitions in the tail of the spectrum of the gain curve. In the framework of diffusion, the time and position-dependent energy density of amplified spontaneous emission can be described by a diffusion equation with absorption and gain terms. In particular, the diffusive system is represented by a set of three diffusion equations (which are coupled differential equations) accounting for the absorption of pump light, seed light, amplified spontaneous emission, and the rate equations for the concentration of dye particles in the excited state.

Solution of the coupled diffusion equations allows to predict the behavior of the lasing photons for different values of inversion and scattering strength, the transient behavior of random lasing emission and also the spiking.

Diffusing light will spend more time interacting with active material in a characteristic volume than ballistically propagating light producing na avalanche-like intensity bursts, induced by incoherent feedback [307]. Moreover, it has been found recently that the photon statistics of a random laser is very similar to that of a regular laser [69, 136, 372].

3.2.2 Mode structures in random lasers

Although the simplified model of diffusion with gain, originally discussed by Letokhov [292], is very powerful in predicting certain emission properties of a random laser, it also neglects some important aspects. In particular light rays in a random laser are subject to interference while undergoing a random walk and may build up random modes similar to those present in a standard laser. In a macroscopic cavity a mode is a stationary solution of the Maxwell equations which possess a well-defined resonance wavelength and spatial distribution of the field, such for example Gaussian modes. Even if in a random laser scattering events may be as close as a few micrometers, nothing forbids the presence of such standing waves also in random lasers. However, in contrast to standard lasers, the spatial distribution of the field may not be predicted a priori because it is practically impossible to access information about the exact location of all of the involved scatterers. It is well known, for example, that a single micrometer-sized spherical particle may sustain Mie modes whose efficiency depends on the index contrast with its surroundings. The presence of resonant phenomena in random lasers was firstly reported in reference [71]. Sharp peaks were reported in the random lasing spectra (close to or below a nanometer in width) on top of the line-narrowed fluorescence in thin zinc oxide samples pumped with picosecond pulses; moreover, the area from which the sharp emission originated was localized in a sub-micrometer portion of the sample [70]. Various hypotheses have been advanced to explain such intriguing phenomena; here we will discuss some of them.

3.2.3 Anderson localization

When the scattering is weak, the transport of light may be described as a diffusion process. If the scattering is sufficiently strong, recurrent light scattering events are probable. Interference between counter propagating waves in a disordered structure gives rise to enhanced backscattering (see previous chapters). However, with increasing disorder, the wave transport in three-dimensional media should show a phase transition to a state without transport – the Anderson localization. Experimentally, distinguishing between the effects of absorption and localization is a challenge. It is well known that optical absorption destroys photon localization because it suppresses the interference of scattered light. Optical gain, which has the opposite effect to absorption should, in principle, enhance photon localization. Cao *et al.* (see Fig. 3.3) demonstrated in their experiments that sharp peaks, present in the spectrum, originate from a submicroscopic area of

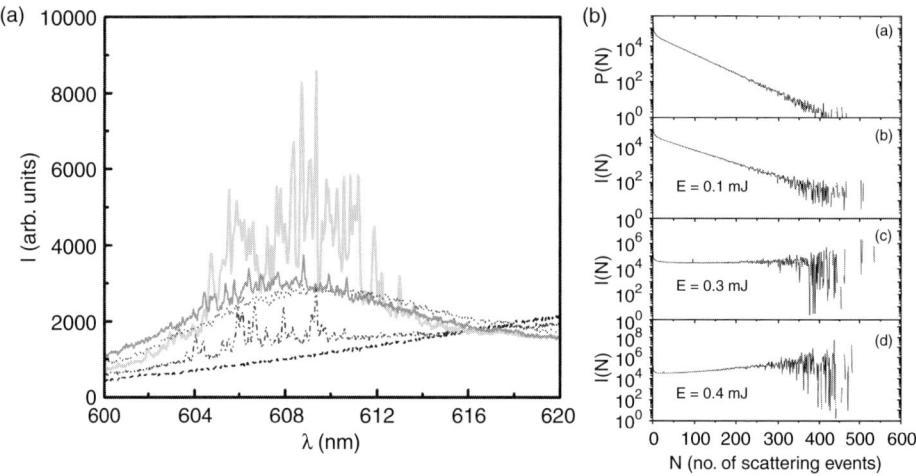

Figure 3.3 (a) Random lasing spectrum obtained from different samples in the weak scattering regime. Reproduced with permission from [544] © 2006 APS. (b) Calculated path length distribution of the spontaneously emitted photons for different values of pump strength. Reproduced with permission from [348] © 2004 APS.

the change, however, this alone does not demonstrate an overall suppression of the diffusion process. In fact, it has been demonstrated by numerical simulation [13, 425, 503, 504] that localized and sharp modes are possible both in the localized regime as well as in the weakly scattering regime (see Fig. 3.3). Moreover, the presence of sharp peaks has been demonstrated in samples with weak scattering [388, 544].

Two different experiments may be proposed to confirm definitively the onset of Anderson localized random lasing. The first is to find a system possessing a tunable scattering strength, in which the presence of sharp peaks is revealed only for very small mean free paths. The second is measurement of the transport properties, performed simultaneously with the acquisition of the random lasing spectrum, as done by [453] for linear systems.

3.2.4 Lucky photons

Sharp peaks in the spectrum have been measured in a large variety of systems, in both the weak and strong scattering regimes, and in many different pumping geometries and conditions [137, 495, 544]. Thus these spikes may be originated from the diffusive nature of light propagation that is not necessarily connected with the degree of localization. Chabanov *et al.* studied the decay rate statistics of light trapped in a large diffusive system [79], demonstrating that an extremely small set of light rays can perform very long trajectories (and thus possess very long decay times) with respect to average paths. In a system with gain, these very rare but very long light paths may be the cause of the spikes in the spectrum because, as a result of the competitive nature of stimulated emission, long-lived diffusive photons have a greater chance of catching more energy from the system, and stealing it from less efficient ones. Mujumdar and co-workers [348] calculated that there exists a small subset of modes with very long lifetimes and thus characterized by very long paths. These modes result in spikes in the spectrum. Since the total gain is proportional to the length traveled in the gain medium, these "lucky photons" can acquire a huge gain and give rise to spikes in the emission. The path traveled by such lucky photons depends strongly on the igniting photons that are randomly selected by the spontaneous emission process and as well affects the spikes which occur at random frequencies. In practice, the lucky photon model predicts, for static systems, the occurrence of sharp spikes at random frequencies which are not repeatable in different shots. Lepri *et al.* [291] demonstrated that in the framework of the diffusion approximation, different fluctuation regimes exist and are determined by the rate of population inversion. Their theoretical model predicts intensity fluctuations with "infinite variance" which have been measured close to the threshold regime.

3.2.5 Localized modes

Being of an intrinsically random nature, spikes predicted by the "lucky photon" model do not coincide with the steady peaks that are found in some experiments [71]. These are two different phenomena: "spikes", that are random in wavelength position and intensity, and "peaks", that may show a certain degree of intensity fluctuation but possess a fixed wavelength. Being repeatable, peaks have to be connected with the permanent disorder realization of the scattering structure, and often a specific area of the sample with a variable degree of localization may be identified as the source of the sharp emission. Figure 3.4 shows some localized mode shapes from different systems. Many advances have been made in the experimental study of these peaks. In the their seminal experiment, Van Der Molen *et al.* [496] analyzed modes' spatial extent using confocal microscopy. The same optical apparatus was used both to pump the system and to collect the random lasing emissions, resulting in a measurement in which just local modes had been excited, eventually suppressing extended modes. This approach has been successively refined, allowing retrieval of the emission of a random laser extensively pumped with micrometer spatial resolution and nanometric spectral resolution in both zinc oxide [129] and in titanium dioxide–rhodamine [123] systems, allowing

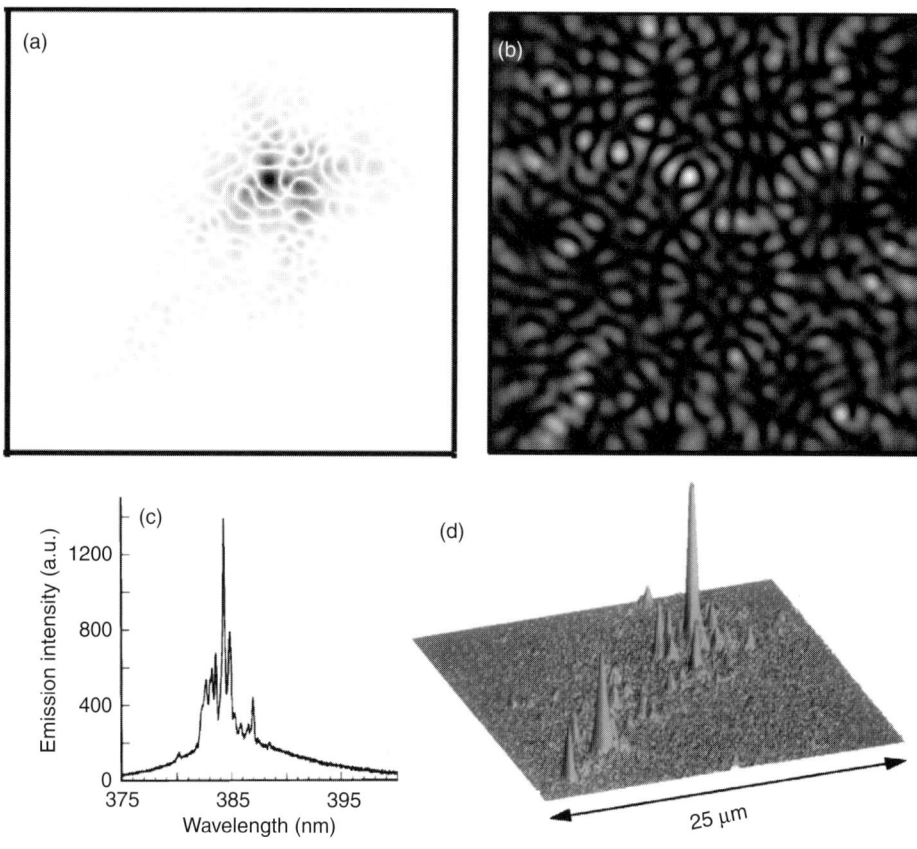

Figure 3.4 Localization in random lasers. Finite difference time domain simulations in the strong scattering (a) and weak scattering (b) regimes. Emission spectrum from a strongly localized sample (c) and the intensity distribution on the sample (d). Reprinted with permission from [70, 425, 504] © 2000, 2002, and 2007 APS.

retrieval of both localized and extended modes in the same system (see Fig. 3.5 and [285]).

In both cases, to access the finest spatial features of a random laser, tiny samples were studied, that is, samples comparable in size to those for a random lasing mode, i.e. tens of micrometers because local spatial and spectral features of a random laser are hidden in larger samples. This may just be due to a mere averaging out of an enormous number of modes, or it may be due to a form of collective emission (see below). Indeed, lasers may appear in various forms: emission may come from very large samples, where local properties are "averaged out" and become invisible to an experiment, or from very small structures, of size comparable to modes, where local properties play a role. In a poorly scattering sample just extended modes may exist [544], even if they can give rise to spectrally narrow modes. Whether spikes or

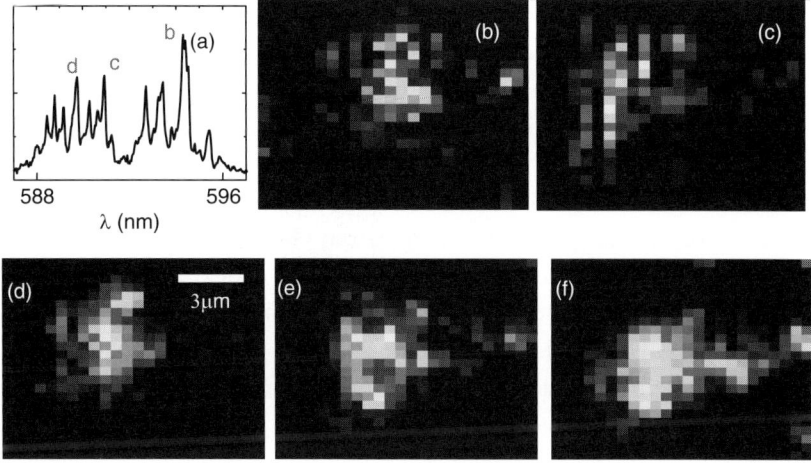

Figure 3.5 Emission from a micrometer-sized random lasing TiO_2 cluster. In (a) the total spectrum is reported. In (b) and (c) spatial distributions of the intensity are shown from the modes identified in (a). In (d), (e), and (f), extended modes activated in different pumping conditions are shown. Reprinted with permission from [285] © 2012 AIP.

peaks appear, random lasers are defined as resonant or coherent feedback random lasers, while if the spectrum is smooth, they are named intensity or incoherent feedback random lasers. Also other parameters, such as the pump duration, the gain strength etc., are relevant to define the emission properties: random lasing is a diverse system in which a multitude of regimes is possible. In the following, a set representative cases are analyzed.

3.3 Tuning emission from a macroscopic random laser

Most random lasers are formed by a haphazard assembly of nondescript optical scatterers with optical gain. Often, the random structure consists of irregularly shaped scatterers of varied size with some average scattering strength. Usually, the macroscopic transport properties of such samples are constant over the wavelength range in which the typical emission appears (a spectral window of breadth typically ~ 100 nm). In other words, the total transmission through the disordered structure is spectrally flat, and photons of all frequencies spend the same average time in the sample; that is, the average lifetime of the involved modes is the same at all frequencies. For such a system the position of the random lasing emission depends on the active medium, which selects for lasing the wavelength with the highest gain.

However, tailoring the average properties of the medium (such as, for example, making the scattering strength for modes at certain frequencies stronger than the average) permits to effectively tune the random lasing emission wavelength. It is the case for the photonic glasses (disordered assemblies of monodisperse, spherical particles), which can sustain resonances at controlled frequencies despite the spheres arrangement, which is completely random. Since they are formed by monodisperse spheres, all possessing the same Mie resonances, a doping of the glass with a gain material enables it to lase at a predefined frequency [417]. The unique resonance properties of this material allow control of the lasing color via the diameter of the particles and their refractive index. Thus a random laser with an a priori set lasing peak can be designed. In Fig. 3.6(a) an SEM image of a photonic glass is shown. The total transmission from three photonic glasses with different sphere diameters are also shown: panels (b) and (c) show resonant spectra from photonic glasses with particles of size (1000 nm and 900 nm) comparable with the wavelengths in the spectrum, whereas panel (d) shows a featureless total transmission for the random medium composed of particles (200 nm), which are too small to resonate at these frequencies. A measurement of the total transmittance is a direct probe of the wavelength dependence of the transport mean free path ℓ. When traveling through the photonic glass certain photons suffer a stronger scattering by being trapped into a Mie resonance, reducing their transport mean free path and enhancing time spent inside the gain medium (obtained by doping with a dye). Since the lasing process is competitive, the resulting emission is centered at the most efficient frequency, and a direct tuning of the emission is thus possible. Lasing from the same gain medium results in frequency-shifted spectra by using monodisperse particles with different diameters [173]. The minimum in the transmission corresponds to a maximum in scattering strength, which tends to attract the lasing curve, pulling it towards

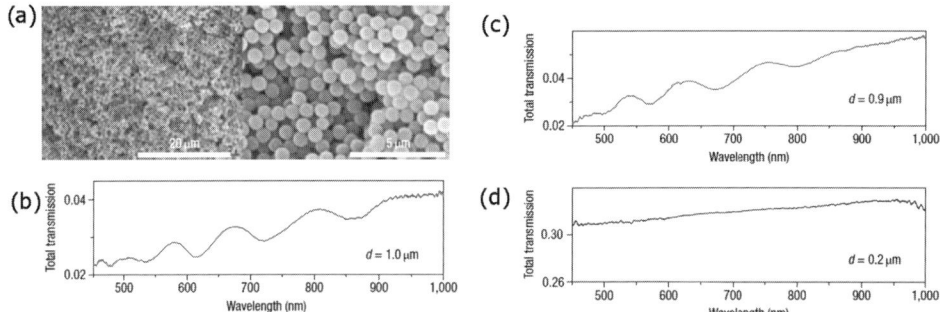

Figure 3.6 Transport properties of a photonic glass: a dielectric composed of monodisperse, disordered spherical particles. Panel (a), two SEM images of a photonic glass. Panels (b), (c), and (d) show total transmission for photonic glasses with different particle diameters. Reprinted with permission from [173] © 2008 NPG.

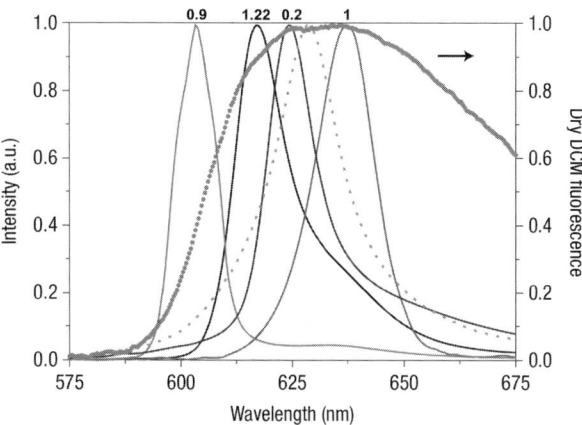

Figure 3.7 Random lasing action in photonic glasses. Continuous lines show random lasing emission for photonic glasses with different sphere diameters (sphere diameter is indicated in microns above each curve peak). The dotted curve shows the pure dry dye fluorescence. A reference sample is made with doped polydisperse TiO_2 powder. The pump energy for all samples is the same. Reprinted with permission from [173] © 2008 NPG.

lower wavelengths within the gain curve of the dye. It is possible, in this way, to select a frequency for the intensity feedback that drives the random lasing emission, as shown in Fig. 3.7. When TiO_2 polydisperse particles are used as scattering material, the lasing frequency is dictated by the dye gain (gain curve maximum).

This phenomenon is the result of the averaging of many scattering events taking place in an extended region of the sample, while properties of the individual resonator are not addressable. Different approaches have been proposed to tuning the intensity feedback emission, either by manipulating the average properties of the scattering medium or of the gain material. For example, Wiersma *et al.* [532] have been able to control emission bandwidth by directly controlling the scatterers' efficiency using temperature-dependent liquid crystals. The authors of [122] used a different approach (engineering the absorption by introducing absorbing, non-fluorescent dye) with similar results: they demonstrated tunability of the average spectral position of the emission due to repulsion from the absorption window.

In a diffusive random laser, the fundamental magnitudes defining the spectrum and the threshold of a random laser are the gain length ℓ_g, the transport mean free path ℓ, and the system volume. However, there may be systems where scattering is subject to different statistics as is light propagation. Non-Gaussian distributions of scattering step lengths lead to an altogether new class of materials in which a scattering mean free path may not be defined, and the system becomes sub- or super-diffusive, depending on whether the average square displacement grows slower or faster than linear with the number of steps. An example is light propagation in systems with Levy statistics [35, 61], where scattering probability is not Gaussian because photons have a nonzero probability of performing very long jumps (the so-called Levy flights). Even if the effects of Levy type diffusion have been measured experimentally in passive systems, in random lasers this is more difficult because effects of linear properties are mixed with the amplification. Evidence of the effects of the scatteres statistics on the emission is given by the authors in [288]. In this paper, systems, possessing the same scattering and gain strength, have been prepared in a way in which individual scattering particles are aggregated in superstructures (clusters) of different sizes and with different inter-cluster distances such that light travels without scattering (mean free path ℓ_f) but intensity increases due to gain, since such a space is filled with dye (see Fig. 3.8). The lower graph shows the behavior of the threshold T as a function of inverse intercluster mean free path, demonstrating that even if overall propagation properties (scattering density and gain efficacy) remain unchanged, the fashion in which the aggregation takes place strongly affects the lasing efficiency. In particular, an increase in intercluster mean free path results in a lowering of the lasing threshold.

3.4 Driving modes of a microscopic random laser

In the previous section we described how random lasing emission is affected by the average properties of the material, such as scattering strength, absorption, and gain. This means that, even if local modes may exist, they are effectively unde-tectable with standard spectroscopic techniques: the emission results from the sum of innumerable modes with slightly different frequencies, and located at different positions inside a macroscopic structure. Different papers show [123, 129] that the typical mode extent is of the order of a few micrometers. Therefore, to study the physics of individual random laser modes demands the number of emitting modes to be restricted.

Mode selection may be obtained by pumping locally a macroscopic sample [496]. The drawback of this approach is that using a very small area pump may result in the effective average mode size being underestimated. This problem was solved by exploiting an extended pumping performed together with local detec-tion [123]. Yet another approach was proposed by Fallert *et al.* [129] (see also Fig. 3.5), who performed a selective detection on a reduced size sample (a clus-ter of zinc oxide), demonstrating that mode size spreads from a few to tens of micrometers.

In all of these cases random lasers were studied by exploiting an experimental setup capable to measure a spectrum presenting a variety of peaks. Such peaks

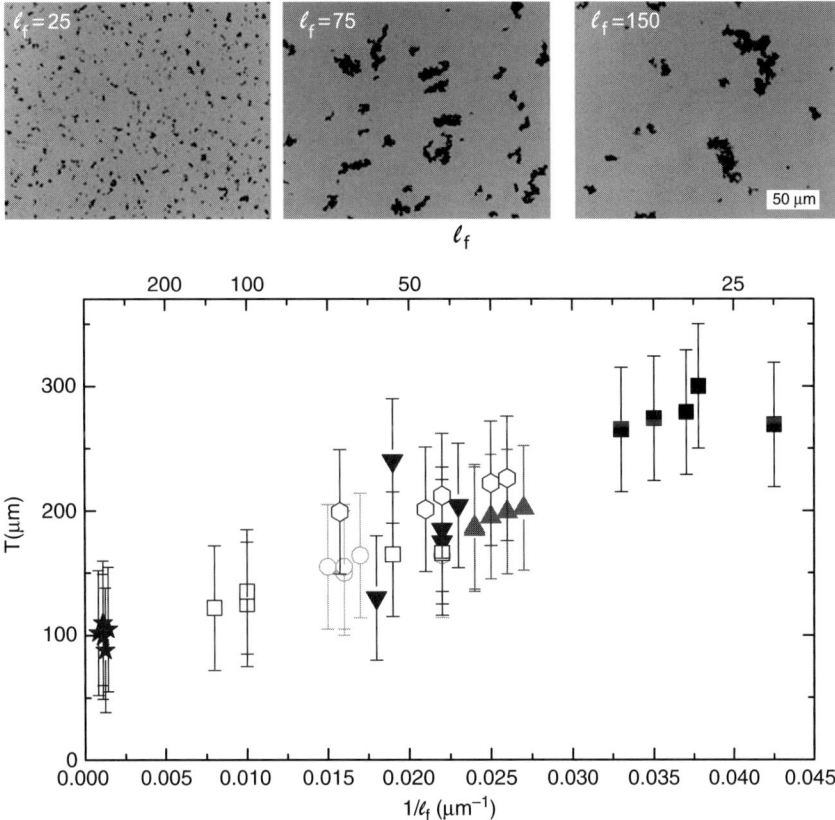

Figure 3.8 Multiscale random lasing. Upper panels show images (from an optical microscope) of samples of different cluster size and relative intercluster mean free path. The lower panel shows the lasing threshold as a function of the intercluster mean free path ℓ_f. Reprinted with permission from [288] © 2013 AIP.

are determined by the particular disorder realization and thus are connected to modes located in a certain area of the disordered structure. A way of distinguishing between stable peaks connected to the structure and the random spikes originated by very long paths (the lucky photons) is to watch the shot-to-shot intensity fluctuations. This is usually done by exploiting pulsed laser pumping with a repetition rate of at least 10 Hz, and monitoring successive realizations of the random lasing emission spectrum from a static sample. By performing an ensemble average, spikes vanish from the mean spectrum while peaks remain visible.

3.4.1 Fluctuations

It is possible to measure the presence of peaks only in a steady sample, that is, in a sample with fixed scatterer positions. Even if their spectral position is stable, peaks are intrinsically subject to intensity fluctuations that originate from various sources [494]. Assuming fixed scatterers, the two main causes of intensity fluctuations are the fluctuation of the gain efficiency (instability, degradation, thermal effects, or displacement of the gain material) and fluctuations in the pump shape and intensity. These two causes may be considerably reduced by implementing technical improvements on the experimental setup (such as finer control on the pump shape) and on the gain medium (such as a stable integration between gain and scattering material like in the case of ZnO). However, there is a feature that cannot be eliminated even in an ideal sample/setup: it is the intrinsically stochastic contribution of spontaneous emission. Spontaneously emitted photons are the first to ignite the stimulated emission, therefore, the modes which lase first are also those that draw energy from their neighbors. Thus, even if the set of active modes is fixed by the disordered structure, the energy distribution between them is inherently stochastic and may fluctuate strongly from shot to shot.

The reason for this energy fluctuation lies in the fact that a random laser is an open system (losses are large for all the modes): being modes not orthogonal they are strongly interacting, and energy may be freely exchanged between them. Such a competition term is well known and is fundamental to explain mode locking of standard lasers in coupled modes theory [186]. It has also been used in different theoretical approaches for random lasers [486]. Strongly connected with mode competition, fluctutations may be exploited to study the effective coupling between two modes. The effective coupling may be extracted from the fluctuation statistics of a large number of spectra taken from a static random lasing system. An example is given in Fig. 3.9(a) in which spectra from five consecutive shots are reported. As can be seen, peak positions are fixed but their intensities are fluctuating. Correlation between two arbitrary modes may be extracted from a plot of the intensity of one mode as a function of the other. For every pair of peaks such a graph provides a number of data points equal to the number of shots. If for two given modes data points spread in a nearly circular area (Fig. 3.9(b)), then they are uncorrelated, meaning that any given intensity in one mode is compatible with any intensity in the other. In Fig. 3.9(c) the correlated case is reported: high intensity in one mode is always associated with high intensity in the other and vice versa. Correlation between modes is due to the fact that overlapping modes share areas of population-inverted emitters so that a single mode is never excited alone. On the other hand, the anti-correlated case is also possible, where the stronger mode eats part of the energy of the weaker one so that high intensity in one coincides with low intensity in the other. This case is more difficult to measure experimentally, since one of the two competing modes usually becomes simply suppressed.

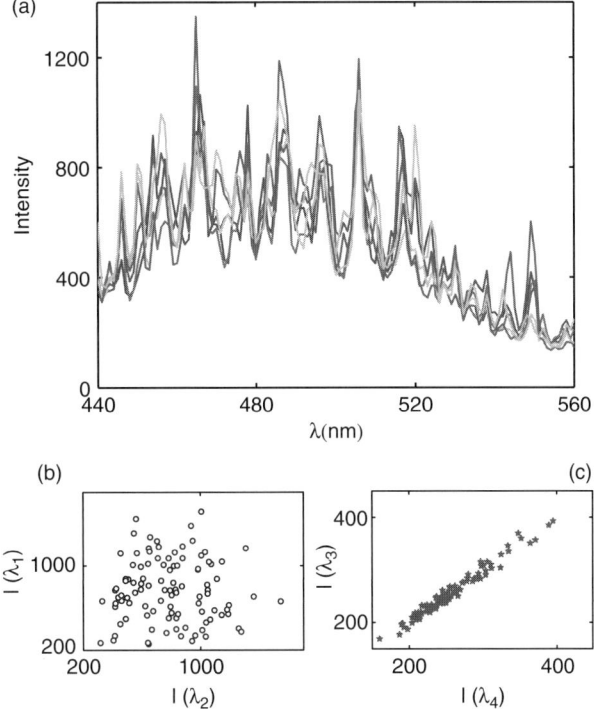

Figure 3.9 Intermode correlation in random lasers. Panel (a) shows five consecutive spectra from a steady random lasing sample. In (b) and (c) the intensities of two modes are plotted against each other. In (b) the pattern of two totally uncorrelated modes is found while in (c) the correlated modes are seen to oscillate in unison.

The intermode correlation $C_{1,2}$ between two modes at λ_1 and λ_2, may be quantified by the correlation in intensity fluctuations. If $I(\lambda_1)$ and $I(\lambda_2)$ are the intensities at these two wavelengths, the pair correlation $C_{1,2}$ is defined as

$$C_{1,2} = \frac{\sum_{i=1}^{N}(I(\lambda_1)_i - \overline{I}(\lambda_1))(I(\lambda_2)_i - \overline{I}(\lambda_2))}{\sqrt{\sum_{i=1}^{N}(I(\lambda_1)_i - \overline{I}(\lambda_1))^2}\sqrt{\sum_{i=1}^{N}(I(\lambda_2)_i - \overline{I}(\lambda_2))^2}}, \qquad (3.5)$$

where $\overline{I}$ represents the average intensity over the N shots at each wavelength. If two modes show a $C_{1,2}$ close to 0, they are uncorrelated (in each mode the intensity oscillates independently around the average value); on the other hand, modes may share or compete for energy, resulting in positive or negative values of $C_{1,2}$.

3.4.2 Local mode selection

Since random laser spectra are composed of a collection of sub-nanometer peaks, an important natural question is whether it is possible to select the emission of a single mode chosen among the many available ones. This procedure has recently been demonstrated [289] and relies on the selective excitation of modes through pumping engineering. The hypothesis behind this achievement, first proposed in a theoretical paper [29], is that pumping configurations can be found that couple optimally to a target mode so that its emission is enhanced, whereas other modes are suppressed. To demonstrate the effect, a random laser based on a single, isolated, small cluster (average diameter between 10 and 25 micrometers) of titanium dioxide was employed furnishing a stable configuration of fixed scatterers. In the proposed configuration (see Fig. 3.10(a)), the cluster lies on the surface of a microscope coverslip and is embedded in the rhodamine solution, which is pumped from the bottom, while the emitted radiation is collected from the top. The numerous laser modes available in the cluster have different coupling abilities with light coming from a given direction and impinging on a given spot of the cluster. Ideally, it would be desirable to be able to send a pump laser beam at any direction and pointed at any desired spot on the cluster. This is achieved indirectly by generating rays of amplified spontaneous emission (ASE) in the surrounding dye with a "green laser" in a manner similar to that used to determine gain in the stripe length method. A green laser excites red ASE from the dye

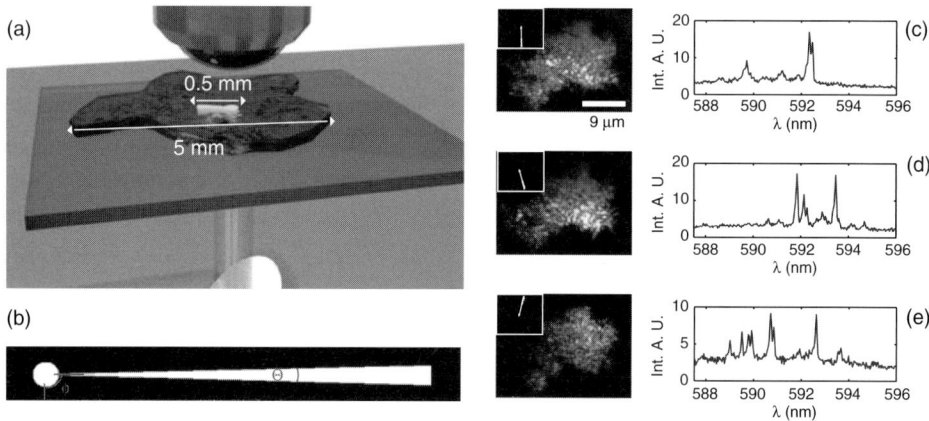

Figure 3.10 Panel (a) shows the experimental setup for controlling emission from a random lasing cluster and panel (b) shows the mask used to produce directional stimulated emission. Panels (c), (d), and (e) report the spectra and spatial distributions of intensity for three different pumping configurations. Reprinted with permission from [285] © 2013 AIP.

and these red beams pump the cluster. The cross-section of the green laser beam determines the direction in which ASE is generated. If additionally a circular area encompassing the whole cluster is illuminated, this prepares the cluster near lasing threshold, so that the rays will trigger lasing only for modes near threshold that couple best with the incoming ASE. This approach is called directionally constrained pumping. At variance with spatial constraint, a directional constraint allows to pump over the threshold modes that are located in distant positions within the cluster that are thus barely interacting. In this way individual excitation may be achieved if the pumping direction is set to favor emission from a target mode chosen from among the many that may be activated in the random lasing cluster.

Since the ASE is directional, a circular sector pumped by the green laser generates an intense ASE beam at the vertex and is the best configuration to control both the angular span of the pumping and the exact point at which the ASE is impinging on the cluster. The wedge is pointing to the center of a circular area which maintains the modes of the cluster barely below threshold, while the wedge feeds the modes that have to be brought to lasing, actively performing the directional selection. The use of a spatial light modulator allows a complete digital control over the green pump beam cross-section, hence the shape of the population-inverted area on the dye bath and, subsequently, over the directions in which ASE is generated. The pointed wedge with angular span (Θ parameter) and orientation (ϕ parameter) is shown in Fig. 3.10(b). In Fig. 3.10(c–e) images of an individual cluster, pumped under three different arbitrary configurations, are presented, together with the respective emitted spectra. In [289] it was demonstrated that through a recursive process it is possible to find the pumping configuration that allows us to generate emission from a single mode.

3.5 Collective random lasing regimes

The measurement of intermode correlations demonstrates that at a microscopic level a random laser is a collection of tiny interacting modes. Even if individual modal features are lost in the multitude at large scale, modes continue to exist also in a macroscopic sample. This evidence has stimulated a theoretical effort aimed at the search for whether this sum of independently oscillating modes is able to give rise to some form of collective state. The first signature of the presence of collective behavior in a random laser was reported by Conti *et al.* [91]. In their original theoretical and experimental work, a novel formulation showed that the emission linewidth of a macroscopic random laser is governed by nonlinear differential equations, formally equivalent to the nonlinear Schrödinger, or Gross–Pitaevskii

(GP), equation governing ultracold atoms. Starting from the lasing condition (gain equals losses), the following equation may be obtained:

$$g_0 \left[a(t) + t_g^2 \frac{d^2 a}{dt^2} - \gamma_s |a|^2 a \right] = [\alpha_0 - \phi_L(t)]a, \qquad (3.6)$$

where $a(t)$ is the temporal shape of the lasing pulse, g_0, t_g, and γ_s are the gain coefficient, the gain lifetime, and the gain saturation coefficient, and $[\alpha_0 - \phi_L(t)]$ is a time-dependent coefficient for losses. Equation (3.6) is identical to the bound-state GP equation for 1D Bose–Einstein condensation with an external potential $\phi_L(t)$. This shows that a spectral region of high-Q modes acts as a trapping potential for the energy levels of the excited photons. Frequencies tend to be concentrated in this spectral range, as Bose-condensed atoms tend to be localized by the external trap [105]. The striking feature is that the equation allows us to predict the energy behavior of the emission bandwidth, as shown in Fig. 3.11. In a nutshell, the spectral condensation results in a line-narrowing of the random laser emission at high mode density (pump energy) in a way similar to that predicted by the Shawlow–Townes law for standard lasers.

A different approach uses spin–glass theory [14] for investigating the behavior of the phases of a multitude of individual modes in the search of collective states in random lasers. This approach [294] allows to draw a phase diagram revealing the interplay between randomness and nonlinearity, and to identify different phases characterized by the not-vanishing complexity Σ which is a measure of the number

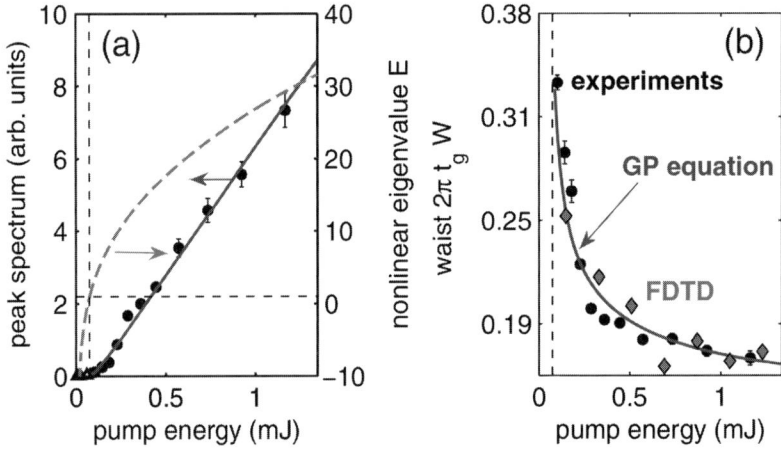

Figure 3.11 Peak intensity (continuous line), nonlinear eigenvalue (dashed line), (a) and spectral width (b) of the random lasing emission as predicted by the Gross–Pitaevskii equation (continuous line), and as retrieved from experiments (full circles). Reprinted with permission from [91] © 2008 APS.

of energetically equivalent mode-locked states. The main finding is that various phases exist for macroscopic random lasing emission and that in certain conditions a self-starting passive mode locking is recognized.

The study of collective regimes in random lasers is difficult because when many modes are activated, the knowledge of the individual behaviour is lost.

Most random laser systems, being large and with many modes, work in intensity feedback conditions: modes cannot be isolated, and any potential signature of collective behavior is diluted. To prove a collective regime for emission it is more advisable to place the system into an individual regime and drive it into a collective oscillation.

This approach [284, 286] demonstrated that a micrometer-sized random lasing structure may be driven from an individual lasing to a phase-locked configuration. The discovery of spontaneous mode-locking of lasers [445], that is, the self-starting synchronous oscillation of electromagnetic modes in a cavity, has been a milestone in photonics, allowing the realization of oscillators delivering ultrashort pulses. Standard multimode lasers without disorder may be driven to a synchronous regime through the so-called mode-locking transition which, so far, has only been shown to occur spontaneously in the presence of a saturable absorber. The authors showed that the same transition occurs in random lasers, allowing the locking of the modes of a resonant feedback random laser, casting its emission in the typical intensity feedback random laser spectrum. The experimental approach used is the same as that used for mode selection and is shown in Fig. 3.10; it consists of an isolated micrometer-sized cluster of titania nanoparticles with static disorder immersed in a rhodamine dye solution. Selected areas surrounding the cluster are pumped optically to generate a directional stimulated emission from the population-inverted areas defined by shaping the beam of a solid-state pump laser, using a reflective spatial light modulator. The result is an inverted area with the shape shown in Fig. 3.12(a).

The transition from an individual to a collective random laser may be studied by varying Θ, that is, the parameter that controls the set of input directions, hence the number of activated modes. At low Θ, the spectrum displays a "resonant feedback" behavior displaying several very narrow (< 0.05 nm) peaks, whereas large values ($\Theta < 100°$) produce a single and smooth RL lineshape (4 nm). The three-dimensional graph in Fig. 3.12(b) reports the behavior of the spectrum as a function of Θ, demonstrating a transition in which a spiky spectrum is found when a small number of modes are active and a smooth spectrum is retrieved at high Θ (large number of modes). Figure 3.12(c) reports a quantitative measure of the spectrum spikiness as a function of Θ. The spikiness S, that is the weight of high-frequency components in the spectrum, is plotted versus Θ

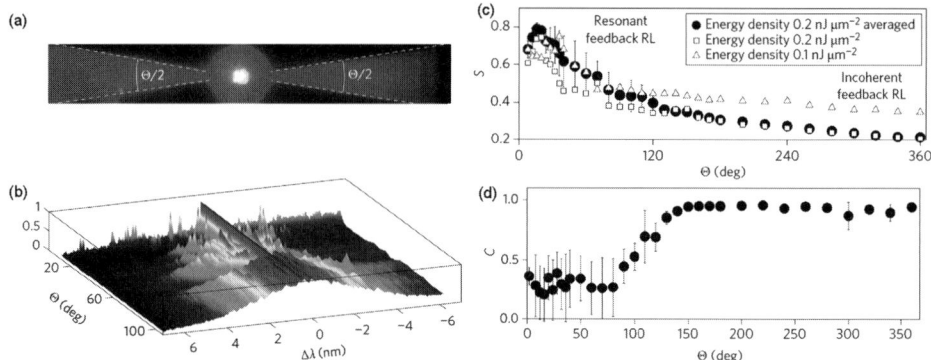

Figure 3.12 (a) Shape of the inverted area exploited to obtain control over the number of activated modes in a random laser. (b) Spectrum from the micro random lasing cluster as a function of the control parameter Θ. (c) and (d) Spectral spikiness S and intermode average correlation C as a function of the control parameter Θ. Reprinted with permission from [284] © 2011 NPG.

at different pump energies (squares and triangles) and averaged over five different disordered structures (filled circles). All curves display the same trend. After rapid initial growth corresponding to an increase in fluence and number of excited modes (appearing on a smooth fluorescence spectrum), S reaches a maximum (resonant regime), followed by a gradual smoothing of the spectrum as Θ grows, until an "intensity feedback" emission is achieved.

The number of activated modes also affects the intermode spectral correlation. In Fig. 3.12(d), the Pearson correlation $\langle C \rangle = \sum_{i \neq j} C_{ij}/2N(N-1)$ averaged over one hundred pairs of modes ($N = 15$) is shown versus Θ. The onset of a strongly correlated regime is obtained for $\Theta > 100°$. The same transition has also been observed in other experiments, revealing a universal trend in which C becomes close to 1 when Θ exceeds $100°$.

The origin of such behavior resides in the onset of a collective state in which the phases of modes retain a fixed relation between themselves. This has been demonstrated by comparing the experiments with the results from coupled mode theory simulations. A cluster of dielectric particles sustains several electromagnetic resonances, which are characterized by a distribution of finite lifetimes and specific frequencies. In such a regime the electromagnetic field can be expressed as a superposition of modes with amplitudes $a_n(t)$ that obey the coupled mode theory equations (see, for example, [14] and references therein),

$$\frac{da_n}{dt} = +i\delta_n a_n - \alpha_n a_n + \sum_{m \neq n} \kappa_{n,m} a_m + g(\delta_n) \frac{a_n}{1 + \gamma_n |a_n|^2}, \qquad (3.7)$$

with $\delta_n = \omega_n - \omega_0$; ω_n the angular frequency of the *mode*, identified by index n; and ω_0 the transition frequency of the resonant system; κ is the inter mode coupling and the last term of the equation reports the gain $(g(\delta_n))$ and its saturation (γ_n). The solution of such a set of differential equations allows us to follow the temporal evolution of the mode's amplitude and phases. As the transition from a spiky to a smooth spectrum depends on the number of activated modes, and thus on their degree of interaction, experimental results may be reproduced by performing a set of simulations in which the coupling strength is varied. In Fig. 3.13(a), the temporal behavior of phases is reported for different degrees of interaction represented by the number of nonzero coupling coefficients n_c. A mode locked behavior (phases synchronized) is evident for the more interacting sample ($n_c = 10$). Simulations also allow us to access the effective degree of spikiness S and intermode correlation C, as depicted in Fig. 3.13(b), in qualitative agreement with results obtained experimentally.

The wedge angle Θ also allows control over the average mode extension, as shown in Fig. 3.5. In practice, extended modes are favored when Θ is large. This

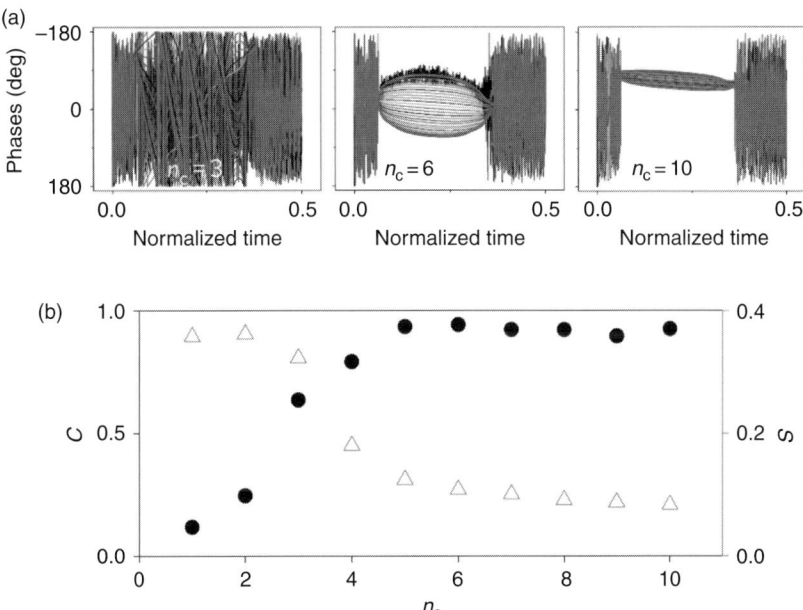

Figure 3.13 Results from coupled modes theory numerical simulations. (a) Present-time evolutions of mode phases for different values of the coupling strength, here implemented as the number n_c of nonzero interaction coefficients. In (b) numerically retrieved behavior of C and S. Reprinted with permission from [284] © 2011 NPG.

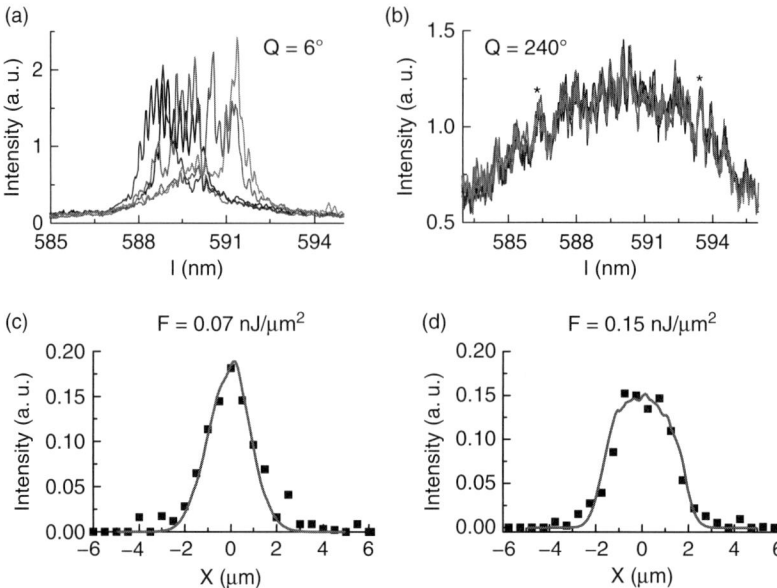

Figure 3.14 Nonlocal effects in a random laser. When few modes are excited their interaction is low and they are allowed to oscillate independently: at their frequency and in their location (panel (a)), but when the interaction is increased by involving many modes, they are forced to involve all the frequencies in the oscillation and spread their location throughout the system (panel (b)). If the energy pumped into the system is increased, the distribution of the condensed state flattens out, adopting the shape of the container, like a fluid (panels (c) and (d)). Reprinted with permission from [283] © 2013.

phenomenon may be explained as a form of condensation, the spatial equivalent of that driving spectral line-narrowing. In the spatial case the interaction is driven by the nonlocality of modes, that is, because modes have a finite extension and energy may flow from one to another. Spectra taken at different locations with low Θ always show very different structures, such as those shown in Fig. 3.14(a). This is a sign of local emission. On the other hand, when Θ is very large, interaction is large and the spectra are identical throughout the cluster (Fig. 3.14(b)). All modes coalesce so that the spectrum is the same regardless of position. To provide a model for this phenomenon, one has to describe the field in the form of a mode density function $\Phi(\mathbf{x}, t)$. Taking into account nonlocality and nonlinearity, we can obtain [287],

$$\frac{d\Phi(\mathbf{x}, t)}{dt} = k_0\Phi(\mathbf{x}, t) + \frac{k_2}{2}\Phi(\mathbf{x}, t) - V\Phi(\mathbf{x}, t) - |\chi||\Phi(\mathbf{x}, t)|^2\Phi(\mathbf{x}, t). \quad (3.8)$$

This equation predicts different behaviors of electric field distribution inside a disordered structure at different energies, which are retrieved in the experiments. In particular, a sharp spatial profile is predicted at low energies (Fig. 3.14(c)), while at high energies a smoothed profile is predicted and retrieved (Fig. 3.14(d)).

3.6 Applications and perspectives

Fabrication of nanostructures for photonics usually requires expensive and time consuming techniques, such as direct laser writing [134], ion beam milling [406], or electron beam lithography [485] for prototyping (which are not usually suitable for large-scale production), or photolithography for static designs. Alternatively, large-scale production of random lasers is relatively cheap and allows access to a world of applications based, for example, on their isotropic 4π angular emission, which may be useful for display applications. In particular, liquid dye based random lasers are fit for the production of flexible screens, while directional emission may be achieved by exploiting liquid crystal droplets in a two-dimensional matrix [172]. Other useful properties are tolerance to fabrication disorder and the possibility of using random lasing structures as lasing paint, for tagging, identification, and coding purposes. Random lasers have been shown to be useful in the field of biology, demonstrating their potential as a tool for identifying cancerous tissue [387], and also in the study of colloidal systems as an instrument able to distinguish between different phases of a shaken granular material [137]. Because of their inherent coherence, standard lasers give rise to a speckled image which effectively limits their use for imaging purposes, thus there is considerable interest in developing light sources with a controllable degree of coherence, which may effectively increase the spatial resolution and image quality [404]. Progress is being made in developing novel polymer/dye materials for biophotonics application, chemically engineered to profit from Förster resonant energy transfer (FRET) [74], and has enabled the first stable near-infrared laser in a colloidal suspension. This may open the way for future imaging techniques mediated by FRET assisted random lasing.

Most of the applications proposed to date exploit the macroscopic emission of the random laser. This is because, when the system is large it is the average properties that are relevant. This avoids succumbing to the fluctuations inherently present when approaching the scale of the individual mode. However, technical advances, allowing selection of individual modes in the disordered structure [285], opened the way to exploiting local properties also in building real micro devices.

An example of single mode control is the application of selective activation of random lasing modes to *identify* coupled resonances located at distant locations, to measure their effective *coupling* in an absolute way, and to demonstrate *switching, routing, and amplification* of signal from one mode to another, realizing in practice a transistor for light. The results are explained in Fig. 3.15, in which two modes of an individual random lasing cluster may be activated independently (spectra in panels (a) and (b)), or at the same time (c). The two modes are named S (Source) and D (Drain) and are activated by exploiting a favorable pumping shape, i.e. activating respectively the source and the gate wedges (c). The cross-talk between modes is evident when the two resonances are activated together. In fact, the dominant frequency in the spectrum becomes the same (593.2 nm in the reported example). This nanophotonic transistor effect in a disordered device is confirmed by panel (d) in

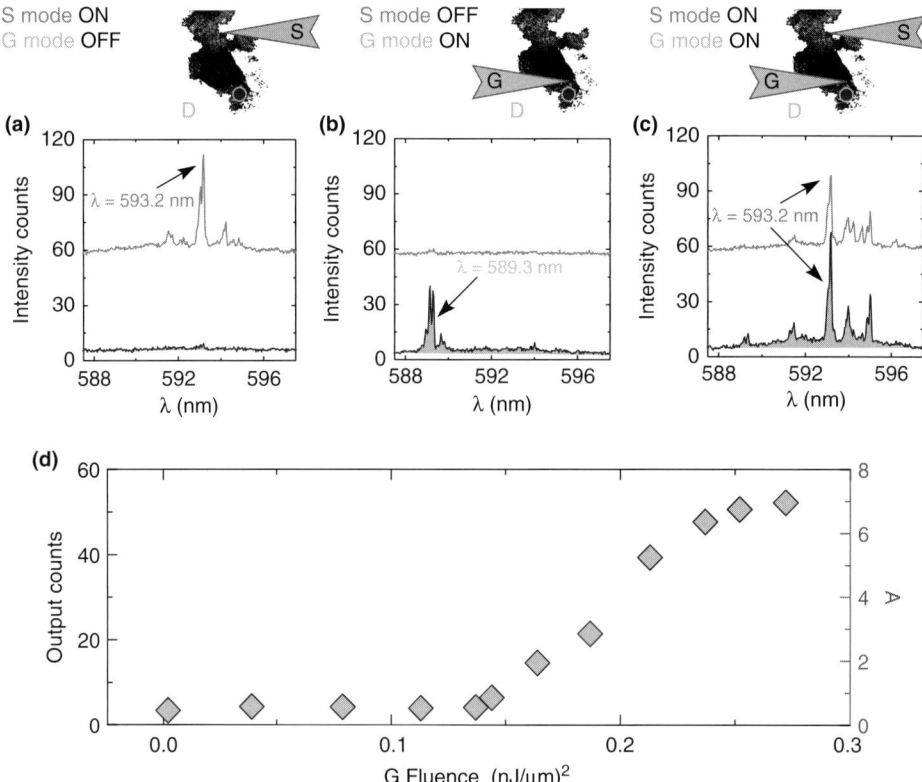

Figure 3.15 Switching and amplification with random lasers. Panels (a), (b), and (c) show three different states of switching when either *source* or *gate* are ON, and when both are (c). Panel (d) shows measurement of the amplification of *drain* signal when the gate is ON as a function of the power injected in the source. Reprinted with permission from [284] © 2011 NPG.

which the amplification at the switched frequency is reported as a function of the gate intensity. In practice, this result demonstrates that it is possible to transmit an optical resonance to a remote point by employing specific control over optical excitations, obtaining a random lasing system which acts both as a switch and as an amplifier, thus opening the way to a novel generation of nano-sized, disordered, and actively controllable photonic devices.

4

Ordered and disordered light transport in coupled microring resonators

SHAYAN MOOKHERJEA

4.1 Introduction

Optical waveguides are used to transport both phase-insensitive and phase-sensitive information in today's communications networks and signal processing devices. Among applications needing high performance waveguides are delay lines and optical phase-locked loops, generating precise phase offsets for waveform sampling, analog-to-digital conversion, and beam-forming. In fundamental terms, high performance waveguides are based on the ballistic propagation of photons [4, 168], which may be capable of reaching the ultimate limits of computational speed and energy efficiency of interconnects [331, 332], since ballistic propagation of electrons [132] is difficult to achieve in practical chips. Therefore, quantifying the impact of disorder on light propagation, and distinguishing between ballistic and non-ballistic transport, is an important and ongoing research endeavor.

Ballistic transport requires that the phase-breaking length be greater than the device length; which is usually satisfied in optical fibers of short length, and in planar glass lightwave circuits, in which the refractive index difference between core and cladding materials is small [260, 261]. But light propagation in high-index contrast waveguides, e.g. in silicon photonics, can be highly susceptible to the effects of disorder. Here, we will focus on periodically structured nanophotonic waveguides, e.g. coupled microring resonators [188, 392, 548, 555] in which light can accumulate significantly more phase and delay per unit length than in conventional waveguides. Such devices can be valuable for many reasons, among which are the facts that on-chip footprint is a precious commodity in real-world devices, and that our ability to increase, via geometrical patterning of waveguiding structures, the optical phase accumulation per unit (structural) length will considerably benefit nonlinear multiwave interactions. However, such devices may also

Light Localisation and Lasing, ed. M. Ghulinyan and L. Pavesi. Published by Cambridge University Press.
© Cambridge University Press 2015.

(a) 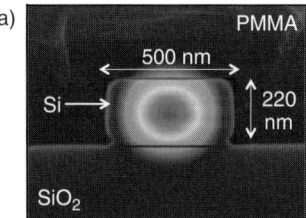 (b)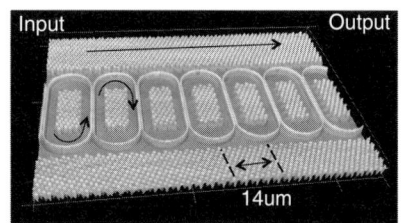

Figure 4.1 (a) Cross-section of a silicon nanophotonic waveguide, used to construct microring resonators in the racetrack configuration. Also shown is the electric field amplitude of the fundamental TE-polarized mode. (b) 3D optical microscope image of a section of a coupled-microring chain, where light propagates down the waveguide by exciting counter-clockwise and clockwise circulating resonant modes, as indicated. Ref. [92]

be more strongly susceptible to the effects of disorder [329], and the issue of whether ballistic propagation can be achieved in coupled-resonator structures is being investigated closely.

From a physics viewpoint, one-dimensional structures are interesting because multiple scattering in transverse dimensions can be restricted, e.g. by using single-transverse-mode waveguides as the constituent building blocks for linear chains of resonators, e.g. microring arrays. Figure 4.1(a) shows the typical waveguide cross-section, on which is overlaid an image of the calculated mode profile of the fundamental (TE-polarized) waveguide mode at the wavelength 1.55 μm. Figure 4.1(b) shows the configuration of the coupled-resonator chain, in which rings are coupled to nearest neighbors via waveguide directional couplers. As the bending radius of our microrings is about 6.5 μm, the TM-polarized mode is quite lossy in propagating around bends and the measured transmission spectrum is only of the TE polarization.

Figure 4.2 shows top-down optical microscope images of coupled microring chains. We have studied chains ranging in length from 5 to 235 resonators, both with and without electrical "control knobs," such as micro-heaters and diodes embedded across the waveguide cross-section, which can tune the refractive index by thermal, carrier injection, or carrier depletion methods. Such structures can be used as filters, delay lines, and for nonlinear and quantum photonics; here, we focus on the more fundamental aspects of light propagation, including a study of non-idealities (e.g. disorder-induced deviations from ballistic transport) in silicon microring resonator chains. Note that experimental studies of one-dimensional transport generally require fabrication using lithographic techniques, unlike three-dimensional random collections of particles and two-dimensional optically induced lattices in nonlinear crystals. However, self-assembled microresonator chains have been achieved using chains of size-selected microparticles; Refs. [20, 554] report on such structures.

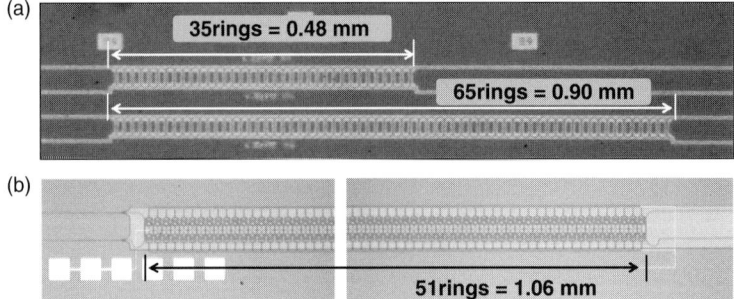

Figure 4.2 Optical microscope images of (a) passive 35-ring and 65-ring devices, and (b) an active 51-ring device, in which a $p - n$ diode is formed across the waveguide cross-section. Light propagation in these structures can be described by a 2×2 transfer matrix formalism, since these devices only propagate the TE-polarized waveguide mode with low loss.

Recently, we reported experimental studies showing that high performance coupled microring devices fabricated using silicon photonics can exhibit ballistic transport, e.g. the absence of disorder-induced localization even in chains of resonators that are 235 microrings long [94]. This level of performance was difficult a few years ago, and is not easy today, but with improving fabrication technology, particularly in the silicon photonics technology platform in which we have been designing our devices, it may become easier to obtain in the near future, if we can understand how disorder plays a role in optical propagation. One aspect of this study is to model the reduction in passband bandwidth that is observed as the length of the chain is increased; this is discussed in Section 4.3.

It is also useful to study the propagation of light in devices with less-than-ideal performance. In Section 4.3, we describe an experimental study of the statistics of photon transport in other devices from the same fabrication batch as that reported in [94], but with different geometric parameters and, consequently, a different impact of disorder on performance. As the length of the periodically structured waveguides was increased, the measurements became suggestive of the regime of diffusive propagation, rather than ballistic transport. Characterized by large relative fluctuations in transmission and propagation time delay, diffusive propagation can imply that the phase of light may be randomized in the longer waveguides. This would be disadvantageous for applications, but the studies provide useful insights into the onset of strong disorder effects in high-index contrast photonic structures.

4.2 Optical microring chains

In a chain of nearest-neighbor coupled microresonators indexed by $n = 1, 2, \ldots, N$, an optical excitation is described by a set of time-dependent coefficients, $\{a_n\}$, which represent the oscillation amplitudes of the resonant field modes of the individual resonators at a particular optical frequency.

The evolution of the amplitudes is given by,

$$i\frac{da_n}{dt} + \Omega a_n + \Omega \kappa (a_{n-1} + a_{n+1}) = 0, \tag{4.1}$$

where Ω and κ are the self-coupling and the (dimensionless) nearest-neighbor coupling coefficients [555].

Labeling the periodicity of the lattice as Λ (14 μm in Fig. 4.1(b)), we take the solution of Eq. (4.1) to have the form,

$$a_n(t) = \frac{1}{\sqrt{N}} e^{i(\omega t - nk_m \Lambda)}, \tag{4.2}$$

where $k_m = m\,2\pi/(N\Lambda)$. Substituting Eq. (4.2) into Eq. (4.1), we find, in the limit of weak coupling, that

$$\frac{\omega_m}{\Omega} = 1 + 2\kappa\,\cos(k_m\Lambda), \quad m = 0, 1, \ldots, N - 1 \tag{4.3}$$

are the N normalized eigenfrequencies of the chain. The eigenvectors corresponding with the eigenfrequencies given by Eq. (4.3) are known as the normal modes of the structure, and allow us to calculate the Green function and the density of states per site [119, Section 5.3.1].

In the limit of large N, Eq. (4.3) defines ω as a continuous function of k, which, in the first Brillouin zone, takes values from $-\pi/\Lambda$ to π/Λ. As $k \to 0$ or $k \to \pm\pi/\Lambda$, the group velocity $v_g = d\omega/dk \to 0$, i.e. slow light is expected at the edges of the band, compared with the speed of light at the band center. However, it is known that this regime is sensitive to disorder [395]: the maximum achievable slowing factor is limited by disorder [337], the propagation loss is also enhanced by the slowing factor [124, 200, 366], and so is the disorder-induced backscattering into the input port [344, 373, 383].

From a practical perspective, there is another form of "slow light", which is more resilient to disorder. From Eq. (4.3), the time taken to propagate from input to output for light of frequency $\omega = \Omega$ (at band center) is $\tau_{min} = N/(2\Omega|\kappa|)$, i.e. the group delay depends on the magnitude of the coupling coefficient. We define a slowing factor as the ratio of the group index *at band center* to the group index of light in a conventional waveguide, i.e. without longitudinal patterning. For microrings in the "racetrack" configuration (see Fig. 4.2), with a bending radius R and directional coupler length L_c, the effective radius of curvature is defined as $R_{eff} = R + L_c/\pi$ and the slowing factor is $S = \pi R_{eff}/(\Lambda|\kappa|)$. Typical values of S range from 10 to 30, which are sufficient for fairly large enhancements in the efficiency of nonlinear interactions (e.g. four-wave mixing scales as the fourth power of S).

Note that we have summarized here a simple theory in terms of the tight-binding model. From a computational perspective, light transport in coupled-resonator optical waveguides is more conveniently calculated by cascaded transfer matrices [95, 391], analogous to computational studies of electronic waveguides.

Slow light results from a large phase accumulation per unit length at the resonant frequencies of the chain of microrings; similarly enhanced scattering is observed in photonic crystal waveguides [373, 482] compared with simple slab or wire wave-guides without periodic patterning [4]. Away from the band-edges, the dispersion relationship of such a waveguide is quite linear, the transmission can be flat, and the inverse of the localization length does not vary significantly with wavelength [233, Fig. 7.9]. Thus, measurements of *in-band* transmission at different wave-lengths of a tunable continuous-wave laser source can yield statistical data, under the assumption of ergodicity, which can be used for insights into the nature of light propagation in slow-wave structures and the transition away from the ballistic regime when disorder is more significant.

4.2.1 Experimental details

Here, we report on microring chains of length $N = 35, 65, 95, 135,$ and 235 unit cells that were fabricated at the IBM Microelectronics Research Laboratory on silicon-on-insulator wafers. The waveguides were single mode, with transverse dimensions approximately $0.50 \, \mu\text{m} \times 0.22 \, \mu\text{m}$. The precise nature of the disorder induced by fabrication is difficult to measure; the fabrication process was charac-terized in an earlier study to result in a line-edge roughness, $\sigma = 1.1$ nm and a correlation length $L_c = 60$ nm [545].

The slowing factor was about 16 for the measured bands, and can conveniently be obtained from measurement of the spectral transmission of light through the device using a tunable laser. Coupled-resonator waveguides are generally char-acterized by a (dimensionless) inter-resonator coupling coefficient κ in the range 10^{-3} to 10^{-1}, which yields a spectral half-bandwidth $\Delta\lambda_{1/2}$ of 3 pm to 0.3 nm for microrings of similar size to those used here; the equation connecting these quantities is

$$|\kappa| = \sin\left(2\pi^2 n_g \frac{R}{\lambda} \frac{\Delta\lambda_{1/2}}{\lambda}\right), \tag{4.4}$$

where R is the ring radius and n_g is the group index of the waveguide. The value of $|\kappa|$ can be used to predict the group delay of light propagating through the N-resonator chain,

$$\tau = \frac{\pi n_g R N}{|\kappa| c}. \tag{4.5}$$

Equation (4.5) predicts, as mentioned earlier, that the group delay increases with the number of resonators, N, and inversely with the bandwidth, $\Delta\lambda_{1/2}$.

Transmission and group delay were measured using an optical vector network analyzer [165], based on the principle of swept-wavelength interferometry. Optical coupling into and out of the waveguides was achieved using tapered and lensed polarization-maintaining fibers, using transverse-electric (TE) polarization. The fiber-to-fiber insertion loss was -21 dB (which has since been improved to about -13 dB for microring chains of similar length and similar racetrack dimensions, using an improved fabrication process). This number includes the loss in coupling from the input fiber and output fiber to the silicon waveguides (estimated to be -8 dB/coupler from measurements on waveguides of various lengths, later improved to -4 dB/coupler), and residual losses from waveguide bends, radiation, etc. Linear fits to the spectrally integrated power transmitted through resonator chains ranging in length from 35 to 235 microrings were obtained for four chips from each wafer, and indicated an average loss per ring of -0.062 dB/ring (including the effects of band-center slow light, as described below). On nearby test sites, the waveguide propagation loss was obtained by measuring transmission through four waveguides (without resonators) of different lengths, but with the same number of bends, and was measured to be -3.5 dB/cm (later improved to about -2 dB/cm).

4.3 From ordered transport to disorder-induced bandwidth collapse

A few groups have demonstrated on-chip resonator chains that are hundreds of resonators long [94, 361, 545], significantly longer than what was achieved through previous efforts: about a dozen resonators [343]. As such long chains become more readily manufacturable, it may be important to understand how long these chains can be made, before the transmission bandwidth collapses from the combined effects of loss and disorder.

As reported in [339], we have used numerical simulations of light transport through a coupled-resonator chain to study this problem. The Lyapunov exponent (which is the inverse of the localization length, ξ) was calculated for the tight-binding Hamiltonian,

$$H = \sum \omega_n (|a_n\rangle\langle a_n| + \kappa_{\omega,(n,n\pm1)} |a_{n\pm1}\rangle\langle a_n|), \qquad (4.6)$$

where ω_n is the on-site energy and $\kappa_{\omega,(n,n\pm1)}$ are the dimensionless (time-domain) nearest-neighbor coupling strengths of the nth oscillator ($n = 1, 2, \ldots, N$). We usually assume that the κ are identical in a disorder-free chain (except for some apodization at the input/output sections of a long chain, which can be ignored here). For a given (finite) value of N, the number of resonators in the chain, the (dimensionless) energy separation between the two eigenmodes closest to the band center is

$$\frac{\Delta\omega}{\omega} = \frac{2\pi\kappa_\omega}{N+1}. \tag{4.7}$$

The time-domain normalized coupling coefficient κ_ω is related to the field amplitude coupling coefficient κ of the waveguide directional coupler through the geometry of the device,

$$|\kappa_\omega| = \frac{\lambda}{4\pi^2 n_g R_{eff}|\kappa|}, \tag{4.8}$$

where n_g is the group index of the waveguide, and R_{eff} is the effective radius of the racetrack resonator (equal to the actual radius of a circular microring).

In our simple model, disorder σ comprises two independent uniformly distributed random variables: variations in round-trip path length of each resonator (standard deviation/mean $= 2 \times 10^{-4}$, typical for today's silicon microfabrication technology) and in the magnitude of the nearest-neighbor field coupling coefficient (standard deviation $= 0.075$). Monte-Carlo simulations of transport were calculated for different realizations of the random coupling coefficients and resonant frequencies. A good fit through the points generated by a direct numerical simulation was found by the parametric equation,

$$\xi^{-1} = b^2 + c^2/(a^2 - \epsilon^2), \tag{4.9}$$

where a, b, and c are fitted constants, and $\epsilon = \omega/(\langle\omega_n\rangle\langle\kappa\rangle)$ is the normalized frequency. The constant a is slightly greater than 2, and describes the shifting of the band-edge outwards from the band-edge of a disorder-free tight-binding chain; b governs the localization length away from the band-edges, such that in the limit $b \to 0$ the standard expression for the localization length of a 1D chain is obtained; and c is proportional to the variance of the disorder. Inverting the relationship between N and ϵ results in an equation which dictates, for a given value of ϵ, over what length of chain $N \le \xi$ propagation is possible. Note that in low-loss microring arrays (here, -0.06 dB/ring), disorder-induced bandwidth collapse occurs before the loss-induced extinction of light transmission; the dynamic range of the measurement typically exceeds 60 dB.

Accordingly, we hypothesized an equation between the usable bandwidth $\Delta\omega$ and the number of resonators N,

$$\Delta\omega(N) = a\left(1 - \frac{Nc^2}{1 - Nb^2}\right)^{1/2}, \tag{4.10}$$

and $\Delta\omega$ collapses to zero at the critical length,

$$N_{crit} = 1/(b^2 + c^2). \tag{4.11}$$

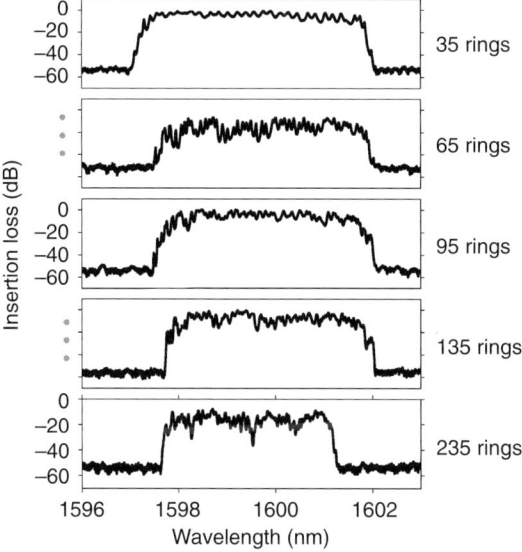

Figure 4.3 Measurements of the transmission intensity of a single passband through coupled microring chains of length 35, 65, 95, 135, and 235 rings, showing the effects of disorder and loss in reducing the usable bandwidth as the chain length increases. In these low-loss structures (approx. 0.06 dB/ring), measured with high dynamic range apparatus, the bandwidth reduction, e.g. collapse of transmission by more than 30 dB at wavelengths not very near the band-edge where the group delay enhancement is relatively small (see Fig. 4.7), cannot be attributed solely to propagation loss.

Monte-Carlo transport simulations using the transfer-matrix method showed that N_{crit} slightly underestimated the number of resonators through which light can propagate, but the small bandwidth in this additional regime was accompanied by very large variations (ripple) in transmission, often exceeding 30 dB over a narrow range of wavelengths, rendering this regime not useful for practical applications. Thus, we are justified in taking N_{crit} as a conservative working estimate of the length of the resonator chain at which bandwidth collapse occurs. In the approach towards this maximum propagation distance, the usable bandwidth narrows with length via two regimes, as shown in Fig. 4.3 and Fig. 4.4: a gradual regime followed by a steeply declining regime as $\Delta\omega \to 0$.

A simple theoretical argument may be used to derive a scaling relationship,

$$N_{crit} = O(|\kappa|^2/\sigma^2), \tag{4.12}$$

between N_{crit} and the disorder strength (variance σ^2) and the average coupling strength (field intensity coupling coefficient $|\kappa|^2$). Disorder-induced phenomena such as localization in the 1D chain can be expected to be governed by the ratio

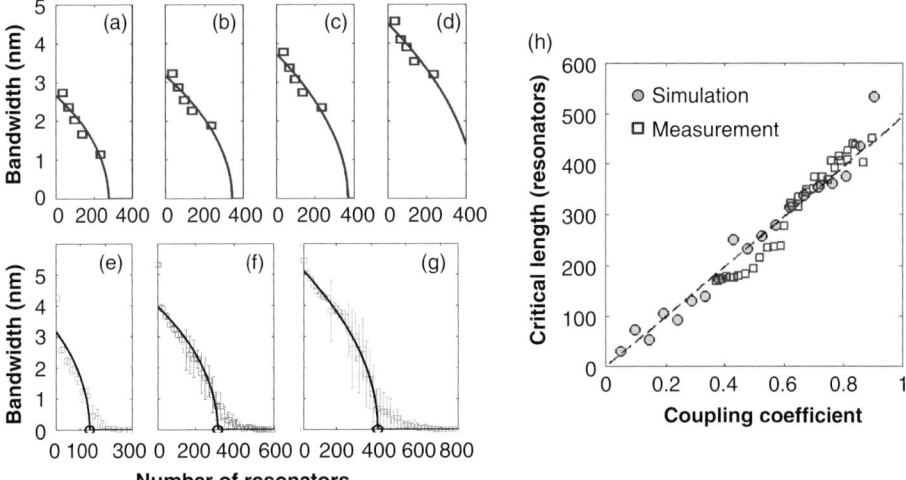

Figure 4.4 (From [337]) Examples of bandwidth scaling with number of resonators. (a)–(d) were experimentally measured at wavelengths 1528.8, 1549.4, 1570.6, and 1599.9 nm, with $|\kappa|^2 = 0.362, 0.454, 0.557$, and 0.692, respectively. (e)–(g) were obtained from simulations with $|\kappa|^2 = 0.0974, 0.334$, and 0.856, and errorbars indicate variations over 16 Monte-Carlo iterations. Solid lines are fitted using Eq. (4.10), and N_{crit} is defined as the number of resonators where the fitted line crosses the horizontal axis, zero bandwidth. (h) $N_{crit} = |2\pi\kappa/\sigma|^2$ on the vertical axis should vary linearly with $|\kappa|^2$, the horizontal axis. Data from measurements are shown with squares, and from *ab-initio* transfer matrix simulations are shown with circles. The fitted dashed line gives disorder strength $\sigma^2 = 0.080$, in close agreement with the assumption made in the simulations, $\sigma^2 = 0.075$.

of disorder to level spacing, $\delta\omega/\Delta\omega$, where $\delta\omega$ is a measure of the disorder on an energy scale and $\Delta\omega$ is the inverse of the density of states. The latter is a function of frequency within the passband, being widest at band center and smallest at the band-edges. Although phenomenological, the criterion $\delta\omega/\Delta\omega = 1$ is useful for estimating the expected critical length; and the implications of this assumption were shown to be consistent with results obtained from direct numerical simulation.

The critical length at which bandwidth collapse occurs is obtained by equating the normalized level spacing $\Delta\omega/\omega$ to the perturbation due to disorder; the latter is taken as equal to $\sigma/N^{1/2}$, where σ represents the standard deviation of the normalized variations (e.g. in the coupling coefficients and/or the phase) and the renormalization by $N^{1/2}$ accounts for exchange narrowing in a chain of N coupled oscillators. For large N, the equality implies that

$$N_{crit} = (2\pi\kappa_\omega/\sigma)^2. \tag{4.13}$$

Figure 4.4 shows a comparison between N_{crit} calculated using Monte-Carlo transfer-matrix simulations of transport in weakly disordered microring chains, and the extrapolation of experimentally measured bandwidth scaling, using Eq. (4.10), in chains of length 35, 65, 95, 135, and 235 rings.

Equation (4.13) predicts that $N_{crit} = O(\kappa/\sigma)^2$ for bandwidth collapse, whereas it has been shown that $\xi = O(\kappa/\sigma)^{2/3}$ for band-edge localization [141, 338]. Therefore, in the comparison between N_{crit} and ξ, the former, although greater in magnitude (i.e. bandwidth collapse occurs after the onset of band-edge localization), exhibits higher sensitivity to the magnitude of disorder, σ.

For practical applications, the regime of near-localized transport is of limited interest, since the transmission has high ripple, group delay is not well defined, and the overall insertion loss of the structure is high. Accordingly, in the next section, we will describe studies of the weakly disordered regime, close to the ideal ballistic regime, where the signatures of disorder first become manifest.

For filtering applications [89, 546], disorder tolerance in coupled resonators can be achieved at larger values of κ, which results in decreasing the finesse, $F = \pi / \sin^{-1}(\kappa)$. Lowering F, where acceptable, also mitigates thermal detuning effects by reducing the buildup of optical power in a few resonators near the input of the chain, which is helpful for nonlinear optics. Out-of-band signal rejection is not necessarily worsened, since the slope of the band-edge is steepened by adding more stages, without incurring a large penalty in the available bandwidth until the localization regime is imminent.

4.4 Statistical measurements of light transport in the intermediate regime

Nearly all optical waveguides are, in practice, used over a wide range of wavelengths, and a routine measurement performed in the laboratory is to measure the spectral variations in the main transport characteristics, e.g. insertion loss and group delay. Such measurements, when carried out with high resolution over a relatively narrow range of wavelengths over which the average transport properties remain constant, yield statistical data which reveal insights into the impact of disorder on light transport. In Fig. 4.5, we show intensity transmission through a set of coupled microring chains with good performance. For different structural parameters (i.e. a different coupling-to-disorder strength ratio), the transmission through a similar set of microrings can be more significantly degraded, as shown in Fig. 4.6.

In fact, measurement of only the average intensity transmission of an optical waveguide constitutes an incomplete story on transport. Absorption plays an important and unavoidable role in practical devices, especially in the slow-light regime, and absorption reduces the average transmission of light in a similar way to localization (exponentially with length). Measuring only the band-averaged group delay is also insufficient, since the average propagation time through the waveguide scales linearly with length in both the localized and non-localized regimes.

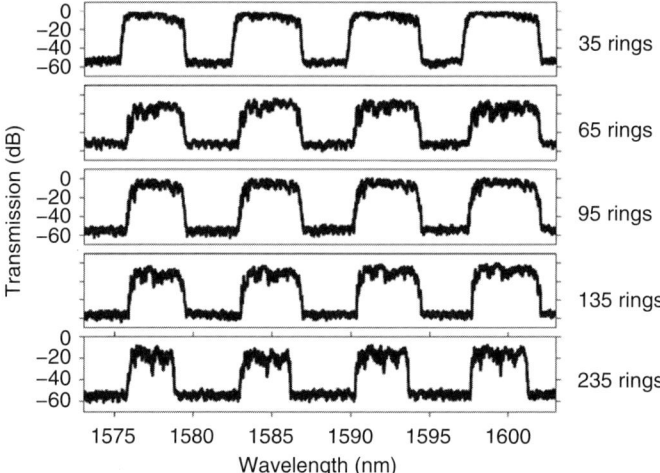

Figure 4.5 (From [92]) Transmission intensity measured for several passbands for coupled microring chains of length 35, 65, 95, 135, and 235 rings (top to bottom), on a chip with relatively wide bandwidths, i.e. large value of inter-resonator coupling coefficient $|\kappa|$ so that the impact of disorder was relatively small. In this case, very long microring chains support transmission with relatively good characteristics

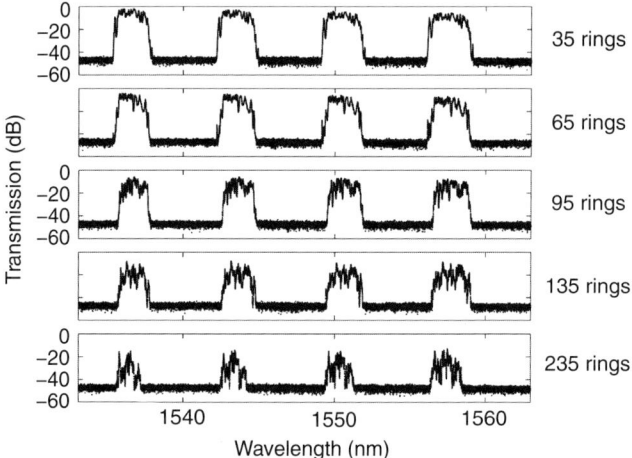

Figure 4.6 Transmission intensity measured for several passbands for coupled microring chains of length 35, 65, 95, 135, and 235 rings (top to bottom), on a chip with smaller bandwidths, i.e. smaller value of inter-resonator coupling coefficient $|\kappa|$ than in Fig. 4.5. Here, the impact of disorder was larger, leading to very large ripples in the transmission of the longer chains, narrower bandwidths, and higher insertion loss.

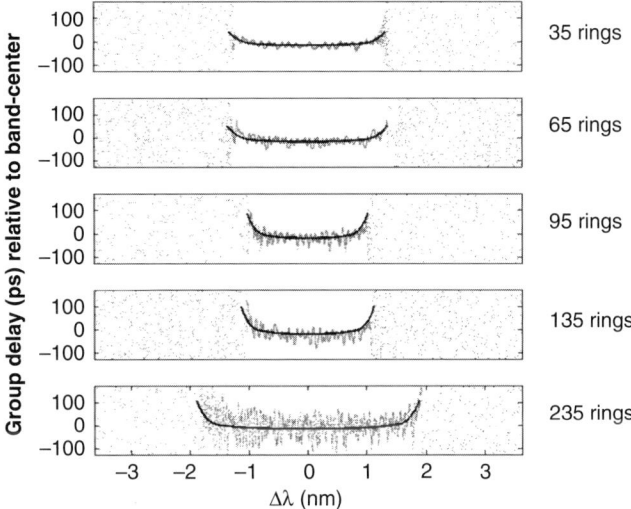

Figure 4.7 Ref [93] Measurements of the group delay (units of picoseconds) of a single passband, for coupled microrings of length 35, 65, 95, 135, and 235 rings (top to bottom). The group delay ripple is defined as the difference between the measured points (dots) and the theoretically expected behavior (solid line). We disregard the band-edges in the analysis described in the text, since this regime is characterized by reduced transmission and greater propagation loss.

However, a measurement of the ripple over a spectral range where, ideally, the transmission should be flat (or described by a parametric equation) can yield useful statistical information. Figure 4.7 shows an example of a high-resolution, flat-top transmission spectrum in microring chains over a single passband, with increasing ripple as the chain length increases. In particular, a more complete distinction between ballistic, diffusive, and localized transport regimes can be obtained through an analysis of the statistical properties of the transmission [59, 241, 318, 413, 466, 498, 501] and propagation time [75, 155, 475].

The main body of experimental evidence in this topic comes from microwave tube waveguides that are about 5 cm wide, filled with 5 mm polystyrene microspheres, or with manually inserted micrometers, and in periodic metallic wire meshes containing metallic scatterers, to enhance scattering of radio-frequency waves [75, 78, 155, 257, 466]. These studies are valuable because the tube can be shaken, or the metallic scatterers rearranged, between successive measurements, so as to achieve a wide range of disorder realizations, which is difficult to do for lithographically fabricated devices. Instead, based on the principle of spectral ergodicity, we measure transport at different frequencies, over which the average value of the nearest-neighbor coupling coefficient and disorder strength (scattering cross-section) may be expected, in disorder-free devices, to be constant, or predictable according to a parametric formula. (According to the principle of spectral ergodicity, an average over an ensemble of random matrices provides the same result as an average over energies for a single element of this ensemble in the limit where the dimension of the matrices is large.)

Optical experiments are also considered difficult because scatterers are usually weak, and the typical number of transverse channels is large [59] (e.g. in slab waveguides). These limitations can be overcome by using high-index contrast single-transverse-mode silicon wire waveguides to form a periodic arrayed waveguide of microresonators. Similarly, enhanced scattering is observed in photonic crystal waveguides [373, 482], compared with simple slab or wire waveguides without periodic patterning [4]. In the absence of self-averaging, i.e. the nanoscale lithographic disorder is "frozen" in time, the variances of transmission quantities normalized to their ensemble averages are not reduced, even as we average over repeated measurements; thus, such averaging can be used to reduce the effects of instrumental noise on the amplitude and phase measurements.

Here, we focus on light transport in low-loss coupled microring chains; the silicon photonic waveguides reported here are nearly 20 times longer than their earlier counterparts made using other material systems [343]. High propagation loss in earlier waveguides inhibited transport over more than a few unit cells, and transmission characteristics in these cases can be dominated by finite-size or coupling effects. By measuring chains up to 235 rings in length, edge effects are relatively minor.

Intensity ripple was calculated from a measurement of $I(\lambda)$ over the central portion of a passband, i.e. away from the band-edges. Similarly, based on a high-resolution measurement of the group delay $\tau(\lambda)$ over the central portion of a passband, we calculated the group delay ripple (GDR) $\Delta\tau(\lambda) \equiv \tau(\lambda) - \langle\tau(\lambda)\rangle$, where the averaging is done over the spectral bandwidth of interest (e.g. pulse bandwidth in data communications). In optical communications, the GDR is usually modeled as a sinusoidal variation [121],

$$\Delta\tau = A \cos\left[2\pi(\delta\lambda/\lambda_p) + \psi\right], \qquad (4.14)$$

where A is the peak-to-peak group delay ripple amplitude, $\delta\lambda$ is the offset from a reference wavelength, λ_p is the period of the ripple, and ψ is the phase of the ripple. Components of the group delay ripple that are on the order of the signal bandwidth affect the signal spectrum by imposing an average chromatic dispersion, which can be compensated at the receiver. Therefore, in the study of fiber Bragg gratings, the average dispersion over the (narrow) band of interest is usually subtracted out from the phase variations, and the residual phase variation is used as the noise statistic which determines performance degradation. The averaging bandwidth depends on the spectral width of the pulses used in data transmission, e.g. 40 Gbps data streams will average over ripple with spectral components less than 100 pm, effectively sensing them as a constant group delay.

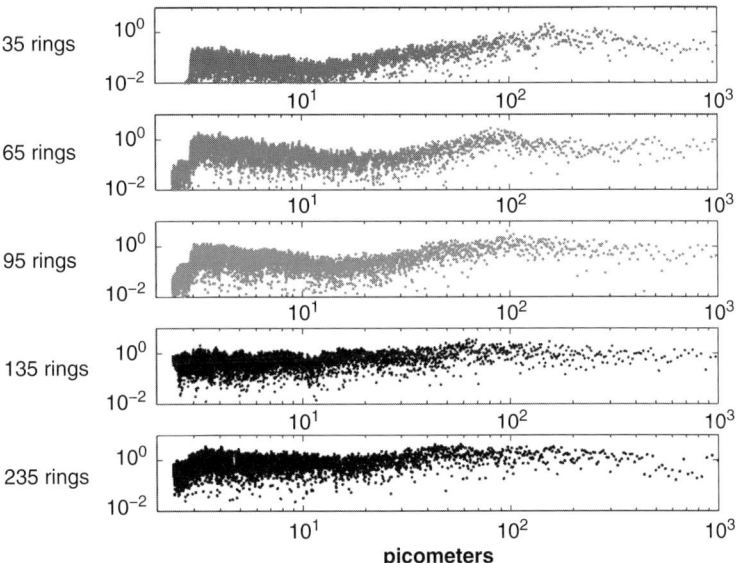

Figure 4.8 Fourier analysis of the group delay ripple (GDR) shows the absence of any specific ripple periodicities in longer chains of resonators. In contrast, a significant fabrication defect would create well-demarcated Fabry–Perot type reflections in the measured spectrum, which would manifest here as sharp and well-defined spikes. In their absence, we use GDR as a disorder statistic in analyzing the transport properties of coupled microring chains.

Here, we use ripple statistics as a disorder metric. Spectral component analysis of GDR, i.e. the Fourier transform of $\Delta \tau |\kappa|/L$, shown in Fig. 4.8, reveals that no significant spatial component dominates the measured GDR distribution. This observation supports our usage of the normalized ripple metrics as a random statistic, rather than an artifact of fabrication, since a significant dominant Fabry–Perot type resonance would result in a sharp spectral component. The high-frequency cutoff at a ripple period of approximately 1.4 pm comes from the resolution of the instrument used to measure group delay.

4.4.1 Intensity statistics

The distribution of the normalized intensity of a classical wave follows the same statistics as the channel, or "angular," transmission T_{ab} of an electron [155, 318, 335, 467]. As shown in Fig. 4.9(a), we observed that $\langle \ln I \rangle$ scales linearly with length, $\langle \ln I \rangle = -L/l$, where the extinction length is $l = 73$ unit cells, based on a measured propagation loss of -0.06 dB/unit cell (each unit cell is 14.3 μm long). In Fig. 4.9(b), we plot the distributions of the logarithm of the intensity, $\ln I$. The lateral shift of the distributions follows from inspection of Fig. 4.9(a).

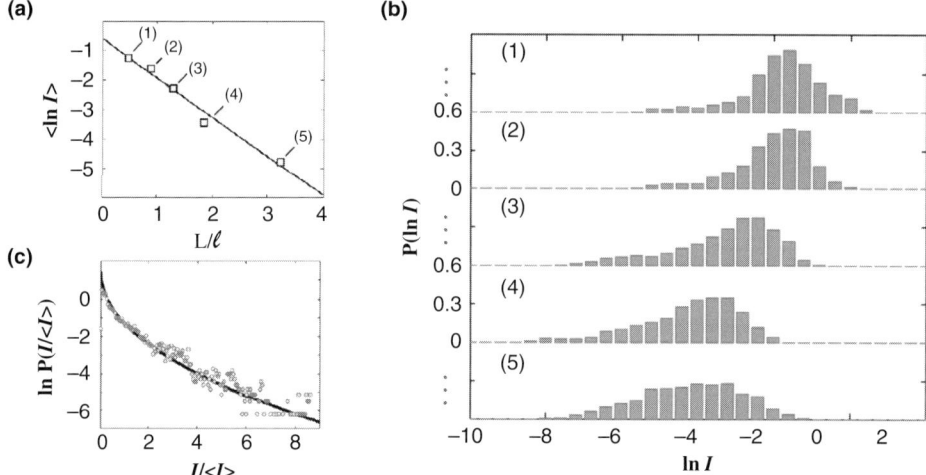

Figure 4.9 (a) The average of the logarithm of the measured intensity, $\langle \ln I \rangle$, decreased exponentially with length (in number of unit cells, L, divided by the extinction length, l). Approximately 5000 measurements were carried out for each waveguide, (1)–(5), at different wavelengths within the transmission band. (b) The probability distributions of $\ln I$ progressively shifted towards lower transmission, in agreement with Fig. 4.9(a), but also became broader as length increases, which suggests significant deviations from ballistic transport. Absorption inhibits the likelihood of higher values of transmission in longer waveguides, thus attenuating the right-side tails of the normal distribution more than the left-side tails [318]. (c) The distribution of normalized intensity, $I/\langle I \rangle$, for the longest waveguide fitted well with the stretched-exponential function, Eq. (4.15) with a fitted conductance value $g = 1.81 \pm 0.04$.

Absorption introduces an asymmetric cutoff to the higher transmission side of the distributions, as predicted by simulations [318]. The measured data show that the mean value and width of the distributions are comparable, i.e. $-\langle \ln I \rangle \simeq \mathrm{var}(\ln I)$, and as such, longer waveguides will show greater fluctuations around the average transmitted intensity.

Further insight can be obtained from the statistics of $\hat{I} \equiv I/\langle I \rangle$, i.e. intensity normalized to its mean, as shown in Fig. 4.9(c), for the longest waveguide. The distribution was fitted by a stretched-exponential function [241],

$$P(\hat{I}) = \exp\left(-2\sqrt{g\hat{I}}\right), \tag{4.15}$$

where $g = 1.81 \pm 0.04$. We inferred a localization length $\xi = gL \approx 425$ unit cells ($L = 235$ unit cells), and localization was neither expected nor observed in these waveguides at the measured wavelengths near the center of the passband.

The stretched-exponential behavior indicated by Eq. (4.15) may be explained by the quasi-discrete nature of the frequency spectrum of coupled-resonator waveguides: at a given excitation frequency, only one spatially extended eigenmode (whose frequency most closely coincides with the tunable laser) determines the summation-of-modes expression for the Green function, thereby leading to Eq. (4.15) [335]. In contrast, a Rayleigh distribution would be obtained if the sum of a large number of modes contributed to the Green function. As such, stretched-exponential behavior is expected in the prelocalized transport regime of few-mode waveguides. In the localized regime, a log-normal distribution of $\hat{I}$ would be expected, with a strong attenuation of the likelihood of measuring higher values of transmittance at longer lengths [241, 498].

Although Eq. (4.15) is characteristic of the diffusive transport regime, which occurs between the ballistic and localized regimes [78, 413, 501], evidence of diffusive behavior is unexpected in one-dimensional waveguides since a single transverse-mode waveguide restricts the possibility for scattering in transverse directions [59, 335]. To examine further whether a diffusive regime was present, we examined the dynamical aspects of transport, which were obtained from our measurements of the statistical distributions of the time delay.

4.4.2 Time statistics

As shown in Fig. 4.10(a), the average measured value of the time delay scaled linearly with the number of unit cells. However, since the average delay scales linearly with length in each of the ballistic, diffusive, and localized regimes [318], we further examined the distributions of the normalized time delay of propagation

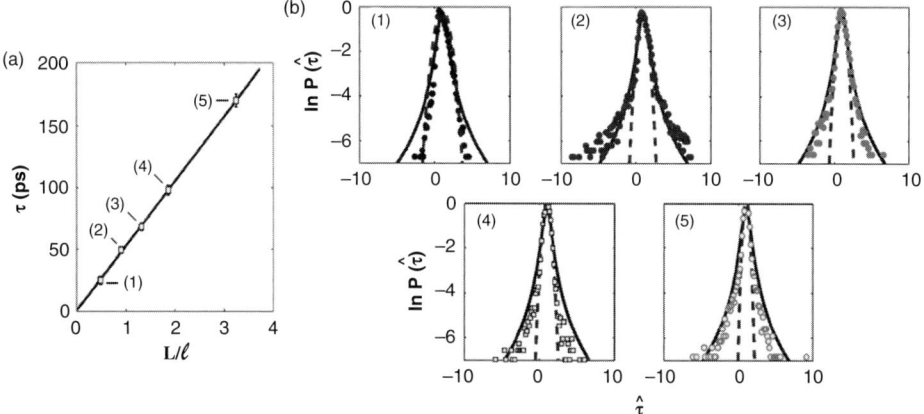

Figure 4.10 (a) The measured propagation delay (τ, units of picoseconds) increased linearly with length ($L = 35, 65, \ldots, 235$ unit cells, divided by the extinction length, $l = 73$ unit cells). (b) The probability distributions of $\tau/\langle\tau\rangle$ are shown using dots. The dashed line is a Gaussian fit to the data, which indicates ballistic propagation statistics, and the solid line is the prediction of Eq. (4.16), i.e. the distribution of single-channel time delay based on a diffusion model, without any fitting parameters, since the parameter Q in Eq. (4.16) is calculated from L and l [502].

($\hat{\tau} \equiv \tau/\langle\tau\rangle$), which are plotted in Fig. 4.10(b), using a logarithmic scale for the vertical axis for clarity.

Whereas the delay time distribution for the shortest waveguide was well described by a Gaussian fit, which is characteristic of the ballistic propagation regime [475], the distributions for the longer waveguides were more closely described by the functional form [155, 335]:

$$P(\tau) = \frac{Q}{2}\left(Q + \hat{\tau}^2\right)^{-3/2}. \tag{4.16}$$

The constant Q in Eq. (4.16) is an experimentally determined quantity based on the extinction length and the length of the waveguide [502] with the values $Q = (0.395, 0.383, 0.365, 0.336, 0.261)$ in the five waveguides. Note that there are no free parameters in Eq. (4.16). This equation was derived for the distribution of delay times under the Gaussian approximation for complex field amplitudes in the diffusive regime [155].

As the length of the waveguide was increased, the statistics of transport evolved from ballistic to diffusive, as shown by the change from a Gaussian fit of the data in the first panel of Fig. 4.10(b) to agreement with the prediction of Eq. (4.16) in the longer waveguides. Equation (4.16) decays polynomially, i.e. more slowly than a Gaussian, at large values of τ and, therefore, significant fluctuations in propagation delay may be measured in the longer waveguides. This is in contrast with the measurements made on a different chip, reported in [94], which was more disorder-resilient (larger value of $|\kappa|/\delta\omega$ in terms of the discussion in the earlier sections), and ballistic statistics were observed even in the longest waveguide.

The product of intensity and delay defines the weighted delay time, $W = I\tau$, which relates to the stored electromagnetic energy in the medium and, upon summation over all modes, is the Wigner delay time [155]. The average value $\langle W \rangle$ obeys the simple equation $\langle W \rangle = \langle I \rangle \langle \tau \rangle$ [502] and the length scaling of the two quantities on the right-hand side of this equation are reported in Figs. 4.9 and 4.10, respectively.

Distributions of $\hat{W} \equiv W/\langle W \rangle$ are shown in Fig. 4.11. According to diffusion theory, these distributions have the functional form,

$$P(\hat{W}) = \frac{1}{\sqrt{Q+1}} \exp\left[\frac{-2|\hat{W}|}{\theta(\hat{W}) + \sqrt{Q+1}}\right], \tag{4.17}$$

where $\theta(x)$ is the Heaviside function ($\theta(x) = 0$ for $x < 0$, and $\theta(x) = 1$ for $x > 1$). Equation (4.17) is shown by the solid black lines in Fig. 4.11, and similarly to Eq. (4.16), there are no fitting parameters. Although scattering theory allows for negative delay times, such measurements are usually associated with low transmission intensity, e.g. dark spots in a speckle measurement [502], and were not seen here.

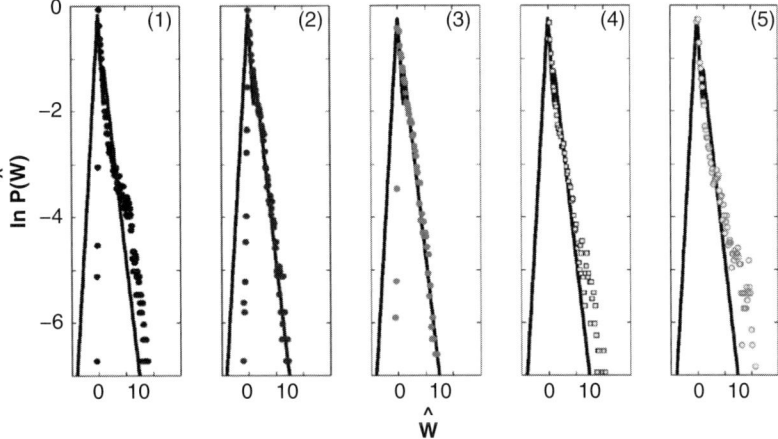

Figure 4.11 The probability distributions of Wigner delay time, $\hat{W} = W/\langle W \rangle$, where W is the weighted delay time, $W = I\tau$, for the coupled-microring lengths of 35, 65, 95, 135, and 235 rings, labeled (1)–(5), respectively, generally agreed with Eq. (4.17), shown by the solid black line, for positive delays. Although allowed by scattering theory, negative weighted time delays were not measured in our experiments.

In comparison with [94], these results demonstrate a transition away from ballistic to diffusive propagation in these measurements, even when the possibility of scattering into multiple transverse modes was restricted by using single transverse mode waveguides. Since the values on the horizontal axis in both Figs. 4.9(b) and 4.10(b) are normalized to their respective mean values (which increase linearly with length), these measurements indicate that the longer waveguides in this batch showed considerable fluctuations in absolute units. As an example, for the longest waveguide, fluctuations measured between $\pm\hat{\tau}$ evaluate to a variability window of about 350 μs in the propagation delay time. The diffusive model of propagation is based on the assumption that the propagating phase is randomized by a large number of collisions with scatterers. Here, the significantly higher scattering in periodically patterned waveguides, compared with conventional waveguides, enabled us to observe experimentally a transition from the Rayleigh regime, where ballistic transport dominates, to one where the transport is described by the stretched-exponential distribution of the normalized intensity, and non-Gaussian statistics of the normalized propagation time.

4.5 Summary

Long coupled microring chains are interesting for physics studies because the availability of measurable statistical variables makes it possible to observe experimentally the transition away from the ballistic transport regime even in the presence of absorption. A next step could be the realization of devices in which this transition could be experimentally controlled, e.g. by control knobs on the inter-resonator coupling coefficients.

As far as applications are concerned, although periodically patterned structures such as photonic crystal waveguides and coupled-resonator optical waveguides accumulate significantly more phase or delay per unit length than in conventional waveguides, the higher degree of scattering may lead to diffusive rather than fully ballistic transport, which randomizes the phase and inhibits some important applications in e.g. coherent optics, optical interconnects, waveguide quantum light circuits, and slow light. As may be expected, the relevant parameter of interest is the ratio of the magnitude of the nearest-neighbor coupling coefficient to the disorder strength; devices in which this ratio is a large value tend to lie well within the ballistic transport regime.

With improvements in fabrication technology, leading to lower waveguide loss, greater uniformity among unit cells, and electronic control over the pole-zero locations of each unit cell, it is possible to envision, in the not too distant future, several applications of coupled microring devices, beyond optical delay lines and filtering [545]. Already, relatively modest lengths of coupled ring chains (a few dozen microrings) have shown efficient, low-power wavelength conversion [343], based on a very high four-wave mixing nonlinearity ($\gamma_{eff} = 4000$ W.m^{-1} at band center), and have been used in the first report of heralded single photons from a silicon chip [107].

Acknowledgments

The author is grateful to Jung S. Park, Michael L. Cooper, Greeshma Gupta, Mark Schneider, Junrong Ong, Ryan Aguinaldo, and Ranjeet Kumar for contributions from their work in the Micro/Nano-Photonics Group at the University of California, San Diego. He is also grateful to William M. J. Green and colleagues at IBM Thomas J. Watson Laboratory (Yorktown Heights, NY), and Dawn Gifford of Luna Technologies (Blacksburg, VA) for their invaluable assistance with this work. This work was funded in part with support from the National Science Foundation.

5

One-dimensional photonic quasicrystals

MHER GHULINYAN

5.1 Introduction

Self-organization is one of the most extraordinary tools in Nature which assembles its elementary building blocks, atoms and molecules, in the form of solid substances. Crystals, in this sense, represent the class of a vast number of solid materials in which atoms (molecules) are brought together in a perfect, spatially periodic manner. A macroscopic crystal, therefore, is made by repeating its smallest microscopic period – the crystal *unit cell* – infinitely in space (Fig. 5.1(a)). Therefore, when translating the crystal by an integer number of unit cell size, the crystal coincides with itself. This property is in the basis of modern solid-state physics and crystallography and is known as the *discrete translational symmetry*. Crystals are thus ordered materials in which atoms show both short- (within a couple of nearest neighbors) and long-range order. Because of the periodic lattice potential, the energy states within a crystal are extended, and the electrons can diffuse freely through it.

At the opposite extreme of crystalline order there are the disordered or amorphous solid materials, in which the short-range order may still exist, while the long-range order is completely missing (Fig. 5.1(c)). In these kind of solids, e.g. in heavily doped semiconductors or glasses, periodicity is lacking and the atomic potential varies in a random manner. In certain cases, when the degree of randomness is large enough, electron wavefunctions localize exponentially near individual atomic sites and the diffusion of charge carriers vanishes (Anderson localization) [11].

With the advances in X-ray diffraction tools at the beginning of the twentieth century, scientists learned to image the signatures of long-range order of crystals by mapping the symmetries of their reciprocal space lattices. The crystal symmetries

Light Localisation and Lasing, ed. M. Ghulinyan and L. Pavesi. Published by Cambridge University Press.

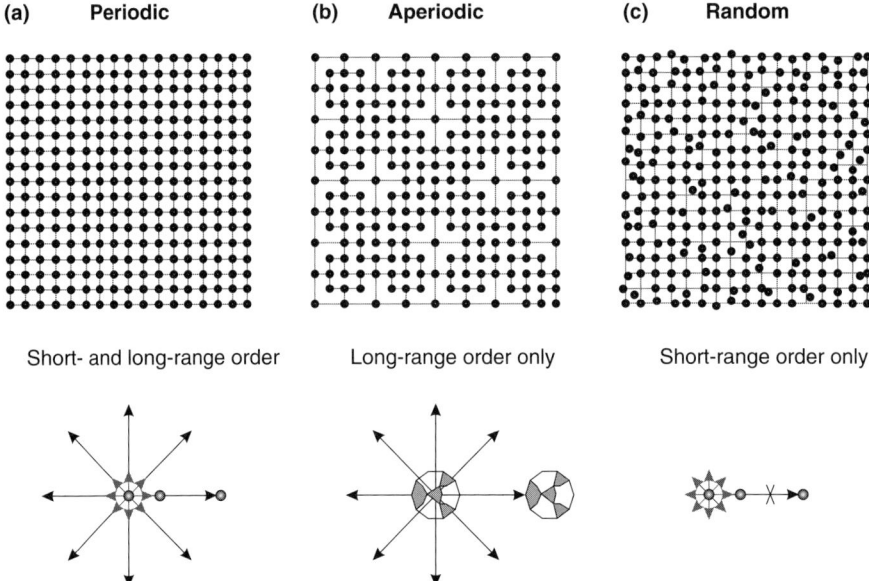

Figure 5.1 Schematic views of a two-dimensional (a) periodic, (b) quasiperiodic and (c) random lattice. Below each panel the corresponding situation for the atomic order is shown.

for all possible arrangements of perfect periodicity are classified in 230 space groups. Contrary to this, the diffraction form, a disordered solid, shows cloud-like concentric rings indicating diffraction from a totally random atomic lattice and, consequently, a lack of any symmetry or, equivalently, any preferential direction in the solid.

5.1.1 In between perfect periodicity and complete randomness

In 1982 scientists found that certain "crystalline" solids were showing point-like X-ray diffraction patterns with symmetries, e.g. five-fold and ten-fold, that did not fit any of the known 230 space groups [428].[1] These solids were named *quasicrystals* and represent a large class of materials in between crystalline and disordered solids. A quasicrystal shows a perfect long-range order but does not possess a periodic lattice (Fig. 5.1(b)). Equivalently, the quasicrystalline lattice can fill the space densely, however, it does not possess translational symmetry.

From a mathematical point of view, two- (2D) and three-dimensional (3D) quasicrystals are formed following aperiodic tiling rules. Discovered by

---

[1] According to the crystallographic restriction theorem the crystals are limited to two-fold, three-fold, four-fold and six-fold rotational symmetries.

mathematicians in the early 1960s,[2] the aperiodic tilings can fill the plane densely by a combination of scaling, rotating, and repeating procedures of compound patterns. Similar rules allow compound forms to fill the space [178].

So far, on our way towards an understanding of the surrounding physical world and natural phenomena, scientists have adopted mathematical approaches which point to reduction of an existing problem into a one-dimensional (1D) one. This is because 1D mathematical models can be described analytically and solved exactly. Higher-dimensional problems, in many cases, are then constructed based on a generalization of 1D models and their solution, while in other cases, numerical methods and laborious computer simulations are required.

In the next section, we will address the principal aspects of 1D quasiperiodicity with a particular focus on 1D Fibonacci chains. The rest of the chapter will be dedicated to the electromagnetic counterpart of 1D Fibonacci structures as relatively the simplest case of the large class of photonic quasicrystals.

5.2 1D quasiperiodicity: Fibonacci chain

Perhaps, the most well known and studied example of an aperiodic sequence of numbers is the Fibonacci sequence [440]. The first two numbers, called the *seed*, are two ones, and the following numbers,

$$2, \ 3, \ 5, \ 8, \ 13, \ 21, \ \dots \tag{5.1}$$

are obtained by summing each time the last two of the list.[3] Mathematically, the recurrence rule defines the nth Fibonacci number as

$$F_n = F_{n-1} + F_{n-2}, \tag{5.2}$$

with the seeds $F_0 = 1$ and $F_1 = 1$. The successive Fibonacci numbers, as shown in Fig. 5.2(a), are intimately connected through the golden ratio,

$$\varphi = \lim_{n \to \infty} \frac{F_n}{F_{n-1}} = \frac{1 + \sqrt{5}}{2} \approx 1.61803 \dots \tag{5.3}$$

The deterministic nature, which follows from recurrence rules used to create aperiodic sequences, and, in particular, those of the Fibonacci type, allows us to construct a variety of many different "forms" – strings or patterns. A nice visual example is the 2D tiling of squares, which are scaled such that their sides are equal to Fibonacci numbers in length, as shown in Fig. 5.2(b).

Scientists have observed that in many plants, such as sunflower and cactus, the florets and stickers are organized on composite spiral-like patterns following

---

[2] Girih tilings were used in the late 1200s for decoration of buildings in medieval Islamic architecture.

[3] In an alternative definition, the seed numbers are 0 and 1, but the recurrence rule is the same.

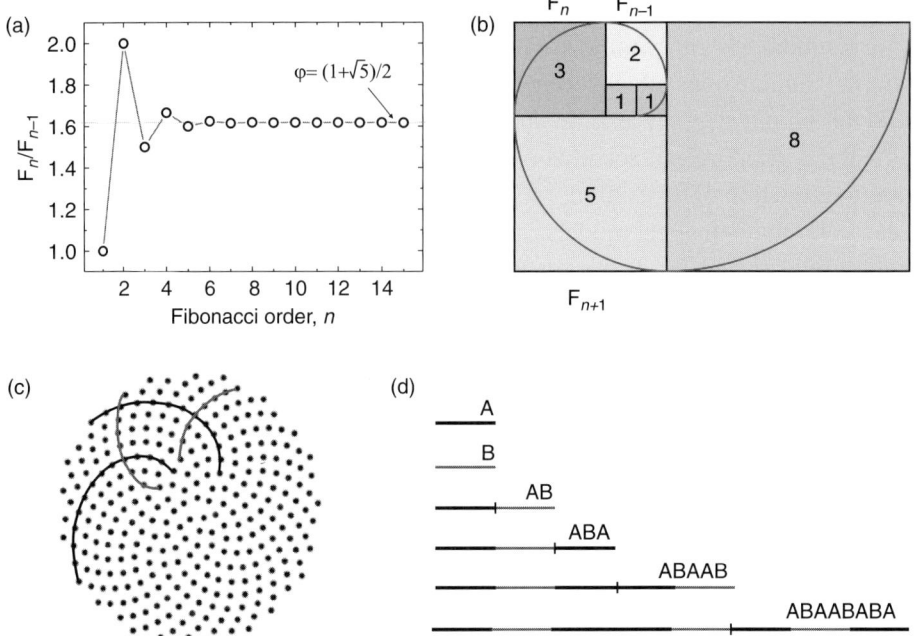

Figure 5.2 (a) The Fibonacci numbers and the golden ratio. (b) Tiling of squares with sides increasing as Fibonacci numbers and the golden spiral. (c) The spiral pattern of Vogel, reproducing sunflower florets. (d) The two-component Fibonacci words.

Fibonacci rules [115, 435]. A model reproducing the sunflower floret pattern has been introduced by H. Vogel [512]. It consists of the generation of a scatter plot in polar coordinates $\left(r = a\sqrt{n}, \; \theta = 2\pi n/\varphi^2\right)$, where a is a scaling factor, n is the index of the floret and φ is the golden ratio (Fig. 5.2(c)). In this model, the angle $\theta_G = 2\pi/\varphi^2 = 137.508°$ is the so-called golden angle, at which in a circle of circumference $a + b$ the ratio a/b of arcs a at $2\pi - \theta_G$ and b at θ_G are equal to $(a + b)/a$.

In 1D, by using recurrence rules, one may construct aperiodic sequences starting from seed *strings*. The most famous example is forming so-called "aperiodic words" by starting with seeds A and B and recurrently applying some rule for the next word generations. In particular, in the case of a Fibonacci sequence,[4] an inflation rule is used,

$$A \rightarrow AB \tag{5.4}$$

$$B \rightarrow B. \tag{5.5}$$

---

[4] Another intensively studied aperiodic sequence is the Thue–Morse sequence, which uses an $\{A \rightarrow AB, B \rightarrow BA\}$ inflation [478].

Table 5.1 *The Fibonacci strings and relations between*
their length and counts of components.

String	Length	Counts(A)[a]	Counts(B)
A	1	1	0
B	1	0	1
AB	2	1	1
ABA	3	2	1
$ABAAB$	5	3	2
$ABAABABA$	8	5	3
$ABAABABAABAAB$	13	8	5

[a] Note: length/counts(A)→ φ and counts(A)/counts(B)→ φ for
long strings.

The first six Fibonacci strings, therefore, are A, B, AB, ABA, $ABAAB$, and
$ABAABABA$ (Fig. 5.2(d)). As can be seen, this inflation rule is equivalent to
adding the $(n-2)$th sequence to the end of the $(n-1)$th in order to generate the
nth series, $F_n = F_{n-1}F_{n-2}$. One particularity of the Fibonacci string is that, while
AA is a frequent instance along an F_n for large n's, other instances such as BB
or AAA may never appear (see Table 5.1). As will be discussed in the next section, this particularity appears to have significance in the context of a 1D photonic
Fibonacci type quasicrystal.

5.3 Photons in a 1D optical potential

The energy spectrum of a quantum particle in a semiconductor crystal is described
by extended Bloch modes and consists of allowed and forbidden bands. Electronic
crystals have their analog in the world of electromagnetic waves (photons) called
photonic crystals [220]. Photonic crystals are complex materials, in which a periodic variation of the material dielectric constant, $\epsilon(\vec{r})$, on a length scale comparable
with the wavelength leads to the formation of energy bands where the propagation
of photons is allowed or forbidden.

Likewise, photonic crystals, solid materials, can be periodic, quasiperiodic and
random, reflecting the variation of $\epsilon(\vec{r})$ in space. This variation, in turn, can be
either continuous (Fig. 5.3(a), top) or abrupt with the space coordinate (Fig. 5.3(a),
bottom). In the first case a smooth variation of $\epsilon(\vec{r})$ between ϵ_1 and ϵ_2 may reflect a
continuous change in material composition (an $Al_xGa_{1-x}As$ alloy, for example). In
the second case an abrupt change in $\epsilon(\vec{r})$ is achieved when two different materials
with dielectric constants ϵ_1 and ϵ_2 are stacked together.

Throughout this chapter, we will focus on the optical properties of 1D photonic quasicrystals of a Fibonacci type which are formed by stacking two different

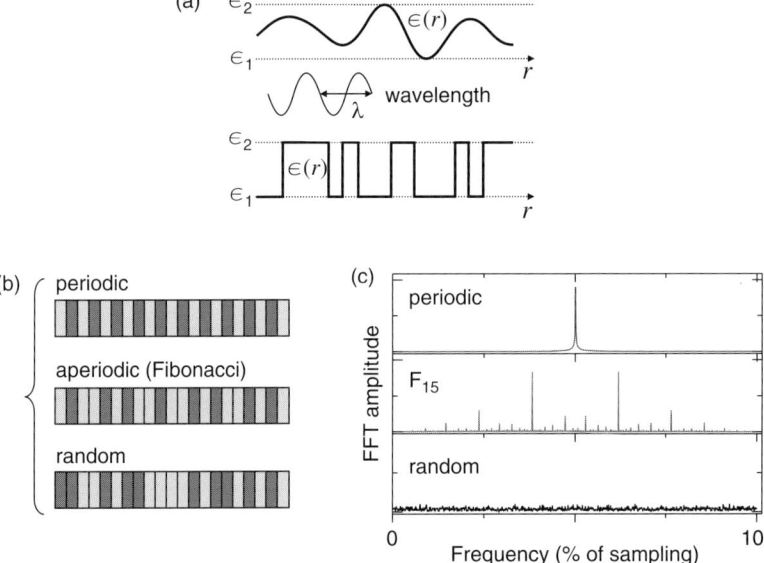

Figure 5.3 (a) The variation of the dielectric constant in space on a length scale
comparable with the light wavelength lies in the basis of photonic crystals. $\epsilon(\vec{r})$
may vary either in a smooth or a step-like manner. (b) With a step-like change of
$\epsilon(\vec{r})$, using two different materials A and B results in the formation of a periodic (top), a quasiperiodic (middle) or a completely random (bottom) optical
potential for photons. (c) The Fourier transformation spectra of the corresponding bi-component layer sequences show a single sampling frequency (periodic), a
self-similar frequency distribution pattern (quasiperiodic) and an uncorrelated flat
spectrum with no specific frequency (random).

materials, say A and B, following the inflation rule discussed above. Prior to this,
we will introduce the basic properties of 1D photonic multilayered structures on
the simple examples of a distributed Bragg reflector (DBR) and a Fabry–Pérot
microcavity.

Distributed Bragg reflector – Otherwise known as a dielectric mirror, the functionality of a DBR is based on the interference between electromagnetic waves
(EMW) which are multiply reflected and refracted at each boundary of the multilayer structure. An EMW of frequency ω_0 is efficiently reflected[5] from a DBR
if the multilayer is constructed by alternating quarter-wave layers of two different
materials A and B with refractive indices $n_A = \sqrt{\epsilon_A}$ and $n_B = \sqrt{\epsilon_B}$, respectively. The quarter-wave condition implies that $n_A d_A = n_B d_B = \lambda_0/4$, where d_A
and d_B are the layer thicknesses, $\lambda_0 = 2\pi c/\omega_0$ is the wavelength and c is the

---

[5] For simplicity, we will consider the case of a normal incidence in the following.

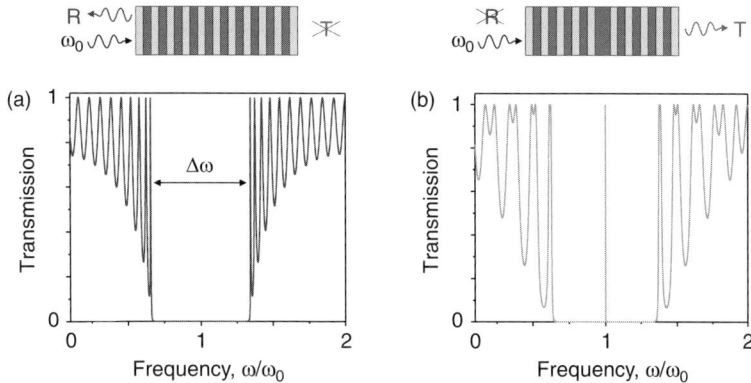

Figure 5.4 The multilayer structure and the transmission spectra of (a) a distributed Bragg reflector and (b) a Fabry–Pérot microcavity.

EMW velocity in free space. The portion of the EMW intensity reflected from the multilayer embedded in air from both sides is given by

$$R = \left[\frac{n_A^{2N} - n_B^{2N}}{n_A^{2N} + n_B^{2N}} \right]^2, \tag{5.6}$$

where N is the number of periods (AB pairs) of the DBR. In absence of losses (scattering, absorption, etc.) the transmission of the mirror is $T = 1 - R$. As shown in the example in Fig. 5.4(a), the transmission vanishes within a bandwidth of $\Delta\omega$ around the frequency ω_0, which means that in this frequency window – *the photonic band gap* – EMWs cannot propagate through the structure. The relative width of the photonic band gap is given approximately by

$$\frac{\Delta\omega}{\omega_0} = \frac{4}{\pi} \arcsin\left(\frac{|n_A - n_B|}{n_A + n_B} \right). \tag{5.7}$$

From Eq. (5.6) we see that the reflectivity of the dielectric mirror grows with either increasing the number of periods N or increasing the refractive index contrast $\frac{|n_A - n_B|}{n_A}$. This last is also crucial for making a wider photonic band gap, as follows from Eq. (5.7), while increasing N affects only the steepness of the band-edges.

A common mathematical approach to calculating the spectral properties of 1D multilayered photonic structures is based on the transfer matrix method [376]. This method consists of relating the electromagnetic fields on two boundaries (left and right, hereafter) of the dielectric layer by means of the transfer matrix, T. In the case of the electric field E this relation is

$$E_l = M \times E_r = \begin{pmatrix} \cos\delta & i\frac{c}{n}\sin\delta \\ i\frac{n}{c}\sin\delta & \cos\delta \end{pmatrix} E_r, \tag{5.8}$$

where n, d, and $\delta = \frac{2\pi nd}{\lambda}$ are the refractive index, the thickness of the layer and the phase change across it, respectively; $E_{l,r}$ are the electric fields at the left and right boundaries of the layer. Equation (5.8) in the same form holds also for the magnetic field.

From continuity requirements it follows that the transfer matrix M for a system of N layers is obtained by simple matrix multiplication of M's of single layers, $M = \prod M_j$ with M_j being the transfer matrix of the jth layer. The complex reflectance, $r(\omega)$ and transmittance, $t(\omega)$ coefficients of the multilayer can thus be calculated from the total transfer matrix M by assuming a unity (normalized to the input) intensity impinging on the structure from one side (left) and no input from the other (right):

$$r(\omega) = \frac{\gamma_l m_{11} + \gamma_l \gamma_r m_{12} - m_{21} - \gamma_r m_{22}}{\gamma_l m_{11} + \gamma_l \gamma_r m_{12} + m_{21} + \gamma_r m_{22}}, \tag{5.9}$$

$$t(\omega) = \frac{2\gamma_l}{\gamma_l m_{11} + \gamma_l \gamma_r m_{12} + m_{21} + \gamma_r m_{22}}, \tag{5.10}$$

where γ_l and γ_m are the inverse of light velocities in the input and output media, respectively.

The complex quantities contain important information about the photonic properties of a multilayer. For example, the power reflection and transmission coefficients are calculated as $R = |r(\omega)|^2$ and $T = |t(\omega)|^2$, respectively. Moreover, the amplitude and the phase of the transmitted light can be retrieved from $t(\omega)$. The phase ϕ, in particular, which is the argument of the complex transmittance, $\arg[t(\omega)]$, allows one to calculate both the density of states (DOS) of the photonic structure, $\rho(\omega)$, and the group velocity of propagating light, $v_g(\omega)$, as

$$\rho(\omega) = v_g(\omega)^{-1} = L_{tot}^{-1} \frac{d\phi}{d\omega}, \tag{5.11}$$

with L_{tot} being the total physical thickness of the multilayer. Figure 5.5(a) plots the DOS of the same dielectric mirror as in Fig. 5.4(a). It shows, in particular, that the DOS, defined typically as the number of states (photonic modes in this case) per unit energetic (frequency) interval, is largely enhanced at certain frequencies adjacent to the photonic band gap; known as *band-edge states*.

Following the definition of Eq. (5.11), the group velocity of a wavepacket centered at the band-edge frequencies and propagating through the multilayer is strongly reduced for those spectral components that lie within the narrow high-DOS lines (see Fig. 5.5(b)). As a consequence, the temporal evolution of a transmitted through DBR light pulse shows important wavepacket broadening as a result of a pronounced delay of these spectral components (mode localization).

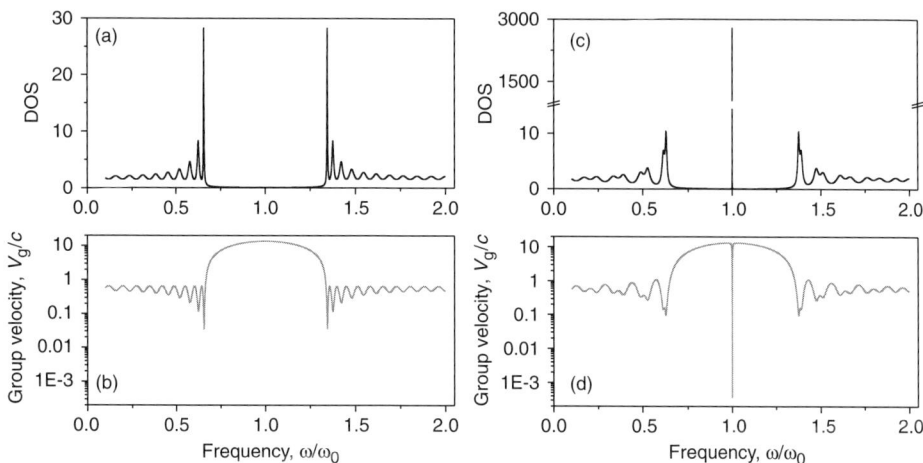

Figure 5.5 The density of states (DOS) and the group velocity for the cases of (a) a dielectric mirror DBR and (b) a Fabry–Pérot microcavity. In the last case a strong enhancement of the DOS and significantly reduced group velocity can be observed at the resonant frequency, $\omega/\omega_0 = 1$.

Fabry–Pérot microcavity – The Fabry–Pérot (FP) microcavity is a dielectric mirror with a defect. The simplest way to introduce a defect in a dielectric multilayer is to eliminate one of the $\lambda_0/4$ layers of either type A or B materials. By doing this, two quarter-wave layers of the same material appear next to each other, thus forming a $\lambda_0/2$ layer, which leads to perfect constructive interference of transmittion at ω_o EMW. As a result, a sharp (unity) transmission channel is formed at the center of the photonic band gap (Fig. 5.4(b)). The narrowness of the transmission line depends on the reflectivity (i.e. the strength) of the dielectric mirrors sandwiching the half-wavelength defect layer. As for the reflectance of the FP microcavity, this vanishes exactly at the resonant frequency, while for the remaining frequencies it shows all the spectral features of a DBR.

It is important to note the dramatic increase of the DOS (Fig. 5.5(c)) and, consequently, the suppression of group velocity at the microcavity resonance frequency (Fig. 5.5(d)). These are signatures of a strongly localized defect state for which the mode wavefunction decays exponentially outside the defect. The ability to confine EMWs in a tiny volume such as that of a defect layer, makes microresonator devices excellent tools for studying light–matter interactions at a nanometric scale. Lasing and Purcell enhancement of the radiative rates of cavity-embedded emitters are typical phenomena which widely exploit the confining properties of microresonators.

Finally, after an introduction to the physics of a dielectric mirror and a microcavity, we now have all the basic ingredients necessary to start with the optical properties of a 1D photonic quasicrystal. In the last part of this chapter we will see how a 1D Fibonacci quasicrystal is formed and in what way the quasiperiodic order modifies the optical properties of an otherwise ordered photonic structure.

5.4 Photons in a 1D quasiperiodic potential

As briefly cited at the beginning of the previous section, an optical potential for electromagnetic waves can be realized by properly stacking thin quarter-wave layers made of two different dielectric materials A and B. In the two extremes, a strictly periodic alternation of these layers will form a perfectly ordered 1D photonic crystal (DBR), while a complete randomization of the stacking order will give rise to a disordered photonic structure (Fig. 5.3(b)).

In terms of the positional order of A and B layers, a 1D quasicrystal stands in between the perfect DBR and a random photonic structure. As already introduced in Section 5.3, following the aperiodic string generation rule, we can now construct the photonic quasicrystal structure of a Fibonacci type (FQ, hereafter) in which the constituent $\lambda/4$ layers of A- and B-type materials are stacked together obeying $F_n = F_{n-1}F_{n-2}$ for the nth Fibonacci generation [149, 184]. Thus, the number of $\lambda/4$ layers of an nth-generation FQ will correspond to the nth-order Fibonacci number (see Fig. 5.6).

The differences between the periodic, aperiodic and completely random arrangements of two-component 1D systems of the same length (all A's and B's together) can be recognized by looking at the Fourier transform spectra of these sequences. Figure 5.3(c) reports such spectra for a DBR, a fifteenth-order FQ and a positionally random multilayer stack, all containing 987 layers. We see immediately that in the case of the DBR (top panel) a single occurrence frequency is present, indicating the perfect alternation of A and B layers. Instead, the flat spectrum of the random structure (bottom panel) is strong evidence that no correlation is present in the sequence of alternation of layer types. More interesting and rich is

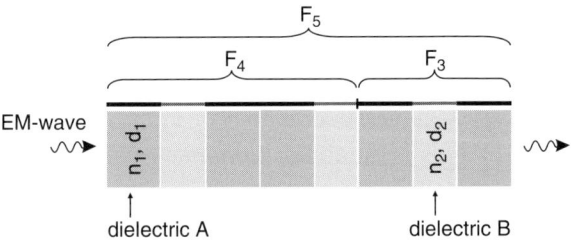

Figure 5.6 A photonic quasiperiodic potential for electromagnetic waves is formed by alternating, according to the Fibonacci inflation rule, two different dielectric layers of appropriately chosen optical thickness $n_i d_i$.

the Fourier transformation spectrum of the FQ (middle panel). We notice that this *A/B*-appearance spectrum represents a self-similar structure full of intense and weaker peaks – signatures of long-range order and absence of perfect short-range periodicity.

5.4.1 Electronic energy spectrum of 1D Fibonacci quasicrystals

Various authors have studied the energy spectrum and the properties of 1D quasi-crystals. Early studies were mainly focused on electronic quasicrystals and much work has been done there. In particular, it has been reported that the typical structure of the energy spectrum, $T(\omega)$, of a Fibonacci system has a self-similar nature [242, 243]. More technically, it is a Cantor set with zero Lebesgue measure [242]. In other words, an arbitrary chosen frequency lies within the gap with a unity probability, and the numerous gaps are densely filling the spectrum. In these spectral gaps, known as *pseudo band gaps*, the DOS tends to zero [72, 359].

Further, it has been shown that as a consequence of the multi-fractal nature of the energy spectrum, the wavefunctions of a Fibonacci quasicrystal show certain exotic localization properties [243]. Namely, the wavefunctions are neither local-ized nor extended – it is said that they are *critically localized* – and thus they decay at a rate slower than exponential ones [140, 242, 450]. These statements have been multiply confirmed through numerical calculations and experiments [199, 313, 384, 447, 461, 518]. Starting from the mid 1990s different studies with their optical (photonic) counterparts confirmed major analogies with the electronic case [40, 149].

5.4.2 Energy spectrum of 1D photonic Fibonacci quasicrystals

From electronic to photonic – Interference is a wave phenomenon [430]. The wave nature of both electrons and photons is the basis of important analogies between electronic charge transport phenomena in solids and light propagation in complex dielectric systems [441]. The spectrum of these analogies includes but is not limited to phenomena such as the optical counterpart of weak localization [256, 488, 543], Anderson localization [106, 153, 531], short- and long-range correlations [144, 487], and universal conductance fluctuations [419], photonic Bloch oscillations [416] and Zener tunneling of light waves [164].

Since photons are uncharged particles and they interact extremely weakly with each other at relatively low light intensities, a light wavepacket that propagates through a dielectric system remains coherent for much longer times than charged particles. This means that dynamic interference effects could be isolated and studied more easily with light than with electrons. In view of this, 1D photonic

Fibonacci quasicrystals were not only realized and spectrally characterized in static experiments but, more importantly, photon wavepacket propagation through optical states around the pseudo band gap of the structure was studied thoroughly [102, 162].

Transmission spectrum – In general, the energy spectrum, $E(\omega)$, of a photonic structure is often identified as its transmission spectrum $T(\omega)$. We will use this last definition throughout the rest of the chapter. As mentioned in Section 5.3, the transmission spectrum of a multilayered FQ can be easily calculated according to the transfer-matrix formalism (Eqs. (5.8) and (5.10)).

Figure 5.7(a) shows the transmission spectrum of a 377-layer thirteenth-order Fibonacci sequence with chosen refractive indices of $n_A = 3$ and $n_B = 1$. This spectrum is different from that of a perfect DBR, but it represents a mixture of general spectral features of a dielectric mirror and an FP-microcavity in terms of

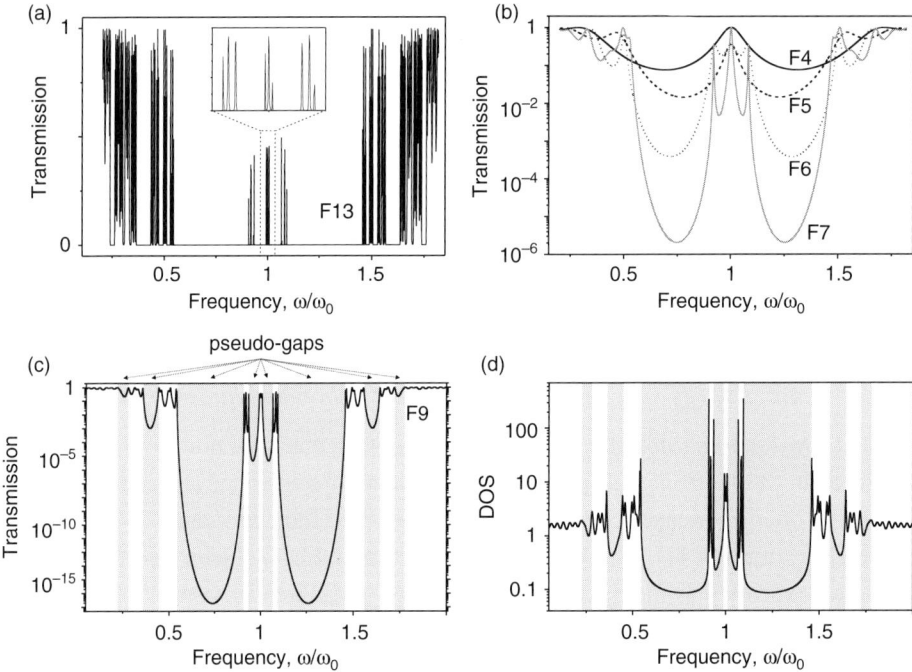

Figure 5.7 (a) The transmission spectrum of a thirteenth-order FQ with 377 layers. The inset shows the zoom around the central frequency revealing the fine modal features. (b) The evolution of the transmission spectra with increasing Fibonacci order: the pronounced deepening of low-T regions is accompanied by the appearance of a richer resonant structure. (c) The shadowed regions in the spectrum of an F9 structure indicate the so-called pseudo band gaps. (d) The same is illustrated on the DOS spectrum.

clearly visible gap regions and narrow transmission lines. The calculated $T(\omega)$ spectra for four successive Fibonacci series are plotted in Fig. 5.7(b). The logarithmic scale of transmission shows how the spectra are evolved from F_4 to F_7, reflecting the deepening of prohibited spectral regions and the appearance of new narrow spectral features.

Experiments with 1D photonic FQs – One of the first realizations of 1D FQs and the spectral characterization of samples was reported by Gellermann *et al.* in 1994 [149]. An electron-gun evaporation technique was used to grow up to 55-layer (F9) samples composed of silica and titanium dioxide layers alternating, following the Fibonacci rule. The quarter-wave ($\lambda_0/4$) layers were centered at a wavelength of $\lambda_0 = 700$ nm and static transmission spectra of the various Fibonacci samples were recorded at around λ_0. The experimental results were successfully compared with the theory based on the transfer-matrix approach. As a typical drawback from the control over growth parameters, in particular, a variation of layer refractive index with an increase in structure size, a certain deviation from the nominal optical path ($n_i \times d_i$) was also observed. By fitting the experimental spectrum a 5% refractive index drift was estimated by the authors for F9 samples.

In works by Dal Negro *et al.* [102] and Ghulinyan *et al.* [162] dielectric Fibonacci quasicrystals were realized by electrochemical etching of silicon. Porous silicon has interesting optical properties [97] and offers an excellent platform for testing ideas experimentally in 1D multilayered photonic structures [102, 161, 162, 163, 164, 416]. When grown in highly doped p-type silicon, porous silicon behaves as an optically homogeneous dielectric material with an effective refractive index n, determined by its porosity. Since it is fabricated by a self-limiting electrochemical process, one can control the refractive index of each layer by varying the electrochemical current during the fabrication. Typically, this last follows a step variation between two constant values, and therefore results in a step-like variation of the effective refractive index profile through the whole structure (the bottom panel case of Fig. 5.3(a)). The thickness d of an individual porous layer is then determined by the etching duration under a constant current density. Hence, the control of etching current and the time guarantee careful control over the optical path throughout the multilayered sample.

This growth technique was used successfully to realize porous silicon-based 1D FQs on silicon substrates and to characterize them both in static transmission and ultrashort-pulse interferometric experiments at near-infrared (NIR) wavelengths [102]. In particular, mode beating, sizable field enhancement, strong pulse stretching, and a strongly reduced group velocity in the band-edge region around a Fibonacci pseudo band gap were all observed and interpreted in terms of Fibonacci band-edge resonances.

A step forward in this direction was performed by further improving the electrochemical growth technique [162], which allowed in particular realization of free-standing multilayered FQs (without Si substrate). By using free-standing structures, in general, natural drifts of the nominal optical path can be controlled at a higher precision permitting realization of thick structures with high optical quality [161, 163]. Moreover, with such samples not only do simpler static transmission experiments become accessible, but also one avoids possible large time offsets due to the silicon substrate during pulse propagation studies. Along with observation of all the various effects reported in the previous study [102], experiments with the new samples provided further insight into the physics of photonic 1D FQs.

In Figs. 5.8(a) and (b) the measured transmission spectra of F9 and F12 (233 layers) are shown. The transfer-matrix calculations, considering the presence of possible drifts in growth parameters, revealed 1% and 4% optical path gradient (linear to a first approximation) for F9 and F12 samples, respectively (solid lines Figs. 5.8(a),(b)). The central wavelength of the F9 sample was chosen to be

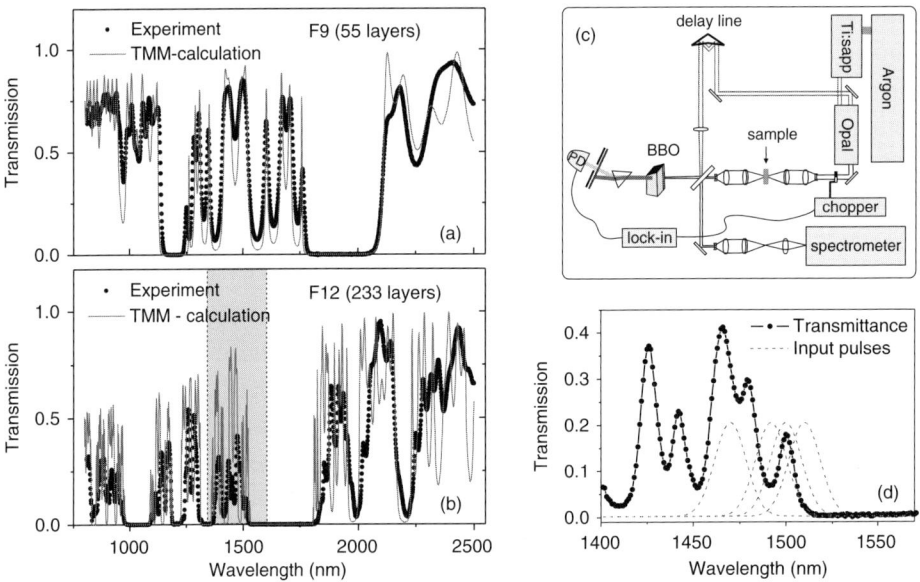

Figure 5.8 (a) The transmission spectrum of the porous silicon-based F9 sample is compared with the transfer-matrix calculations with 1% optical path gradient. (b) The same for the F12 sample. The gradient is 4% in this case. (c) The ultrafast pulse transmission setup based on an optical gating technique. (d) Pulse propagation experiments were performed over the spectral range, which corresponds with the shaded area of panel (b). Dotted lines show examples of Gaussian profiles of the ultrashort pulses centered at different wavelengths [162].

approximately 1500 nm. Instead, the sample F12 was intentionally realized such that the edge states of one of the pseudo band gaps was spectrally positioned at around 1500 nm (see the shaded area in Fig. 5.8(b)), where ultrashort pulse propagation studies could be performed (available range for sweeping the pulse wavelength).

Ultrashort pulse propagation through 1D FQs – Ultrafast time-resolved transmission measurements were performed using an optical gating technique based on signal upconversion (Fig. 5.8(c)). While being simpler compared with an interferometric system [102], it is sensitive to the transmitted intensity only and not the phase. In experiments wavelength-tunable pulses between 1400 and 1570 nm and of 220 fs duration (bandwidth, 14 nm) were sent through the F12 sample, thus probing the band-edge states (Fig. 5.8(d)). The transmitted signal was then mixed in a nonlinear BBO crystal with a time-delayed reference pulse in order to generate a sum frequency signal. This last, selected with a prism and detected with a photodiode, is proportional to the temporal overlap between the signal and the reference pulse. Therefore, the time profile of the transmitted signal was mapped by varying the delay of the reference (temporal resolution of 260 fs).

The left panel of Fig. 5.9 reports examples of the temporal evolution of the four selected transmitted pulses centered at different wavelengths of the band-edge. The top panel shows the free-space pulse (with no sample) the maximum of which is taken to be the zero on the time axis. An analysis of the transmitted signals at different wavelengths provides us with rich information about the nature of the band-edge states.

(a) *Mode beating* – Almost all of the transmitted pulses show an oscillating in time behavior (Fig. 5.9, left panel). This is related to the so-called mode beating phenomenon and manifests generally in a situation where more than one spectral feature (mode) lies within the spectral width of the pulse. The time period, t, of these beatings is related to the frequency separation, $\Delta\omega$ of the involved spectral features as $t = 2\pi/\Delta\omega$.

(b) *Pulse delay* – The pulse delay time, Δt, is considered to be the delay of the center of the mass of the transmitted signal with respect to the zero of the time axis, as shown on the left panel of Fig. 5.9. In [162] Δt was corrected for the trivial delay of the pulse due to the additional optical path of the sample given by $\sum_{i=1}^{N} n_i d_i$, where d_i is the physical thickness of the ith layer in the structure. In particular, the sample F12 ($N = 233$ layers) had a physical thickness of 44 μm and an optical thickness of 76 μm. With an average (weighted) refractive index of $\bar{n} \approx 1.88$, calculated as $\bar{n} = \sum n_i d_i / \sum d_i$, the time offset due to this effect is 99 fs for all pulses that pass through the twelfth-order

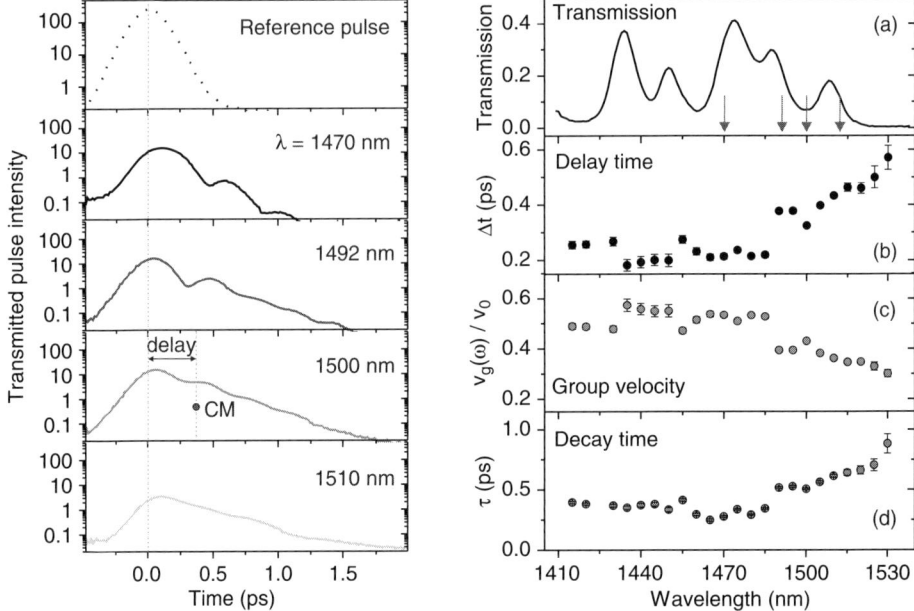

Figure 5.9 Left panels show examples of the time-resolved signals transmitted through the F12 sample. In the top panel the reference (undisturbed) pulse profile is shown, whereas the rest of the panels plot the transmitted pulses at wavelengths of 1470 nm, 1492 nm, 1500 nm and 1510 nm, respectively. The concept of the center of the mass of transmitted pulses and the corresponding delay time are explained too. Right panels report (a) the sample's transmission spectrum, (b) the delay time, (c) the relative group velocity derived from the delay time and (d) the decay time of the signals as a function of the probe pulse wavelength [162].

Fibonacci sample and all the extracted delay times were corrected for this. As a result, all the delays in Fig. 5.9(b) are purely due to the internal modal structure of the sample. Note that the availability of free-standing porous silicon Fibonacci structures avoids correcting for large time offsets due to a silicon substrate.

(c) *Group velocity* – From the delay time, the group velocity v_g = of the wavepacket with respect to light velocity v_m in a medium with an average refractive index of $\bar{n}$ was deduced (Fig. 5.9(c)). The authors have found a maximum group velocity reduction of a factor of three ($v_g/v_m \approx 0.3$) at the band-edge frequencies.

(d) *Decay time* – The intensity of the transmitted pulses decays almost exponentially with time, however, this kind of analysis in many cases may be affected by the time-oscillatory nature of the signal. Figure 5.9(d) shows the obtained decay time constants, τ, extracted from the raw data.

A typical trend that can be observed from the above-described quantities characterizing the pulse propagation (Fig. 5.9(b)–(d)) is that both the delay and the decay times increase as one approaches the pseudo band gap. As expected, these should be accompanied by narrow peaks in the transmission spectrum at the edge of the band gap, corresponding with intense peaks in the DOS spectrum. The reason for not observing narrow peaks in the experimental transmission spectrum (Fig. 5.8(d)) is related to the technical limitations of the measurement apparatus.

Using the transfer-matrix approach, it is possible to model the pulse propagation through the 1D FQ using the structure parameters of studied samples. The temporal response of the system can be obtained by inverse fast Fourier transformation (iFFT) of the product of the incoming pulse shape and the sample's transmission spectrum:

$$T(t) = \mathcal{F}^{-1}[G(\omega)T(\omega)] = \int_{-\infty}^{\infty} G(\omega)T(\omega)\exp(i\omega t)d\omega, \qquad (5.12)$$

where $T(t)$ is the transmitted time-resolved signal, $G(\omega)$ is the (Gaussian) spectral profile of the incoming pulse, and $T(\omega)$ is the transmission spectrum of the Fibonacci sample. The results are summarized in Fig. 5.10(a) and (b). In particular, Fig. 5.10(a) shows the calculated DOS (dashed line) and the spectrum $T(\omega)$ (solid line) of the sample. The temporal evolution of the transmitted pulse intensity was calculated by sweeping the pulse wavelengths through the band-edge states. An example of such a calculation is shown in the inset of Fig. 5.10(b). The extracted delay times (solid line Fig. 5.10(b)) showed satisfactory good agreement with the experimental data.

Disorder-related effects in 1D FQs – From a quick look at the calculated DOS and the transmission shown in Fig. 5.10(a) it is quite evident that while the first shows peaks up to a wavelength of 1550 nm, the second appears to vanish at above 1520 nm. This indicates that states do exist in the spectral range between 1520 nm and 1550 nm, but they barely transmit light for some reason. The origin of this phenomenon is due to the presence of a certain degree of disorder in the quasicrystal and will be discussed in the following.

A comparison between transmission spectra, DOS, and group velocity of a perfect DBR, an ideal Fibonacci, and a Fibonacci structure with certain disorder has been reported and discussed in [369]. In an ideal non-absorbing structure, in which the optical path $n_i d_i$ is exactly the same for each individual layer, whenever the DOS shows a peak, it has unity transmission. Consequently, both the DOS and $T(\omega)$ vanish in the band gap region.

The disorder inside a 1D photonic structure can either be uncorrelated to the position inside the structure or it can manifest as a continuous drift of the optical

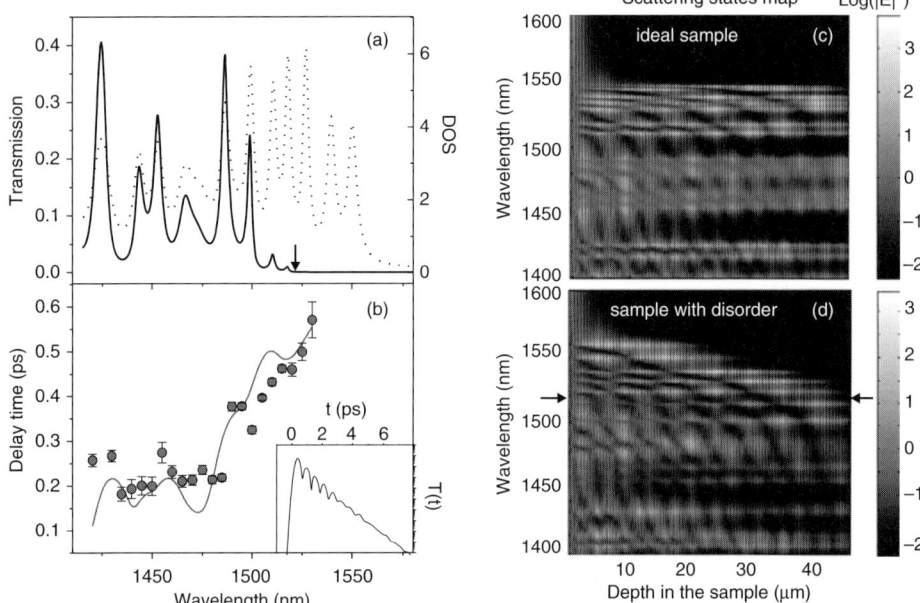

Figure 5.10 (a) The transfer-matrix-calculated transmission spectrum (solid line) of the F12 sample and the corresponding DOS (dotted line). (b) The experimentally measured delay times of the transmitted pulses (scatter data) are compared with the calculated ones (line). An example of the time-resolved transmitted intensity of a probe pulse centered at 1497 nm is plotted in the inset. (c) and (d) show the light intensity distribution inside the F12 sample for the ideal case (no optical path gradients) and for a 4% gradient, respectively. The arrows in the lower graph indicate the wavelength above which the sample transmission becomes almost zero, while internal modal structure can be clearly identified [162].

path (a linear gradient to first approximation). This last is the typical situation when during the preparation of the multilayered samples the growth rate (layer thickness) or the material composition (refractive index) is slowly varying through the final structure [102, 149, 162, 163]. We will therefore examine the effect of this type of disorder.

When a drift is present, the low transmission regime apparently extends for a wider wavelength range than the DOS gap. The reason for this can be found by examining the calculated scattering states maps of the Fibonacci sample, reported in Fig. 5.10(c) and (d). The scattering states map for the ideal case with flat photonic band structure is shown in the top panel. The linear drift in the optical path, $d(n_i d_i)/dx$, along the depth x in the multilayered structure acts as a built-in bias that tilts the band-edge (see Fig. 5.10(d)). This band tilting behaves as the analog of an external static electric field applied to an electronic superlattice structure. Such controlled realization of optical path gradients was exploited recently to observe time-resolved photonic Bloch oscillations [416] and Zener tunneling of light in optical superlattice structures with tilted photonic band structure [164]. Moreover, researchers have also managed to change the degree of band tilting by means of vapor flows through porous multilayers [34, 159].

Generally, and, in particular, in the twelfth-order Fibonacci sample, the drift-induced band tilting reduces the spatial extension of the optical modes, which are closely situated at the edge of the pseudo band gap. As a consequence, the first band-edge states do not extend over the whole sample and, therefore, their contribution to high transmission channels reduces significantly. In other words, a photon may be injected easily into such a spatially squeezed state from one side of the sample, however, in order to be transmitted it needs to tunnel through the band gap which extends for the rest of the sample (see Fig. 5.11). This explains the fact that the transmission spectra and density of modes differ in the region of the pseudo band gap. Equivalently, the long delay times observed around the edge of the pseudo band gap can be associated with the time needed to propagate through the band-edge states. The long decay times are directly related to the spatial confinement of these states.

Further numerical calculations in this direction have also been performed. In particular, the intensity of the electric field distribution inside the Fibonacci structure for few band-edge states was investigated for both an ideal sample and one with optical path gradient. The results of these calculations are shown in Fig. 5.12, in

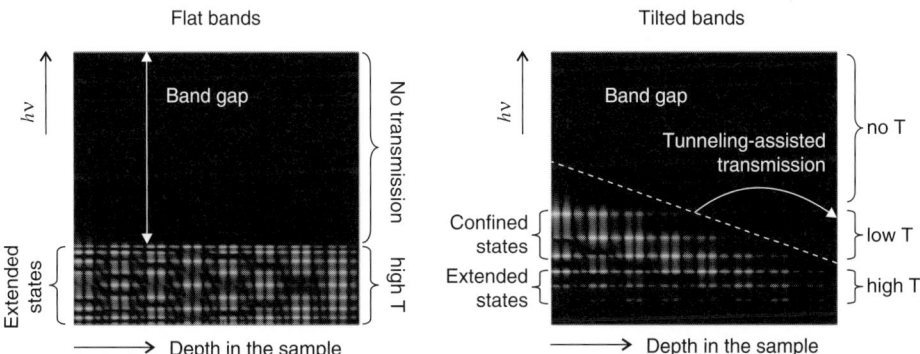

Figure 5.11 (Left graph) In the absence of optical path gradients the photonic bands are flat and the photon states are extended throughout the whole structure. (Right graph) The effect of an optical path gradient on photonic bands is in close analogy with the effect of a static electric field on energy bands in an electronic crystal. In the photonic case the band-edge states are no longer extended Bloch states and are confined spatially between the physical edge of the sample and the photonic band gap. Thus, photon transmission through these states is largely suppressed.

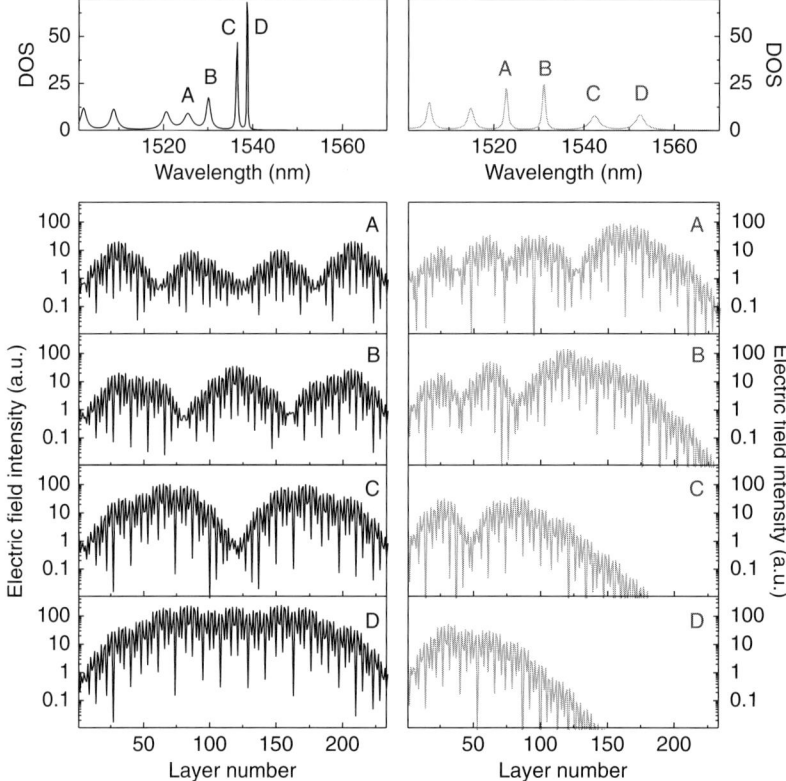

Figure 5.12 The electric field intensity distribution inside the multilayered sample for the first four band-edge states in an ideal Fibonacci structure (left panels) and a 4% drifted structure (right panels). The corresponding DOS spectra around the band-edge are shown in the upper panels.

which the DOS at the band-edge is also reported in the top panels. In the absence of optical path gradients (left panels), the distribution of the electric field intensity shows the characteristic self-similar structure of the band-edge resonances of a Fibonacci quasicrystal.

It can be seen from the right panels of Fig. 5.12 that while the main features of the intensity distribution are preserved when a drift is introduced (4% in this case as in the experiment with F12 samples), the profiles become asymmetric with respect to the center of the sample. Quite visibly, the intensity pattern squeezes more and more towards one side of the quasicrystal as the mode wavelength approaches the pseudo band gap. In accordance with the situation shown in Fig. 5.11 for the tilted band case, the states closest to the gap are spatially more confined between the sample surface and the tilted band gap. Accordingly, when approaching the band-edge (from panel A to panel B of the tilted band case), the intensity profile distorts heavily and the electric field decays efficiently through the rest of the structure. Thus, these calculations show that the natural drifts, occurring during the growth of the 1D photonic FQ, alter the spatial extension of the first few band-edge modes, but do not destroy their specific structure. Higher-order band-edge modes are less affected by the optical path gradients.

5.4.3 Origin of band-edge states and pseudo band gaps

In the previous subsection we reviewed the recent progress in photonic 1D Fibonacci quasicrystals in terms of their realization, spectral characterization and ultrashort pulse propagation studies. We learnt that these photonic structures are somewhat in between ideally periodic photonic crystals and disordered complex dielectric structures. When describing them, terms such as *pseudo band gaps* and *band-edge states* with *exotic (critical) localization* properties are often used. Even so, the reader might still be unclear as to why "pseudo" is used or what we mean by saying "critical localization" and why it happens? The scientific literature is rather sparse in this sense and rigorous mathematical descriptions are not easy to interpret.

Through the rest of this chapter, we will offer the reader a new interpretation of these properties for the 1D photonic Fibonacci quasicrystals. We will show, in particular, how the particularity of the inflation rule of the Fibonacci type affects the properties of a 1D photonic FQ and in which way the so-called pseudo band gaps and the band-edge states with non-exponential decay are formed. Perhaps, you will be rather surprised to realize that the Fibonacci quasicrystal is a particular arrangement of simple photonic building blocks – DBRs and FP microcavities – arranged all together to form a coupled microcavity system.

Coupled microcavity optical superlattice – In a close analogy with a semiconductor double-quantum well, the optical coupling between identical cavities raises the mode degeneracy inducing a repulsion, which splits the single state at a frequency ω_0 into two new frequencies at $\omega_0 \pm \Delta\omega$ [18, 374, 377, 455]. Degenerate mode coupling between N cavities therefore leads to the formation of a miniband of N new optical states, which are densely packed around the resonant frequency. This way, in the very same analogy with electronic superlattices (Fig. 5.13(a)) [125], an optical superlattice can be made by coupling degenerate optical resonators (cavities) within the same photonic structure (Fig. 5.13(b)) [160, 161, 163, 328, 340].

In one dimension, an optical superlattice can be realized by stacking two different quarter-wave dielectric layers A and B in such a way as to form identical cavities separated by dielectric Bragg mirrors (Fig. 5.13(c)). In a generic form, the layer sequence of such a multilayer stack looks like $[(BA)_n B] (AA)_1$

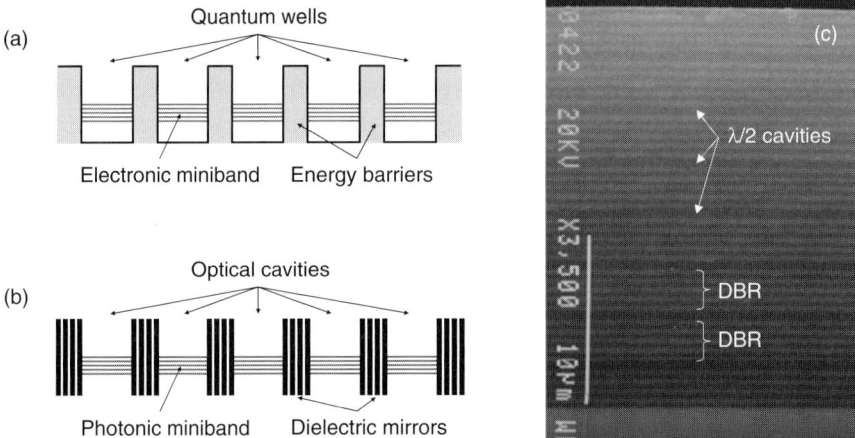

Figure 5.13 In analogy with the electronic coupling of separate quantum wells in a semiconductor superlattice, (a) an optical superlattice can be realized when optical cavities are brought together, (b) which results in the formation of a miniband of extended photonic states. (c) A scanning electron micrograph of a one-dimensional porous silicon optical superlattice with seven coupled microcavities.

$[(BA)_n B] (AA)_2 \dots (AA)_m [(BA)_n B]$, where a series of m microcavities $(AA)_m$ are coupled to each other through the $[(BA)_n B]$ DBRs of $(n + 1/2)$ periods. The amount of mode splitting of cavity resonances is given by the strength of the coupling mirrors. As described in Section 5.3, for a given refractive index contrast, the reflectance of the DBR improves with increasing number of periods. Similarly, for a fixed number of periods, the mirror reflectance grows with an increase in the refractive index contrast, i.e. the stronger the mirrors reflect, the weaker is the mode repulsion.

Figure 5.14 shows the formation of a characteristic transmission spectrum of an optical superlattice of five coupled microcavities. First, let us look at the spectrum of a single DBR, which has a layer sequence $[(BA)_n B]$ with $n = 5$ (top panel).[6] As discussed previously it is characterized by a photonic band gap centered at a relative frequency $\omega_0/\omega = 1$ and extends to a certain frequency range around it (indicated by the shaded area).

Next, we introduce a defect in the DBR layer sequence thus creating a $\lambda/2$ cavity sandwiched between two DBRs: this creates a sharp transmission peak – the Fabry–Pérot resonance – around the central frequency and splits the DBR band gap into two forbidden spectral regions (Fig. 5.14(b)). It is important to underline that the new band gaps on the left and right sides of the FP peak are part of the *same and*

---

[6] The exact number of DBR periods n is not essential throughout this analysis.

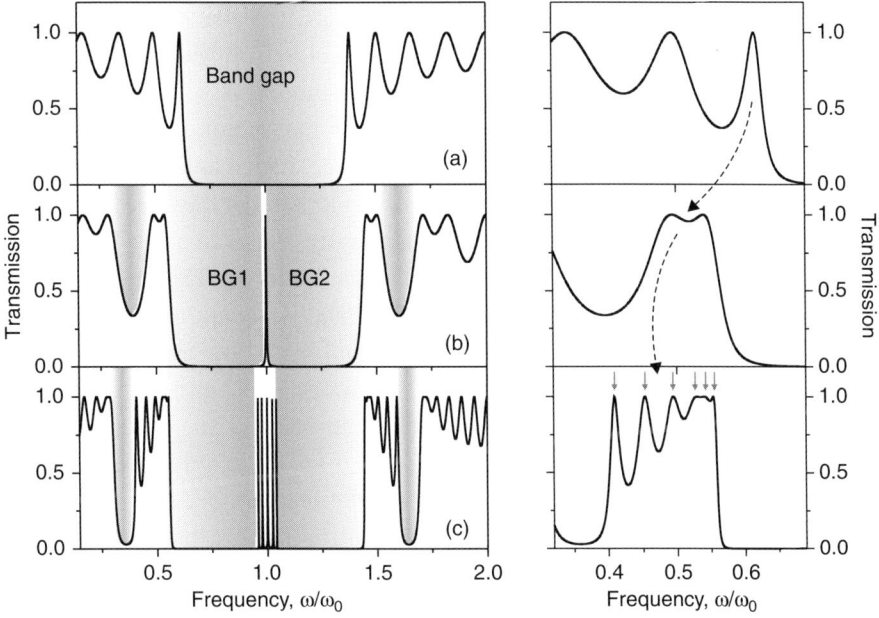

Figure 5.14 Broad range transmission spectra of (a) a dielectric mirror, (b) a Fabry–Pérot microcavity and (c) five coupled microcavities. The zooms of spectra around the band-edge states are shown in related panels on the right.

unique band gap of the DBR sequence $[BABA \dots B]$ and therefore have the center frequency ω_0/ω.

Moreover, the FP resonator's spectrum contains other details which are absent in the DBR's spectrum. While these details are not evident at first glance, they can be revealed by looking at the blow-up of the corresponding $T(\omega)$ curves in the vicinity of the band-edge. From the right-hand panels of Fig. 5.14 we notice, in particular, that with respect to the DBR's spectrum the band-edge states of the FP structure have undergone spectral reshaping. Namely, a paired (double-peaked) band-edge state is now appearing. Understanding the origin of these doublets is important as it will be a crucial point when analyzing the origin of Fibonacci pseudo band gaps.

Recalling the layer sequence of the 1D FP microcavity:

$$[BABA \dots B] \, AA \, [BABA \dots B],$$

and remembering that by definition they have the same central frequency, we notice that two identical DBRs are separated through a layer AA, which can be interpreted as a system of coupled DBRs! As already discussed above, such a coupling should induce degenerate mode splitting. Thus the origin of the band-edge doublet comes from a coupling, sandwiching the DBRs in the FP structure sequence. This coupling is valid for all DBR states. In fact, by looking back at the broad range transmission spectrum of the FP microcavity (Fig. 5.14(b)), we notice that the doublets are present for all other modes far from the band-edge. Moreover, the formation of these doublets induces relatively deep low-transmission (gap-like) regions between neighboring doublets. The successive gap-like regions become shallower when moving further from the band-edge (shallower shaded areas in Fig. 5.14(b)).

These observations become more evident when looking at the spectrum of a five coupled microcavity structure (Fig. 5.14(c)). Firstly, we notice the formation of photonic miniband states, five in total, which are densely packed around ω_0. As in the case of the single FP structure, this miniband is separating the fundamental DBR band gap in two regions at $\omega < \omega_0$ and $\omega > \omega_0$. Secondly, from the corresponding zoom of the band gap edge we notice that now we have a multi-peaked modal structure (right bottom panel). Importantly, one counts six peaks which correspond exactly with the total number of DBRs in a five coupled microcavity structure. Moreover, the secondary gap-like features become much deeper and well defined. Finally, these observations confirm the fact that in analogy with the coupling between identical $\lambda/2$ AA layers the dielectric mirrors also couple within the same photonic crystal.

It is now the right moment to look at the formation of the band structure of a 1D photonic Fibonacci quasicrystal.

1D photonic Fibonacci quasicrystal – Let us first examine the layer sequence of the FQ. As an example we consider the seventh-order Fibonacci sequence, which reads,

$$ABAABABAABAABABAABABA.$$

Now, remembering that in the photonic case (i) A and B layers have a quarter-wave thickness and, hence, (ii) an AA layer forms a $\lambda/2$ cavity, we can write this Fibonacci sequence in the form,

$$[AB]\,AA\,[BAB]\,AA\,[B]\,AA\,[BAB]\,AA\,[BABA].$$

We notice immediately, that this Fibonacci multilayer is nothing other than **a system of four coupled microcavities with unequal dielectric mirrors** (see Fig. 5.15(a) and (b)). It is important to again note here the fact that due to the recurrence rule of the Fibonacci type, the AA occurs frequently along an F_n for large n's, while other sequences such as BB or AAA may never appear (see Table 5.2). In view of this, we can say now that a 1D photonic Fibonacci quasicrystal, independently of its recurrence order n, will be a coupled $\lambda/2$ microcavity system with unbalanced DBRs.[7]

---

[7] In order to form coupled cavities (three at least, since two cannot be formed), n should be larger than 6. F4 and F5 have only one cavity, while no cavity is formed in lower orders F1, F2 and F3.

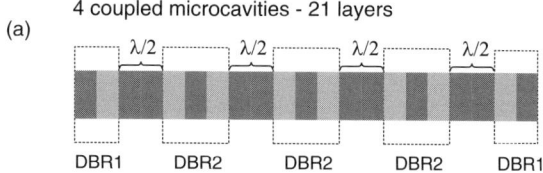

(a) 4 coupled microcavities - 21 layers

DBR1   DBR2   DBR2   DBR2   DBR1

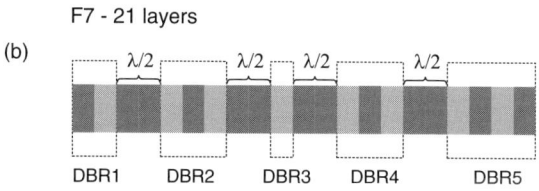

(b) F7 - 21 layers

DBR1   DBR2   DBR3   DBR4   DBR5

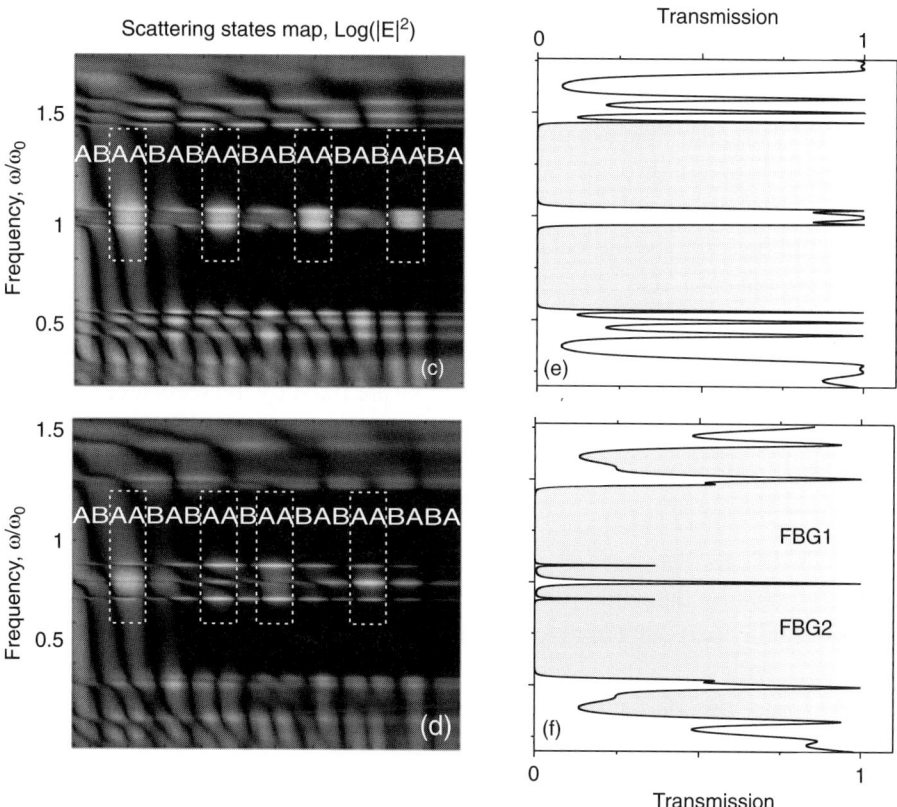

Figure 5.15 (a), (b) The schematics of layer alternation in a four coupled micro-cavity and in a seventh-order FQ structure. (c) The scattering states map and (e) the corresponding transmission spectrum of the coupled microcavity structure. (d) and (f) show the same quantities for the 1D FQ.

Table 5.2 *The occurrence of AA cavities in layer sequences of 1D FQs (from third to seventh Fibonacci orders).*

1D FQ	counts(A)	counts(B)	counts(AA)
ABA	2	1	0
$AB\|AA\|B$	3	2	1
$AB\|AA\|BABA$	5	3	1
$AB\|AA\|BAB\|AA\|B\|AA\|B$	8	5	3
$AB\|AA\|BAB\|AA\|B\|AA\|BAB\|AA\|BABA$	13	8	4

For comparison, in Fig. 5.15(c) and (d) the light intensity distributions inside a four coupled microcavity (4CMC) structure with balanced DBRs [160] and an F7 1D photonic quasicrystal are shown. Their respective transmission spectra are plotted in Fig. 5.15(e) and (f). We note that the weaker external mirrors in the case of the 4CMC structure are chosen intentionally so as to achieve a total of 21 layers as in a F7 structure. It is however important that the intra-cavity (coupling) mirrors are all identical. These intensity maps visualize nicely the layer positions of the $\lambda/2$ cavities. In particular, one can observe that in the case of the 4CMC structure they are evenly distributed through the multilayer (see the bright intensity spots around $\omega/\omega_0 = 1$) in contrast with their inhomogeneous distribution in an F7 quasicrystal.

It thus follows that the optical coupling of cavities inside a Fibonacci multilayer is depth dependent. Also, it is clear that this coupling is not changing in a random manner with the position inside the multilayer, but it is related to the exact DBR sequences which strictly reflect the Fibonacci string generation rule. In other words, in a 1D FQ the sequences of the DBRs are fixed by the recurrence rule. For example, the first five DBRs are AB, BAB, B, BAB and $BABA$.

In view of the above discussion, certain aspects of FQ *pseudo band gap* formation and the nature of the *band-edge states* are seen under a new light. In particular,

- The so-called fundamental pseudo band gaps (FBG1 and FBG2 in Fig. 5.15(f)) are essentially parts of a unique band gap separated by the photonic miniband of coupled cavities.
- Higher-order pseudo band gaps are low-transmission regions formed by neighboring multi-peaked states of the coupled DBRs (more external shaded regions in Fig. 5.15(f)). The origin of their formation is exactly the same as in the case of coupled microcavity structures with balanced mirrors (see Fig. 5.14(c)).
- Fundamental band-edge states are of two different types. The first are coupled cavity miniband states and originate from AA layers. The band-edge states of the second type are essentially coupled DBR states and, with respect to the coupled cavity states, are situated on the opposite side of the gap.

We now examine the properties of the band-edge states of the first type, i.e. the AA-layers-related coupled cavity miniband of the Fibonacci quasicrystal. As already mentioned above, the coupling of cavities inside the 1D FQ takes place via unbalanced mirrors. This results in the formation of photonic states which are not homogeneously distributed within the spectral width of the miniband, as opposed to the case of a coupled microcavity system with uniform DBRs [161, 163]. For a more detailed analysis we consider an eighth-order FQ which counts eight occurrences of AA cavities. For simplicity, we rename the AA cavities to C_m ($m = 1, 2, \ldots, 8$) and rewrite the Fibonacci sequence as

$$AB\ C_1\ BAB\ C_2\ B\ C_3\ BAB\ C_4\ BAB\ C_5\ B\ C_6\ BAB\ C_7\ B\ C_8\ B.$$

The typical broad range transmission spectrum as well as the zoom of the miniband region are plotted respectively in panels (a) and (b) of Fig. 5.16. As we can see, the miniband is composed of three regions of relatively high transmission states. While a total of six marked peaks can be counted, the number of states involved in the miniband is eight, where two states are less pronounced and are

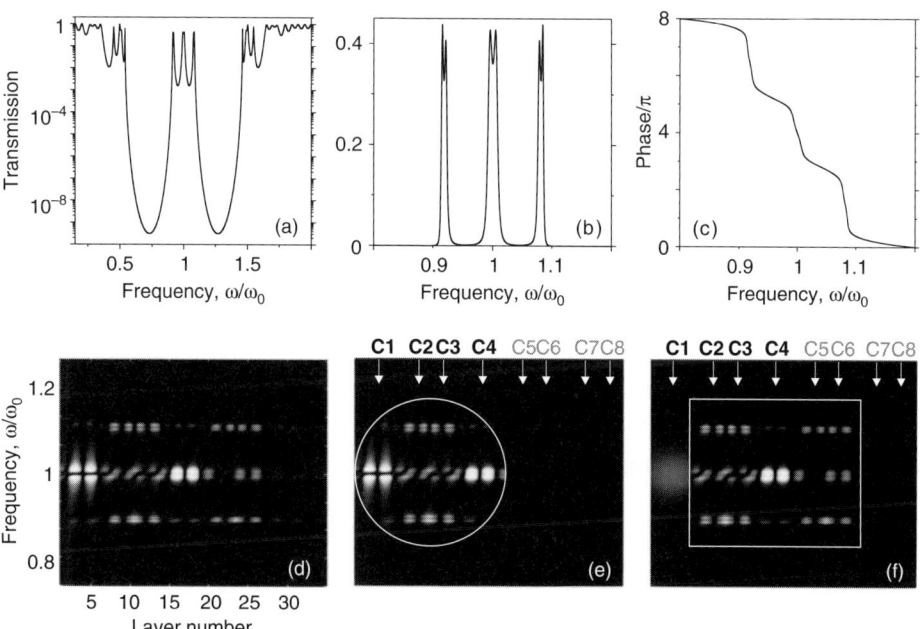

Figure 5.16 (a) The broad range transmission spectrum of an eighth-order FQ. (b) The zoom of the spectrum and (c) the accumulated phase around the miniband region. (d) The scattering states map of the miniband states. Panels (e) and (f) explain the degenerate mode coupling between various cavities distributed through the Fibonacci sequence.

not visible in the $T(\omega)$ spectrum. The accumulated phase, which is a very useful tool when analyzing unknown resonant structures [42, 157, 158, 160], is plotted in Fig. 5.16(c) and counts a total of exactly 8π when sweeping the frequency through the miniband region. The three high-T channels are separated by two regions of very low transmission. These last are often attributed to the "category" of higher-order pseudo band gaps, while their origin is clearly related to the cavity miniband structure.

In order to understand better the formation of the inhomogeneously split miniband, we refer to the electric field intensity distribution inside the quasicrystal around the miniband central frequency ω_0 (Fig. 5.16(d)). The positions of the bright spots, which indicate the large filled intensities accumulated in cavities, show certain symmetric patterns both through the depth inside the structure (x-axis) as well as over the frequency range (y-axis) around the $\omega/\omega_0 = 1$ point.

As an example, we analyze two of the symmetric patterns, which are shown in panels (e) and (f) of Fig. 5.16. Firstly, we notice that cavities C_1 and C_4 are weakly coupled through the large mirror $BABC_2BC_3BAB$, therefore the resulting mode splitting is small (y-axis). More precisely, this double-peaked state is formed by a symmetric coupling of three cavities C_1, C_4 and C_7, since an identical mirror $BABC_5BC_6BAB$ is present between C_4 and C_7:

$$AB\ C_1\ [BAB\ C_2\ B\ C_3\ BAB]\ C_4\ [BAB\ C_5\ B\ C_6\ BAB]\ C_7\ B\ C_8\ B.$$

We also notice the cavities C_2 and C_3, which are strongly coupled through the very weak mirror B, therefore, the resulting mode splitting pushes the new states far from $\omega/\omega_0 = 1$ to the extremes of the miniband [157, 158]. Moreover, each of these new states is itself a doublet state, which can be explained by further analysing the FQ structure. Namely, as in the case of C_2-C_3 coupling, we notice that the same scenario of coupling also takes place for C_5 and C_6 cavities, which pushes the new split states exactly to the same frequencies as for the C_2-C_3 case thus forming a new couple of degenerate states at the extremes of the miniband. In this case, however, the coupling between cavity pairs C_2-C_3 and C_5-C_6 is much weaker:

$$AB\ C_1\ BAB\ C_2\ B\ C_3\ [BAB\ C_4\ BAB]\ C_5\ B\ C_6\ BAB\ C_7\ B\ C_8\ B,$$

therefore, only a small amount of mode splitting is observed.

Thus, while this kind of analysis can be taken further by considering other cavity pairings (we leave this as an exercise for the reader), the formation of inhomogeneously distributed miniband states in a 1D photonic quasicrystal of Fibonacci type is now comprehensible.

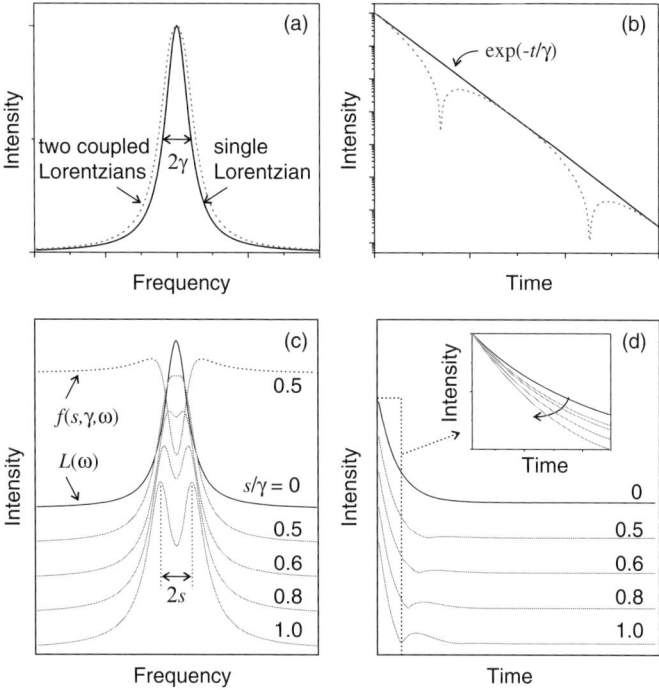

Figure 5.17 (a) Spectral lineshape of a single (solid line) and two weakly coupled Lorentzians (dashed line). (b) The corresponding inverse FFT transformations of the single and coupled states. (c) The evolution of the doublet splitting with increasing coupling, s/γ. (d) The corresponding time decays of doublet states showing the oscillatory time behavior. The inset is the zoom of the time-spectra at the initial part of the decays.

Localization properties of band-edge states – It is well known that the energy spectrum of a localized state at a frequency ω_0 is described by a complex function,

$$t(\omega) \sim \frac{\gamma}{\omega - \omega_0 + i\gamma},\qquad(5.13)$$

and its transmission spectrum has a Lorentzian lineshape (Fig. 5.17(a)),

$$L(\omega) = |t(\omega)|^2 \sim \frac{\gamma^2}{(\omega - \omega_0)^2 + \gamma^2}.\qquad(5.14)$$

Here, 2γ is the full width at the half-maximum (FWHM) of the resonance. In other words, γ describes how well the cavity confines the electromagnetic energy or, equivalently, how lossy it is. Thus, the FWHM in the frequency domain represents the decay rate, proportional to γ^{-1}, of the resonance in the time domain. In fact, the inverse Fourier transformation of a Lorentzian lineshape is an exponentially decaying in time function (solid line in Fig. 5.17(b)).

In the case when two identical resonances (same γ) are degenerately coupled, the resulting lineshape is described through a coherent sum of two Lorentzians as

$$D(\omega) \sim \left| \frac{\gamma}{\omega - \omega_0 + s + i\gamma} + \frac{\gamma}{\omega - \omega_0 - s + i\gamma} \right|^2, \qquad (5.15)$$

where $2s$ is the coupling induced splitting of the doublet resonance. This lineshape can be also presented as

$$D(\omega) \sim \left| \frac{\gamma}{\omega - \omega_0 - i\gamma} \right|^2 \times \left| \frac{2\left[(\omega - \omega_0)^2 + \gamma^2 \right]}{(\omega - \omega_0)^2 - \gamma^2 - s^2 + 2i\gamma(\omega - \omega_0)} \right|^2. \qquad (5.16)$$

This result has the form $L(\omega) \times f(s, \gamma, \omega)$, in which we recognize the single-resonance Lorentzian $L(\omega)$ convoluted with the function $f(s, \gamma, \omega)$. It is easy to see that $\lim_{s \to 0} D(\omega) = L(\omega)$, i.e. when the coupling is infinitely weak, one ends up with a single Lorentzian lineshape.

In a weak coupling situation (e.g. a strong intra-cavity mirror) the splitting is small and a clear doublet is not formed (dotted line in Fig. 5.17(a)). While this lineshape appears as an effectively broader "single-peaked" resonance without any evident fine spectral features, its Fourier spectrum reveals a clearly faster decay (dotted line in Fig. 5.17(b)) than the pure exponential decay of the Lorentzian resonance. Moreover, the composite nature of this double-state resonance is brought to light nicely in the time domain; it can be observed at a relatively longer time scale that the state decay is not monotonic but it performs regular oscillations. The time period of these oscillations is defined by the inverse of the doublet splitting, s^{-1}. This is what is typically observed in time-resolved pulse propagation experiments [102, 162, 164, 416], which essentially reflects the convolution of the ultrashort pulse lineshape with that of the composite resonance; regular beatings occur between two coupled states which manifest as oscillations in the transmitted intensity (see Fig. 5.9). In addition, the transmitted pulse is also characterized by a larger delay which is caused by the transient time necessary to build up the double resonance.

Figure 5.17(c) shows the evolution of the spectral lineshape of two coupled Lorentzian states at increasing coupling strength s/γ. The corresponding iFFT spectra, plotted in Fig. 5.17(c), show that the oscillations become faster for larger doublet splittings and the intensity decays faster at short times (see the inset of Fig. 5.17(c)).

When more than two resonances are degenerately coupled within a photonic structure, the decay of such a state may show an even more complicated time behavior. In particular, more than one oscillation period can appear in the intensity decay profile, but also, the amplitude of beatings can vary because of unevenly distributed degenerate resonances within the structure. A photonic structure which fits well with this situation is the 1D Fibonacci quasicrystal. As we discussed earlier in this subsection, the band-edge states of a 1D FQ are miniband states originating from AA half-wavelength layers unevenly distributed throughout the photonic quasicrystal. While for their localization properties terms such as "exotic" or quasi-localization are typically used to describe a weaker-than-exponential decay, we see that this property is exactly what is expected for photonic resonances originating from degenerately coupled states.

Acknowledgments

We wish to thank our co-workers Claudio J. Oton, Luca Dal Negro, Lorenzo Pavesi, Riccardo Sapienza, Marcello Colocci and Diederik S. Wiersma for their important contributions to the study of 1D Fibonacci quasicrystals presented in this chapter. We are also thankful to Daniel Navarro-Urrios, Paolo Bettotti, Zeno Gaburro, Georg Pucker, Matteo Galli, Lucio C. Andreani and Andrea Melloni for many fruitful discussions and our common studies on 1D coupled photonic microresonator systems. This work was financially supported by the INFM projects RANDS and Photonic and by MIUR through the Cofin 2002 "Silicon based photonic crystals" and FIRB "Sistemi Miniaturizzati per Elettronica e Fotonica" and "Nanostrutture molecolari ibride organiche-inorganiche per fotonica" projects.

6

2D pseudo-random and deterministic aperiodic lasers

HUI CAO, HEESO NOH, AND LUCA DAL NEGRO

6.1 Introduction

Unlike a conventional laser which utilizes mirrors or periodic structures to trap light, a random laser relies on the multiple scattering of light in a disordered gain medium for optical feedback and light confinement (see Figure 6.1) [67, 70, 528, 529]. Coherent laser emission has been generated from various random structures ranging from semiconductor nanoparticles and nanorods to polymers and organic materials. Over the past decade, random lasers have generated significant interest among researchers because of their unique applications [67, 529]. For example, specifically designed random lasers combine high radiance with low spatial coherence [357, 403], ideally suited for speckle-free, parallel, high-speed imaging, laser projection and ranging [404]. The sensitivity of the random lasing threshold to the amount of scattering has been explored for cancerous tissue mapping [387, 389]. Moreover, the spectral fingerprint of the random structure [70, 71], the micron size [70], the low fabrication cost, and the robustness against surface roughness and shape deformation point to potential applications in optical tagging and identification [528]. Since random laser cavities are formed by scattering and are much easier to fabricate compared with the mirror-based cavities, they may be applied to lasers at wavelengths where high-reflectivity mirrors are either not available or very expensive, e.g. UV laser, X-ray laser, γ-ray laser. A detailed description of random lasers can be found in Chapter 3.

However, a major limitation to device applications of random lasers is the lack of control and reproducibility of lasing modes. Namely, the frequencies and spatial locations of lasing modes are unpredictable, varying randomly from sample to sample. One way to solve this problem is to utilize pseudo-random structures, or more generally, deterministic aperiodic nanostructures (DANS) [98]. Unlike

Light Localisation and Lasing, ed. M. Ghulinyan and L. Pavesi. Published by Cambridge University Press.
© Cambridge University Press 2015.

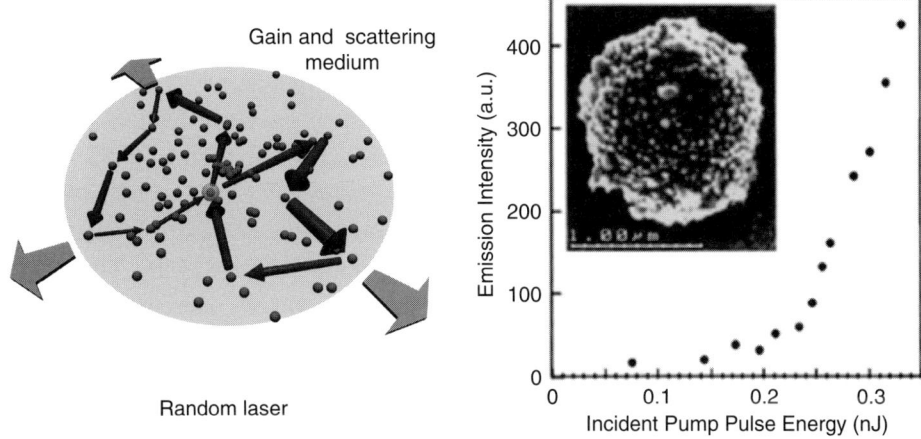

Figure 6.1 Left panel: Sketch of the random laser concept (after Ref. [528]). In a random laser the cavity is absent but multiple scattering between particles in the disordered material keeps the light trapped long enough for the amplification to become efficient, and for laser light to emerge in random directions (figure by Robert Tandy, member of the A. Douglas Stone research group at Yale Univ.). Right panel: First experimental demonstration of random lasing in a micron-sized cluster of ZnO nanoparticles (after Ref. [70])

structures generated by a random process, DANS are generated by the iteration of simple mathematical rules with a controllable degree of spatial complexity that interpolates between periodic and random systems. DANS lasers can combine the advantages of random lasers with the fabrication and design reproducibility required for optoelectronics integration.

A DANS laser is advantageous over a photonic crystal (PhC) laser because it has rich emission spectra and complex resonances. The DANS can support a variety of lasing modes, e.g. spatially localized modes, fractal-like modes and extended modes with distinct and well-defined frequencies and output directionalities [56, 57]. The diversity of modes provides a huge degree of freedom for design and applications. Thus, the DANS laser is a promising candidate for engineering a broadband coherent and miniature light source on-chip.

6.2 Overview

Deterministic aperiodic nanostructures are rooted in symbolic dynamics and prime number theory, and thus possess very rich spectral features [98, 312]. DANS can be implemented in physical systems as diverse as dielectric multilayers [102, 149, 242, 342], nanopillars [56, 566] and nanoparticle arrays [99, 170, 171]. The structural complexity of DANS is measured by their spatial Fourier spectra, which are discrete (singular) for quasiperiodic systems, singular–continuous or absolutely continuous for pseudo-random structures of increasing complexity.

Most studies on DANS are performed on the passive systems. This chapter is focused on the active DANS. The addition of optical gain can lead to a lasing action. In recent years, there have been detailed studies on lasing in quasiperiodic photonic media which have only rotational symmetry but no translational symmetry. Notomi *et al.* reported lasing action due to the two-dimensional (2D) quasiperiodicity of photonic quasicrystals with a Penrose lattice [360]. The lasing modes are extended bulk modes, formed by standing waves coherently extended among bulk quasicrystals, despite the absence of true periodicity or translational symmetry. Baba *et al.* fabricated a seven-hole-missing defect in a twelve-fold symmetric quasiperiodic photonic crystal (QPC) and realized lasing in the defect mode which is a whispering-gallery mode [363]. They also observed lasing in the defect-free QPC due to localization of 2D Bragg modes at the photonic gap edge [364]. A QPC single-cell resonator was proposed and realized by Kim *et al.* [235]. Lasing occurred in a well-defined hexapole-like localized state which has high quality factor and small mode volume. Compared with the photonic crystal defect lasers, the QPC defect lasers can have lower lasing thresholds [279, 280, 281]. This is attributed to the higher rotational symmetry of the QPC, e.g. the twelve-fold symmetry of the dodecagonal QPC, which provides more isotropic confinement of the whispering-gallery modes in the defect region. In addition to the 2D QPC lasers, Mahler *et al.* constructed a one-dimensional (1D) quasiperiodic distributed feedback laser based on a Fibonacci sequence [314]. They showed that engineering of the self-similar spectrum of the grating allows features beyond those possible with traditional periodic resonators, such as directional output independent of the emission frequency and multicolor operation. A three-dimensional (3D) icosahedral QPC was fabricated with a seven-beam optical interference method and multidirectional lasing was demonstrated [244]. Compared with the 3D periodic crystal, the QPCs have higher symmetry and are more favorable for the formation of 3D photonic band gaps. Hence, they are more efficient in providing the feedback mechanism for lasing action.

The above lasing structures have discrete Fourier spectra. On the other hand, there has been little work on lasing in structures with absolutely continuous Fourier spectra, or pseudo-random DANS. Recently, we conducted numerical and experimental studies on lasing in 2D DANS with continuous or singular–continuous Fourier spectra [358, 553], and demonstrated lasing action by localized optical modes in active photonic membranes with pseudo-random deterministic morphologies [553]. We showed that the lasing modes of pseudo-random structures occur at reproducible spatial locations and frequencies, only slightly affected by structural fluctuations in different samples. The ability to engineer a high density of lasing modes and make them reproducible is essential to chip-scale photonic applications. Moreover, we were able to tune the structural aperiodicity in singular–continuous Thue–Morse structures and found an optimal degree of aperiodicity where light confinement is maximal and lasing is strongest [358]. At various degrees of aperiodicity, different types of modes acquire the highest quality factors and may be selected for lasing. Our work opens a way to control lasing characteristics by engineering structural order in deterministic aperiodic systems.

6.3 Pseudo-random laser

The primary example of the pseudo-random DANS is the Rudin–Shapiro (RS) sequence [100, 312]. In a two-letter alphabet, the RS sequence can be obtained by iteration of the inflation: $AA \rightarrow AAAB$, $AB \rightarrow AABA$, $BA \rightarrow BBAB$, $BB \rightarrow BBBA$. This inflation method has recently been generalized from a 1D sequence to 2D by alternating the 1D inflation map along orthogonal directions [99]. According to this rule, we can deterministically assign the positions of scattering objects (spheres, cylinders, etc.) across a 2D array once their minimum separation has been chosen. Following this approach, we have fabricated in a free-standing GaAs membrane 2D arrays of air holes arranged in an RS sequence. A 190 nm-thick GaAs layer and a 1000 nm-thick $Al_{0.75}Ga_{0.25}As$ layer are grown on a GaAs substrate by molecular beam epitaxy. Three InAs quantum wells are embedded in the GaAs layer. The RS pattern was written on a 300 nm-thick ZEP layer with electron beam lithography. Then the pattern was transferred into the $Al_{0.75}Ga_{0.25}As$ layer by chlorine-based inductively coupled plasma reactive ion etching with the ZEP layer as the mask. The ZEP layer was subsequently removed in an oxygen plasma cleaning process. Finally the $Al_{0.75}Ga_{0.25}As$ layer was selectively etched in a dilute HF solution. Figure 6.2(a) shows the scanning electron microscope (SEM) image of a 2D array of air holes arranged in the RS sequence in the free-standing GaAs membrane. The air holes have a square shape with the side length d = 330 nm. The edge to edge separation of adjacent air holes is 50 nm. The total size of pattern is about 25 μm × 25 μm, containing 2048 air holes. 2D Fourier transformation of the structure produces the spectrum shown in Figure 6.2(b), which corresponds with the far-field diffraction pattern. Unlike a quasiperiodic structure with discrete Bragg peaks [360, 390], the RS structure features a large density of spatial frequency components which form nearly continuous bands. As the system size increases, the spectrum approaches the continuous Fourier spectrum of a white-noise random process.

To investigate the resonant modes of the 2D RS structure in a slab geometry, we performed numerical calculation using the 3D finite-difference time domain (FDTD) method. The structural parameters used in the simulations were identical

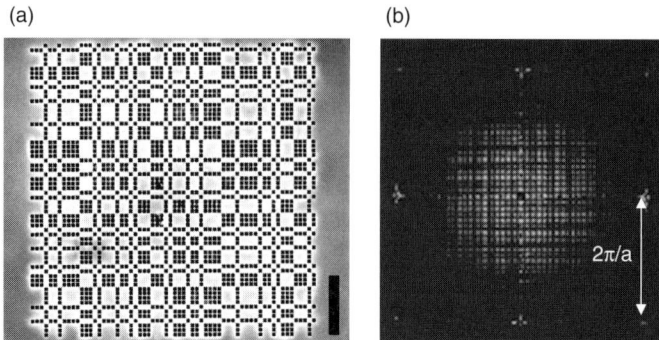

Figure 6.2 (a) SEM image of a 2D array of air holes arranged in the Rudin–Shapiro sequence in a free-standing GaAs membrane. The black vertical bar is 5 μm. (b) Reciprocal space representation of the structure obtained by Fourier transform of the SEM image. The nine red spots correspond with the reciprocal vector of square lattice with the lattice constant $a = 380$ nm [553].

to those of the fabricated sample in Fig. 6.2(a). We considered the transverse electric (TE) field, namely the electric field parallel to the slab, because experimentally the laser emission is TE polarized. Optical pulses, with a center wavelength of 880 nm and spectral width 30 nm, were launched at randomly selected locations across the pattern. Figure 6.3(a) is the calculated spectrum of the electromagnetic field inside the structure after the excitation pulse left. It contains many resonant modes of distinct frequencies. The spatial distribution of electric field intensity, shown in Fig. 6.3(b), illustrates that the resonant modes are localized at different positions of the structure. The top and bottom insets of Fig. 6.3(b) show two strongly localized modes at wavelength $\lambda = 859.2$ nm and 867.1 nm. Their quality (Q) factors are 570 and 200, respectively. The right inset shows a coupled resonance with $\lambda = 867.2$ nm and $Q = 300$. Because of its flat Fourier spectrum, the RS structure can support many localized modes, each having a well-defined frequency and position. The Q values of these localized modes are limited by the vertical leakage of light out of the slab.

In the lasing experiments, the samples were cooled to 10 K in a continuous-flow liquid helium cryostat, and optically pumped by a mode-locked Ti:sapphire laser (pulse width $\sim$ 200 fs, center wavelength $\sim$ 790 nm, and pulse repetition rate $\sim$ 76 MHz). A long working distance objective lens with a numerical aperture of 0.4 was used to focus the pump light to the structure at normal incidence. The diameter of pump spot on the sample was about 2 μm. The pump spot was moved across the sample to excite localized modes at different positions. The emission from the sample was collected by the same objective lens. The emission spectrum was measured by a half-meter spectrometer with a liquid nitrogen cooled coupled charged device (CCD) array detector. Simultaneously the spatial distribution of the emission intensity across the sample surface was imaged by a TE-cooled CCD camera.

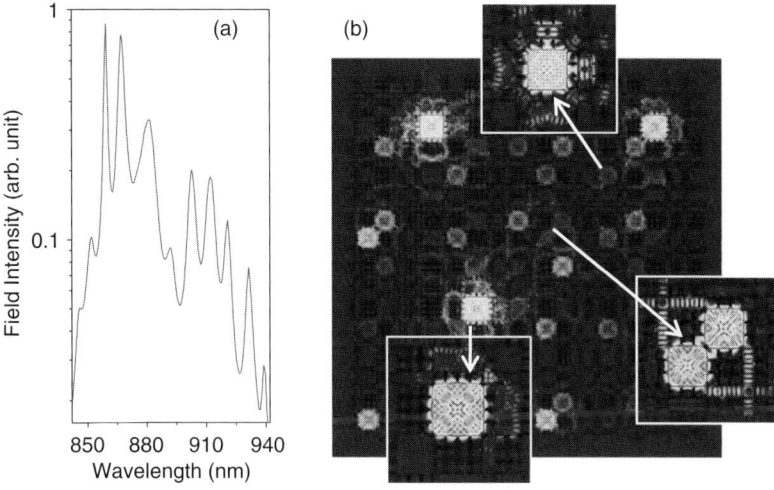

Figure 6.3 (a) Calculated field intensity inside the structure versus wavelength after short excitation pulses are launched at random locations across the structure. (b) Calculated spatial distribution of electric field intensity (in logarithmic scale) across the RS slab. The insets are the magnified view of resonant modes. The top and left insets are two localized modes at $\lambda = 859.2$ nm and 867.1 nm, respectively. The right inset shows a coupled resonance at $\lambda = 867.2$ nm [553].

Figure 6.4(a) shows the evolution of the emission spectrum with increasing pump power. Two spectral peaks were observed at low pump. One of them grew rapidly with increasing pump power above a characteristic threshold. Figure 6.4(b) shows the threshold behavior of the peak intensity versus the incident pump power P. We observed that the spectral width $\Delta\lambda$ of the peak also decreased dramatically with increasing P. Above threshold, the hot carrier effect, produced by the short pump pulse, prevented a further reduction of $\Delta\lambda$ and at higher P it caused a slight increase of $\Delta\lambda$. This is because the carrier distribution kept changing in time during the short lasing period following the pump pulse. Consequently the refractive index changes in time, causing a continuous red-shift of lasing frequency [213, 390]. In our time-integrated measurement of lasing spectrum, the transient frequency shift results in a broadening of the lasing line. Such broadening increases with the hot carrier density and becomes dominant at the highest pump power.

To map the spatial profile of the intensity of this lasing mode, we have used a bandpass filter to block the pump light and imaged the spatial distribution of the emission intensity on the sample surface. The image in Fig. 6.4(c) reveals a hot spot of laser emission on top of a broad background of (amplified) spontaneous emission within the pumping area. It is evident that the lasing mode is localized inside the RS pattern whose boundary is marked by the white line in Fig. 6.4(c). The lateral dimension of the lasing mode is approximately 2 μm, which agrees well with the size of localized modes in our numerical simulation. When we shifted

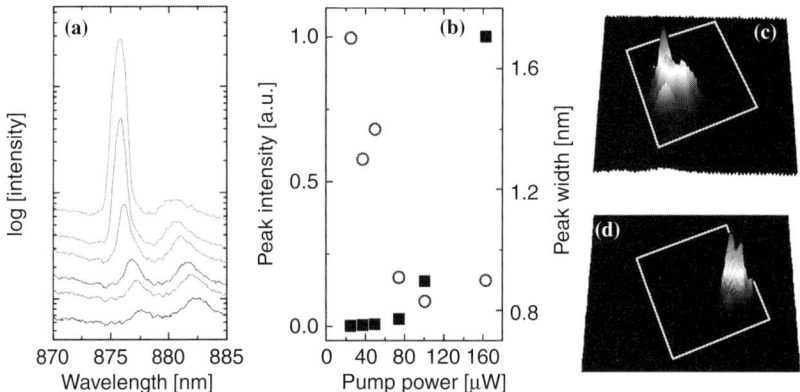

Figure 6.4 (a) Measured emission spectra at different pumping levels for the RS pattern shown in Fig. 6.2. (b) Intensity and width $\Delta\lambda$ of the lasing peak versus the incident pump power. (c) Image of the lasing mode. (d) Image of another lasing mode at a different position of the structure [553].

the pump spot across the sample, the lasing modes changed in terms of location, frequency and spatial profile. Figure 6.4(d) shows the intensity distribution of a lasing mode excited at a different position in the structure. This mode exhibits two regions of intense laser emission, suggesting that it originates from a coupled resonance, in agreement with the numerical calculation shown in the right inset of Fig. 6.3.

To prove the reproducibility of lasing modes in the pseudo-random DANS, we fabricated three identical RS patterns on the same wafer. Figure 6.5(a) shows the lasing spectra of three patterns when the same areas were optically pumped. We observed the same lasing peaks with a slight shift in wavelength. More quantitatively, the wavelength shift is within 0.3% of the lasing wavelength. Although the lasing peak blue-shift with increasing pump power below and near the lasing threshold [Fig. 6.4(a)], it remained at almost the same spectral position once the pump power exceeded the lasing threshold. The wavelength shift was caused by a change in the carrier density, which was clamped above the lasing threshold (a result of gain saturation). The three spectra in Fig. 6.5(a) were taken at pump powers well above the lasing thresholds, thus the shift of lasing wavelengths was caused by a slight structural variation due to limited fabrication accuracy.

To simulate the wavelength shift, we extracted the structures from the top-view SEM images and input them to the 3D FDTD calculation [371]. We launched short excitation pulses within the pump area and found the resonant modes of similar wavelength. As shown in Fig. 6.5(b), the mode wavelength indeed shifts slightly from pattern to pattern. The resonant wavelengths are not exactly equal to those measured, they differ by approximately 1%. This difference is attributed to the

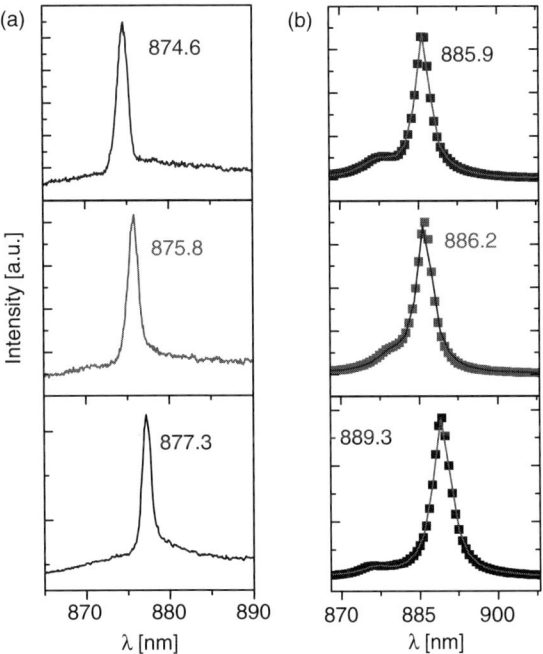

Figure 6.5 (a) Measured lasing spectra of three RS patterns fabricated on the same wafer. The incident pump power is (from bottom to top) 120 μW, 350 μW, and 420 μW. The wavelengths λ of lasing modes are written next to the peaks in the unit of nm. (b) Calculated excitation spectra of those structures which were directly extracted from the SEM images. The mode wavelengths are written in the unit of nm [553].

limited accuracy in determining the value of refractive index due to low temperature and hot carrier effect. Nevertheless, both the overall trend of wavelength shift and the intensity distribution of the modes match well with our experimental data.

Finally we stress that pseudo-random structures are not truly random structures, because they are well defined by the generation rule and have zero configurational entropy [58]. The intrinsic structural correlations in pseudo-random systems may suppress the localization effects [177]. Even in 1D RS structures the previous theoretical studies do not agree on the localization properties, partly due to the absence of rigorous analytical results [254, 255, 308, 381]. Our numerical results, showing spatially localized modes in 2D RS structures, illustrate the nature of resonances in such pseudo-random systems. The presence of a complex hierarchy of structural correlations in the RS structures, originating from the deterministic generation rule, does not suppress the localization character of the resonant modes. Note that numerical simulation is performed on the passive structures (without gain), but in the lasing experiment the samples are pumped and have optical gain. Although light localization has been predicted theoretically in certain types of deterministic pseudo-random scattering potentials (continuous and smooth varying potentials, different from the binary RS sequences) [58, 177], the possibility of lasing by localized modes in binary pseudo-random structures has not been assessed theoretically. Our experimental results demonstrate that localized modes can support lasing oscillation in the presence of gain. We emphasize that it is physically incorrect to consider the complex electromagnetic interactions of our pseudo-random structures as originating from multiple defect perturbations of the underlying periodic systems. It is well known [216] that only quasiperiodic structures such as Fibonacci and Penrose quasicrystals can be regarded as defect perturbations of periodic systems, due to the presence in their Fourier spectra of well-defined discrete peaks. For pseudo-random RS structures, the absolutely continuous nature of the Fourier spectra makes light transport and localization properties fundamentally different from the periodic structures with multiple defects. The possible benefit is that pseudo-random structures can support localized modes at all frequencies just like random structures, while the defect modes in periodic structures can only exist within photonic band gaps.

6.4 Optimization of structural aperiodicity for lasing

In this section, we consider another type of DANS – the Thue–Morse (Th–Mo) structure. It is generated by two symbols, A and B, following the inflation rule: $A \rightarrow AB$ and $B \rightarrow BA$ [245, 303]. The successive Th–Mo strings are A, AB, $ABBA$, $ABBABAAB$, etc. The spatial Fourier power spectrum of an infinitely long Th–Mo structure is singular continuous [312], in contrast with the singular (δ function) Fourier spectra of periodic and quasiperiodic lattices. Consequently, photonic band gaps and light transport in Th–Mo structures display unusual properties, e.g. fractal gaps, anomalous diffusion [179, 204, 217]. Resonant transmission and frequency trifurcation have been reported in Th–Mo multistack layers [303, 398]. Omnidirectional reflection and photoluminescence enhancement are also demonstrated [103, 104, 399].

Recently the Th–Mo inflation method has been generalized from a 1D sequence to 2D [32, 100] as following $A \rightarrow \begin{pmatrix} A & B \\ B & A \end{pmatrix}$, $B \rightarrow \begin{pmatrix} B & A \\ A & B \end{pmatrix}$. A (or B) stands for the presence (or absence) of a dielectric/metallic scatterer on a 2D square lattice. Such a structure is shown to support critical states that lie in between localized states and extended states [56, 341].

Recently, we have investigated, experimentally and numerically, lasing behavior in 2D Th–Mo structures [358]. Instead of removing scatterers from a square lattice following the Th–Mo inflation rule, we modulated the size of scatterers on the square lattice, namely, A and B represent scatterers of different size. Varying their size ratio b induces a gradual transition from aperiodic ($b = 0$) to periodic ($b = 1$) systems. Maximal light confinement in quasi-2D Th–Mo membranes is achieved at an intermediate value $0 < b < 1$, where lasing becomes the strongest.

The 2D Th–Mo structures are fabricated on a GaAs membrane that is free-standing in air. The membrane is 190 nm thick and contains three uncoupled layers of InAs quantum dots (QDs). Circular holes are etched into the layer via e-beam lithography and reactive ion etching. The sample fabrication procedure is similar to that described in the previous section. The two different sized air holes are arranged according to the 2D Th–Mo sequence, corresponding with the building blocks A and B on a square lattice. The lattice constant is $a = 320$ nm. The lateral dimension of the pattern is $20\,\mu m \times 20\,\mu m$, or equivalently $64a \times 64a$. The radius R_1 for A is fixed, while R_2 for B is varied. Figure 6.6 shows two fabricated structures with $b = R_2/R_1 = 0.83$ and 0.67. Their spatial Fourier spectra exhibit primary peaks that correspond with the reciprocal vectors of the square lattice. In addition, there is a large density of spatial frequency components that

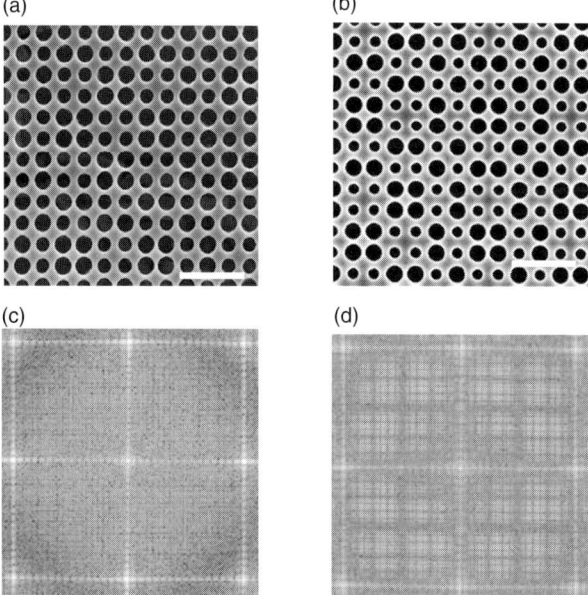

Figure 6.6 (a,b) Top-view scanning electron microscope (SEM) images of parts of Th–Mo patterns fabricated in a GaAs membrane. Air holes of radii R_1 and R_2 are arranged in 2D Th–Mo sequence. $R_1 = 130$ nm, and $R_2 = 108$ nm in (a) and 87 nm in (b). The scale bar is $1\,\mu m$. (c,d) Reciprocal space representations of the structures (a,b) obtained by Fourier transform of the SEM images. Besides the primary peaks which correspond with the square lattice, there are many secondary peaks due to structural aperiodicity, which become weaker as R_2 approaches R_1 [358].

result from the structural Thue–Morse aperiodicity. They form a nearly continuous background when the structure size is large enough. As b increases towards 1, the structural aperiodicity is reduced, and the continuous background of the spatial Fourier spectra diminishes.

The lasing experimental setup is the same as that in [553]. Femtosecond pulses from a mode-locked Ti:sapphire laser are used to excite the InAs QDs. The pump spot has a diameter $\sim 3\,\mu\text{m}$. Figure 6.7(a) shows a series of emission spectra taken at different pump levels. The emission peaks grow with the incident pump power P. Figure 6.7(b) plots the intensity of one peak at wavelength $\lambda = 978$ nm as a function of P. The data (solid circle) display a clear threshold behavior. Above the pump power $\sim 3\,\mu\text{W}$, the peak intensity grows much more rapidly with P. Meanwhile the peak width (full width at half maximum – FWHM) decreases monotonically with increasing P. It experiences a quick drop around the threshold pump power $\sim 3\,\mu\text{W}$, and eventually reaches a value of 0.71 nm at high pumping $P = 20\,\mu\text{W}$. These data demonstrate the onset of lasing at $P \sim 3\,\mu\text{W}$. We have observed numerous lasing modes as we scan the pump spot across the 2D Th–Mo structure. The lasing threshold varies from mode to mode. The frequencies of lasing peaks vary with the pump positions, indicating that lasing modes are spatially localized in the 2D Th–Mo structure.

Next we probe lasing in several Th–Mo patterns of different ratio $b = R_2/R_1$. The lattice constant a is set at 300 nm, and R_1 at 114 nm, while R_2 changes from 93 nm to 69 nm, and 45 nm. Among the structures, those of $R_2 = 69$ nm ($b = 0.61$) demonstrate the strongest lasing, namely, the lasing peaks are the highest under identical pump conditions. We also fabricated samples of $b = 0$, but could not achieve lasing with the maximum power of our pump source.

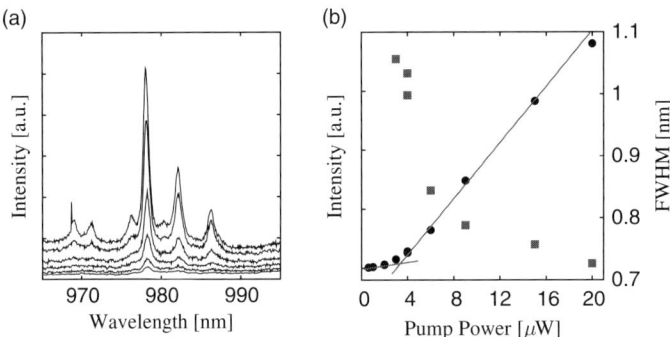

Figure 6.7 (a) Emission spectra taken at the incident pump power $P = 3, 4.1,$ 6, 9, 15, 20 μW. (b) Measured intensity (circle) and spectral width (square) of one emission peak at wavelength $\lambda = 978$ nm versus P showing the onset of lasing at $\sim 3\,\mu\text{W}$. The lines are linear fitting of peak intensity below and above the threshold pump power ($\sim 3\,\mu\text{W}$) [358].

To understand the lasing behavior in Th–Mo structures, we perform 3D numerical simulation with a commercial program (Comsol Multiphysics 3.5). The thickness of the GaAs membrane is identical with the experimental value, and the refractive index of GaAs is 3.4. Since the lasing modes usually correspond with high-quality (Q) resonances of the passive system, we compute such resonances within the frequency range of laser emission. Our simulation is limited to transverse electric (TE) polarization (electric field parallel to the membrane), because the lasing modes are TE polarized due to stronger index guiding and preferential gain of InAs QDs for TE modes.

We fix R_1 at $0.4a$, and change R_2 from 0 to $0.4a$ with a step of $0.05a$. Because of limited computer memory, the lateral dimension of the structure is $16a \times 16a$, significantly smaller than that of the real samples. Hence, our numerical simulation aims to provide qualitative instead of quantitative explanation for the experimental results. For each value of b, we find the highest Q factor for all modes with the frequency range of interest. Its value is plotted against b in Fig. 6.8(a). The global maximum of Q is reached at $b = 0.75$. Hence, there is an optimal value of b where the overall Q factor is maximized. This result explains the experimental observation of strongest lasing at an intermediate value of b between 0 and 1. However, the exact value of optimal b depends on the pattern size. Moreover, the maximal Q increases with the pattern size, thus we expect a much higher Q for the lasing modes in the real structures.

Several types of high-Q modes are identified in the frequency range $0.2 < a/\lambda < 0.5$, each has a distinct field distribution. Figure 6.8(c) shows the spatial distribution of magnetic field for three types of modes, labeled I, II and III. As b increases, different modes become the highest-Q mode within the frequency range of interest. For example, when b is close to 0, Type I (squares in Fig. 6.8(a)) is the highest-Q mode. For $0.37 < b < 0.75$, type II (triangles) takes over. At $b = 0.88$, type III mode (circles) has the maximal Q. We trace the change in Q factor for each type of mode as b varies. Figure 6.8(b) plots the Q values for type II and III modes as a function of b. With increasing b, the Q of type II first increases, reaches the maximum at $b = 0.75$, and then decreases. Type III exhibits a similar trend, but reaches the maximal Q at a larger value of b. Hence, the Q factors for different modes are maximized at different b. Typically the higher Q modes have lower lasing threshold, thus we may choose different types of mode for lasing by varying b.

To understand why Q is maximal at an intermediate value of b, we analyze light leakage from the structure. In a free-standing membrane, light can leak either horizontally from the edge of the pattern or vertically to the air above or below. These two leakage rates are characterized by $(Q_h)^{-1}$ and $(Q_v)^{-1}$, respectively. The

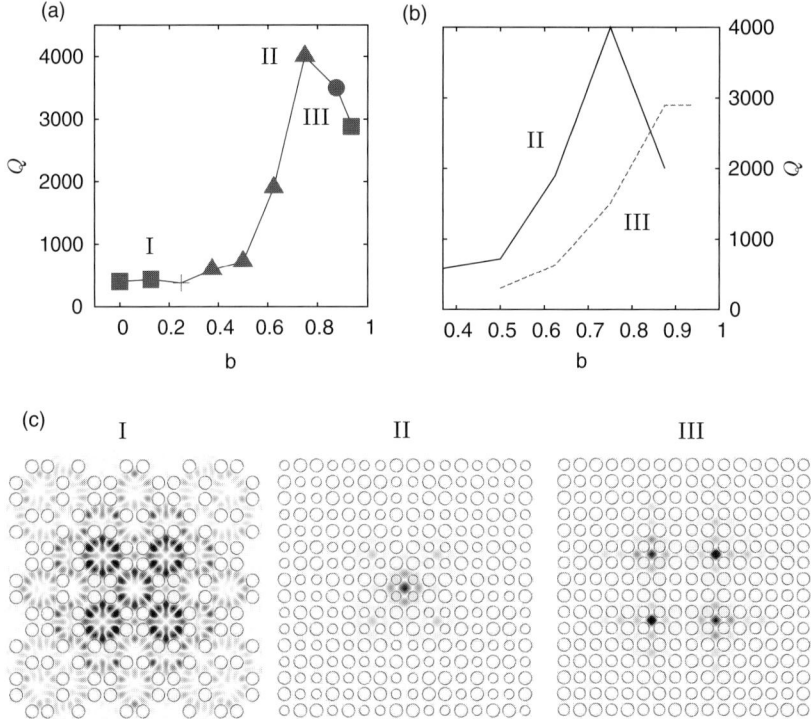

Figure 6.8 (a) Calculated Q values of the highest Q modes for each b. Different symbols represent different types of modes, e.g. square for type I, triangle for type II, circle for type III. (b) Q factor as a function of b for type II (solid curve) and III (dashed curve) modes. (c) Spatial distribution of magnetic field for type I, II and III modes (from left to right). Type II becomes the overall maximal Q mode at $b = 0.75$ [358].

total leakage rate is $Q^{-1} = (Q_h)^{-1} + (Q_v)^{-1}$. In Fig. 6.9(a), we plot Q_h and Q_v for the highest-Q modes as a function of b. While Q_v increases monotonically with b, Q_h is peaked near the optimal value of b. The variation of Q_h is associated with the change of mode size

$$s \equiv L^2 \frac{\int [I(x, y)]^2 dxdy}{[\int [I(x, y)]dxdy]^2}, \tag{6.1}$$

where $I(x, y)$ is the spatial distribution of field intensity and L is the lateral dimension of the structure. As shown in Fig. 6.9(b), s reaches the minimum at $b = 0.75$. This can be seen from the field patterns in Fig. 6.8. The type II mode is more confined horizontally than types I and III, thus it has smaller light leakage through the pattern boundary and higher Q_h. The trend of s with b can be understood qualitatively from the photonic band gap (PBG) effect. In the periodic structure $b = 1$, the high-Q modes are band-edge modes which are spatially extended and have large

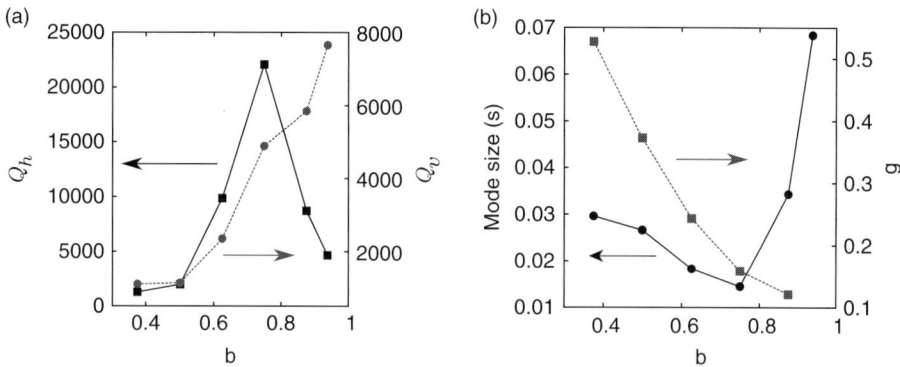

Figure 6.9 (a) Calculated Q_h and Q_v for the highest-Q modes versus b. Q_h becomes maximal at $b = 0.75$ while Q_h increases monotonically with b. (b) Lateral size s of the highest-Q mode and percentage g of the continuous background in the spatial Fourier spectra of Th–Mo structures as a function of b. s is minimal near the optimal value of b where Q_h is maximal. g decreases monotonously as b increases [358].

s. When the structural aperiodicity is introduced $b < 1$, defect modes are formed in the PBG and they become localized. As b decreases, the defect modes move from the edge to the center of the gap. The mode size s decreases, because light confinement is the strongest at the gap center. However, the structural aperiodicity reduces the PBG. When b is small enough, the gap closes, and s increases. Hence, there is an optimal value of b where s is the minimal. The variation of Q_v with b can be explained by the change of spatial Fourier power spectra of the structure. The spatial Fourier spectrum of the Th–Mo structure is singular continuous. The continuous background provides spatial frequencies that allow light of any horizontal k vector outside the light cone being scattered into the light cone and then leaking vertically out of the membrane. As b increases, the continuous background reduces (Fig. 6.6). Consequently the vertical leakage decreases and Q_v increases. This is confirmed by the percentage g of the continuous background computed from the spatial Fourier spectra. As plotted in Fig. 6.9(b), g drops continuously with increasing b.

6.5 Conclusions and future prospects

In summary, we have demonstrated lasing in localized optical resonances of deterministic aperiodic structures with pseudo-random morphologies. The localized lasing modes in the Rudin–Shapiro arrays of air nanoholes in GaAs membranes occur at reproducible spatial locations and their frequencies are only slightly affected by structural fluctuations in different samples. Numerical study on the resonances of the passive systems and optical imaging of lasing modes has enabled us to interpret the observed lasing behavior in terms of distinctive localized resonances in two-dimensional Rudin–Shapiro structures. Although the relatively narrow gain spectrum of InAs quantum wells does not allow broadband lasing in the current samples, our numerical simulations confirmed that the RS structures support many localized modes of distinct frequencies. Once we switch to semiconductor quantum dots, which have a broad gain spectrum, we expect that lasing can occur over a wide frequency range. The deterministic aperiodic Rudin–Shapiro system provides an alternative approach from random media and photonic crystals for the engineering of multi-frequency coherent light sources and complex cavities amenable to predictive theories and technology integration.

In addition, we have realized lasing action in quasi-2D Thue–Morse structures that are fabricated in a GaAs membrane. By changing the relative size of two scatterers which correspond to the building blocks A and B, we gradually varied the degree of aperiodicity and investigated its effect on lasing. We found that there exists an optimal degree of aperiodicity where the quality factor reaches a global maximum and lasing becomes its strongest. This is attributed to an enhancement of horizontal confinement of light in a finite-sized pattern by structural aperiodicity. However, the continuous background in the spatial Fourier spectrum of the Thue–Morse structure facilitates vertical leakage of light out of the membrane. At various degrees of aperiodicity, different types of modes acquire the highest quality factors and may be selected for lasing.

This work opens the way to controlling lasing phenomena at multiple frequencies via the tuning of aperiodic Fourier space. We think that the study of deterministic aperiodic media with gain, which is still in its infancy, can provide many exciting opportunities to engineer novel types of laser structures with distinctive optical properties. In particular, the controlled generation and manipulation of localized optical modes carrying orbital angular momentum, recently demonstrated in aperiodic Vogel spiral arrays with circularly symmetric Fourier space [101, 270, 271, 300, 483, 484] promises to impact a number of device applications in singular optics, secure communication, and optical sensing when combined with gain media. Theoretical as well as experimental activities along these directions are currently being pursued by the authors.

Acknowledgments

We thank our co-workers Jin-Kyu Yang, Seng Fatt Liew, Svetlana V. Boriskina, and Jacob Trevino for their important contributions to the studies presented in this chapter. Professor Glenn Solomon grew the semiconductor samples by molecular beam epitaxy, Dr. Michael Rooks worked with us in fabricating the nanostructures with electron beam lithography, and Dr. Mikhail Guy assisted us in the computer simulation. We acknowledge Professors Eric Akkermans, A. Douglas Stone, and Z. Valy Vardeny for stimulating discussions. Our research program was supported by the NSF grant DMR-1205307, the AFOSR Award FA9550-10-1-0019, and by the NSF Career Award No. ECCS-0846651. Facilities use was supported by YINQE and NSF MRSEC DMR 1119826.

7

Three-dimensional photonic quasicrystals and deterministic aperiodic structures

ALEXANDRA LEDERMANN, MICHAEL RENNER, AND
GEORG VON FREYMANN

This chapter presents an overview of the state-of-the-art of photonic three-dimensional deterministic aperiodic structures with a special emphasis on photonic quasicrystals. Reaching out into the third dimension has its own challenges. As discussed in detail in the preceding chapters, generating a deterministic aperiodic structure in one and two dimensions requires correctly calculated layer thicknesses (one dimension) or positions on a plane, where material is removed (forming holes) or added (forming cylinders, dots or other objects). Because of solid substrates all of these structures are inherently stable and one can concentrate directly on their physical properties. Having a properly calculated point-set does not necessarily help in three dimensions. Without supporting structures holding the building blocks in position, all of these constructs will collapse. Hence, devising proper support structures which do not alter the overall properties under study is one challenge; the second being the actual fabrication. For photonic structures feature sizes have to be on a scale of 100 nm and separation between features should not exceed a few microns – hence, part of this chapter is dedicated to technological questions. Because of these challenges there are only a limited number of groups working on three-dimensional aperiodic structures, a field which is just starting to expand. However, the benefit of reaching into the third dimension is to create a real material showing perhaps surprising properties, like the three-dimensional quasicrystals Shechtman *et al.* [428] discovered in 1984.

In Section 7.1 we will briefly classify the different aperiodic sequences used later on with respect to their Fourier spectra, especially keeping in mind that three-dimensional structures can only be of finite dimensions. Section 7.2 is then dedicated to the generation of three-dimensional point-sets and proper support structures, presenting the cut-and-project method as a powerful tool for generating

Light Localisation and Lasing, ed. M. Ghulinyan and L. Pavesi. Published by Cambridge University Press.
© Cambridge University Press 2015.

quasicrystalline structures with almost any desired symmetry properties. Fabrication technologies with respect to three-dimensional aperiodic structures are briefly summarized in Section 7.3, before Section 7.4 presents three-dimensional realizations of quasicrystals and deterministic aperiodic structures.

7.1 Classification

Wavefunctions of electrons, phonons, or photons propagating in deterministic aperiodic lattices can neither be described with Bloch waves and bandstructures as solutions of the wave equation with a periodic potential, nor with purely statistical tools like those used for completely random structures (e.g. Monte-Carlo simulations). The wavefunctions display a very rich behavior reaching from self-similar to fractal characteristics or power-law localization – with strong wavelength dependencies. Being neither periodic nor truly disordered there exists no theoretical description of deterministic aperiodic structures, which are as extensive as for crystalline or amorphous materials. However, deterministic aperiodic structures can generally be divided into two groups: structures which show quasiperiodicity and structures which do not. This distinction can be made by examining the Fourier transform of the underlying potential or – in the case of photonic structures – of the underlying dielectric density distribution, FT $(\rho(\vec{r})) = \hat{p}(\vec{k})$. For quasiperiodic (and crystalline) systems, the Fourier transform consists of densely spaced but well separated Bragg peaks (pure point). Non quasiperiodic systems are characterized by an absence of Bragg peaks in the Fourier transform. Here, we can further subdivide into structures with absolutely continuous (similar to amorphous systems) and singularly continuous contributions to the Fourier-transform. This is summarized by Lebesgue's decomposition theorem, which states that every Fourier transform can be decomposed into the three above-mentioned contributions [26]:

$$\hat{p}(\vec{k}) = \hat{p}_{pp}(\vec{k}) + \hat{p}_{sc}(\vec{k}) + \hat{p}_{ac}(\vec{k}).$$
(7.1)

Typical examples for quasiperiodic systems are structures distributing their elements relying on the golden mean (e.g. the Fibonacci sequence) or the silver mean. Structures having absolutely continuous components in the Fourier transform can be generated with the help of the Rudin–Shapiro sequence, structures with singularly continuous components primarily based on the Thue–Morse sequence. A comprehensive introduction to deterministic structures can be found in a recent review by Poddubny and Ivchenko [386] and in the preceding chapters.

7.1.1 Quasi-periodic Fourier spectra

Following the strict definition given above, a quasicrystal is a structure showing Bragg spots in its diffraction pattern under kinematic (single) scattering (first Born approximation) without possessing translational invariance – as distinct from periodic crystals. One of its most striking features, namely exhibiting rotational symmetries – forbidden for periodic lattices – in its diffraction pattern, is not a necessary condition as pointed out in [302]. Hence, we can investigate the well-known one-dimensional Fibonacci sequence to explore all relevant properties with respect to Fourier spectra.

The Fibonacci sequence is based on the two letter alphabet (A, B) together with the substitution rule $\sigma(A) = AB$, $\sigma(B) = A$, which can also be written as

$$\sigma : \begin{pmatrix} A \\ B \end{pmatrix} \rightarrow \begin{pmatrix} 1 & 1 \\ 1 & 0 \end{pmatrix} \begin{pmatrix} A \\ B \end{pmatrix} = S \begin{pmatrix} A \\ B \end{pmatrix} = \begin{pmatrix} AB \\ B \end{pmatrix}. \quad (7.2)$$

Using the substitution matrix S one can obtain the absolute frequencies v_n^A, v_n^B of letters A and B in a word w_n after n substitutions:

$$\begin{pmatrix} v_n^A \\ v_n^B \end{pmatrix} = S^{n-1} \begin{pmatrix} 1 \\ 1 \end{pmatrix}. \quad (7.3)$$

In the limit $n \rightarrow \infty$ the ratio of the frequencies of A to B is the golden mean $\tau = (1 + \sqrt{5})/2$. This value also appears in the roots of the characteristic polynomial of the substitution matrix S. The roots also fulfill the so-called Piso–Vijayaraghavan property [309] which is – together with the requirement that $\det S = \pm 1$ – a sufficient condition for pure point components in the diffraction pattern.

This rather abstract mathematical definition can be understood by considering a physical realization of the Fibonacci chain. The two elements A and B are represented by two distances L and S between elements, e.g. gold nanodots on a surface. The positions x_n of the n elements (with $L = S \cdot \tau$ and $S = 1$) forming the one-dimensional (finite) Fibonacci chain are given by

$$x_n = n + \frac{1}{\tau} \left\lfloor \frac{n+1}{\tau} \right\rfloor, \ n \in \mathbb{N}_0. \quad (7.4)$$

Here, $\left\lfloor \frac{n+1}{\tau} \right\rfloor$ presents the integer part or floor of $\frac{n+1}{\tau}$.

The function x_n can be divided into a sum of two functions, each of which describes a periodic spacing, yet with incommensurate periods. The first term defines a periodic spacing equal to one; the second term is also periodic, as it increases by τ^{-1} each time n is increased by τ. This translates into Bragg peaks spaced periodically with two incommensurate periods Q_1 and Q_2 in Fourier (reciprocal) space. Thus, the total diffraction pattern consists of the two sets of peaks from Q_1 and Q_2, plus additional peaks at linear combinations of Q_1 and Q_2. As Q_1 and Q_2 are incommensurate, the result is a set of Bragg peaks that almost densely fills the reciprocal space.

Computing the Fourier components of the distribution of Eq. (7.4), which are nonzero for

$$Q_{h,h'} = \frac{2\pi\tau^2}{\tau^2 + 1}\left(h + h'\frac{1}{\tau}\right), \ h, h' \in \mathbb{Z}, \tag{7.5}$$

finally results in the following Fourier transform of the Fibonacci chain (for derivation see, e.g. Ref. [215]):

$$F(Q) = \sum_{h,h'} F_{h,h'}\delta(Q - Q_{h,h'}), \ h, h' \in \mathbb{Z}, \tag{7.6}$$

where $Q = \Delta\vec{k} = \vec{k}_{\text{incident}} - \vec{k}_{\text{diffracted}}$ and the amplitude factor $F_{h,h'}$ is given by

$$F_{h,h'} = \text{sinc}\left(\frac{\pi\tau}{\tau^2 + 1}(\tau h' - h)\right)\exp\left(i\pi\frac{\tau - 2}{\tau + 2}(\tau h' - h)\right), \ h, h' \in \mathbb{Z}. \tag{7.7}$$

The diffraction peaks of the one-dimensional Fibonacci chain show an intensity distribution according to $|F_{h,h'}|^2$ and are indexed by two integers h and h'. As any h' value can be associated with any h, the diffraction peaks form a very dense pattern. Yet, the individual diffraction peaks are still sharp peaks, just as the peaks obtained from periodic photonic crystals, due to the inherent long-range order. Thus, the diffraction patterns of quasicrystals display a so-called pure point Fourier spectrum.

To demonstrate this, we start with a fictitious realization of the Fibonacci chain, where element A is represented by a slab of material with refractive index $n = 1.5$ and width $d = 0.2a$, while element B corresponds with the background material. Both elements are arranged on a periodic lattice with lattice constant a (see Fig. 7.1), corresponding with the mathematical generation rule at the beginning of this section. Because of the arrangement on a periodic lattice, we expect a clear diffraction spot – besides the characteristic Fourier components – due to the lattice constant a. However, this peak will appear at much larger diffraction angles

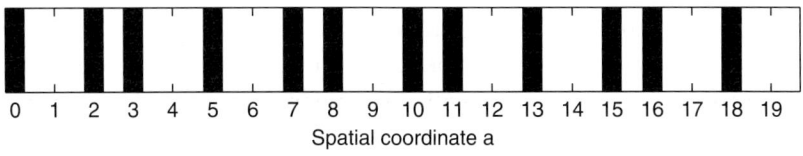

0 1 2 3 4 5 6 7 8 9 10 11 12 13 14 15 16 17 18 19
Spatial coordinate a

Figure 7.1 (a) Dielectric distribution for a 20-element Fibonacci chain. Unit cells to which element 1 of the sequence is assigned are partially filled with a value of $n = 1.5$; the background is $n = 1$.

and, hence, can easily be recognized. The corresponding diffraction patterns for two different lengths of the Fibonacci chain can be seen in Figs. 7.2(a) and 7.3(a). Increasing the number of elements from 100 to 1000 leads to a finer separation of Bragg peaks and, hence, a clearer representation of the pure point Fourier spectrum.

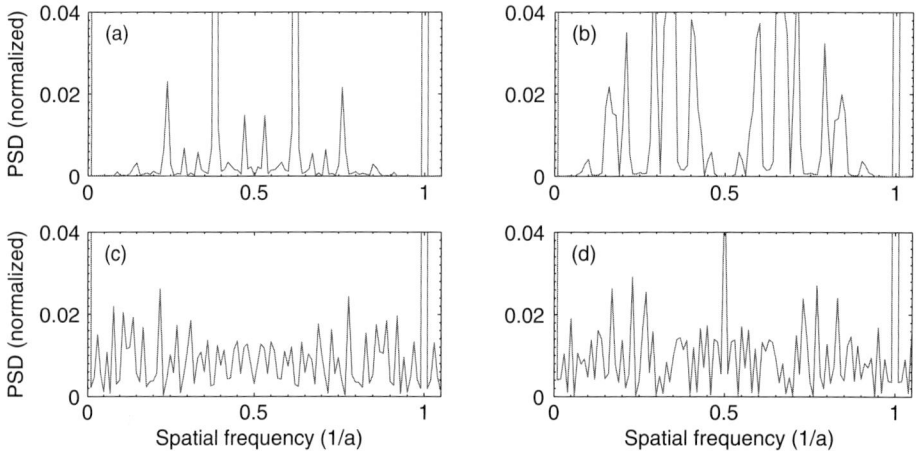

Figure 7.2 Power spectral density (PSD) given by $|F|^2$ of a 100-element-long (a) Fibonacci, (b) Thue–Morse, (c) Rudin–Shapiro and (d) random one-dimensional dielectric distribution. Because of the arrangement on a periodic lattice, a clear diffraction spot from the remaining periodicity can be observed. The PSD is normalized to the height of the periodic Bragg spot.

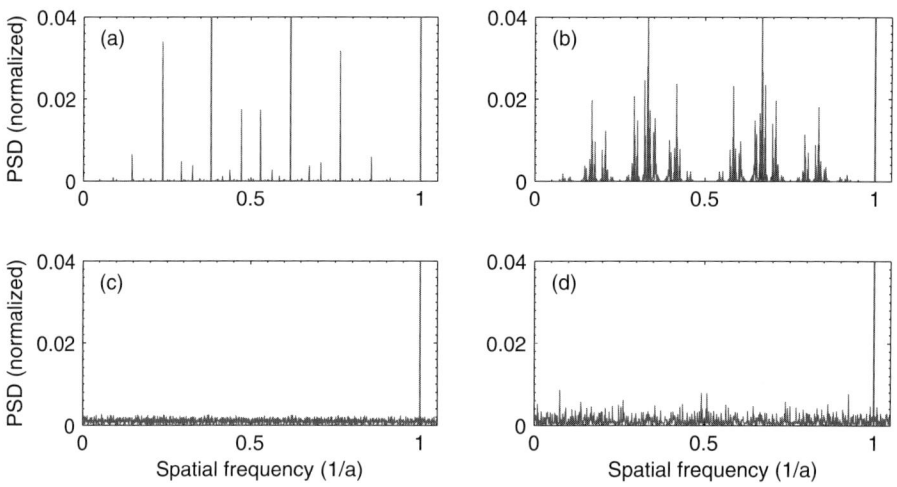

Figure 7.3 Same as Fig. 7.2 but for a 1000-element-long (a) Fibonacci, (b) Thue–Morse, (c) Rudin–Shapiro and (d) random one-dimensional dielectric distribution.

7.1.2 Non quasi-periodic Fourier spectra

Singularly continuous spectra – Thue–Morse sequence

As with the Fibonacci sequence the Thue–Morse sequence consists of a two-letter alphabet (A, B) but with different substitution rules, $\sigma(A) = AB$, $\sigma(B) = BA$ which reads as

$$\sigma : \begin{pmatrix} A \\ B \end{pmatrix} \rightarrow \begin{pmatrix} 1 & 1 \\ 1 & 1 \end{pmatrix} \begin{pmatrix} A \\ B \end{pmatrix} = S \begin{pmatrix} A \\ B \end{pmatrix} = \begin{pmatrix} AB \\ BA \end{pmatrix}. \quad (7.8)$$

The substitution matrix fulfills the Pisot–Vijayaraghavan property but does not have a vanishing determinant. Consequently, the diffraction pattern does not show Bragg peaks in the thermodynamical limit $n \rightarrow \infty$. A more detailed mathematical study reveals that the sequence has singularly continuous Fourier components which behave as a Cantor set [84, 194]. Both elements A and B appear with the same frequency.

Spectral band gaps also exist in one-dimensional Thue–Morse multilayer systems. These band gaps show a fractal splitting when system size is increased [5]. Because of the simultaneous existence of delocalized and critically localized modes it is possible to distinguish between two types of band gaps: one type which is similar to band gaps in periodic structures and another which is characterized by a self-similar splitting into smaller gaps depending on the size of the system. While the first type gets more prominent with increasing system size, leading to less transmission, the latter simply splits up at constant transmission level [217]. Dal Negro *et al.* studied the transmission spectrum through multilayers under varying incident angles and observed angle-insensitive band gaps for both polarizations [103]. A two-dimensional generalization of the Thue–Morse sequence in a rod arrangement was studied by Moretti and Mocella [341].

For the same type of realization as in Fig. 7.1 the Fourier spectra for finite Thue–Morse systems are depicted in Fig. 7.2(b) and Fig. 7.3(b). For 1000 elements the characteristic clustering of Fourier components at certain reciprocal lattice vectors can clearly be seen.

Absolutely continuous spectra – Rudin–Shapiro sequence

The Rudin–Shapiro sequence consists of a four-letter alphabet (A, B, C, D) with the substitution rule, $\sigma(A) = AB$, $\sigma(B) = AC$, $\sigma(C) = DB$, $\sigma(D) = DC$. Putting this into matrix notation one obtains,

$$\sigma : \begin{pmatrix} A \\ B \\ C \\ D \end{pmatrix} \rightarrow \begin{pmatrix} 1 & 1 & 0 & 0 \\ 1 & 0 & 0 & 0 \\ 0 & 1 & 0 & 1 \\ 0 & 0 & 1 & 1 \end{pmatrix} \begin{pmatrix} A \\ B \\ C \\ D \end{pmatrix} = S \begin{pmatrix} A \\ B \\ C \\ D \end{pmatrix} = \begin{pmatrix} AB \\ AC \\ DB \\ DC \end{pmatrix}. \quad (7.9)$$

The eigenvalues of the substitution matrix do not exhibit the Pisot–Vijayaraghavan property. The diffraction pattern of the Rudin–Shapiro sequence is diffuse which is described by an absolutely continuous Fourier spectrum. This means that every interval in reciprocal space has a contribution to the integrated diffraction intensity which is proportional to its length. Hence the integrated scattered intensity is a continuous and differentiable function. Baake and Grimm [24] showed mathematically that the diffraction patterns of the Rudin–Shapiro and purely random structure are identical although both have different entropies. Generally, the Rudin–Shapiro sequence can be reduced to a two-element sequence by assigning $A(C)$ and $B(D)$ to $A(B)$. This resulting two-letter sequence is again realized as a dielectric distribution, as outlined in Fig. 7.1. The resulting Fourier spectra can be seen in Fig. 7.2(c) and Fig. 7.3(c). Furthermore, we compare a true random series of the same length to the Fourier spectrum of the Rudin–Shapiro series (see Fig. 7.2(d) and Fig. 7.3(d)). Note that for small realizations ($<<$ 1000 elements) the Rudin–Shapiro series converges faster to an absolutely continuous spectrum than the random series. Here, several configurations have to be investigated and averaged to reach the same flat Fourier spectrum. It seems that Rudin–Shapiro is, even for shorter chains, already perfectly "disordered."

The Rudin–Shapiro sequence has been studied less often experimentally. Nevertheless, there are some theoretical publications which deal with one- or two-dimensional structures [56, 505]. Agarwal *et al.* report strongly structured transmission spectra of porous silicon multilayers having four different refractive indices [5]. Apart from two-dimensional plasmonic and dielectric nanostructures, there are no experimental higher-dimensional generalizations of the Rudin–Shapiro sequence.

7.1.3 Mode structure

The unique structure of the aperiodic or quasiperiodic patterns themselves is expected to give rise to so-called critically localized modes [242, 336, 463, 521], which do not decrease exponentially but have a power-law dependence with distance. The localization occurs due to the broken translational symmetry (similar to disordered photonic systems), yet the self-similarity of the patterns causes resonances between self-similar lattice configurations. As a consequence, the amplitude of a critical wavefunction has its maximum at a certain lattice site, but has a series of subsidiary nonzero values at other lattice sites related by self-similarity due to "tunneling effects." Such localized modes are predominantly caused by short-range effects, i.e. nearest-neighbor resonances [336, 521]. Theoretical calculations for quasicrystals have shown that the light wave modes at the photonic band gap edges are strongly localized on specific quasilattice sites, namely on those with highest local symmetry. The corresponding bands just below the main photonic gaps are very flat and of very low group velocity, which is consistent with the strong localization of the wavefunctions. As disorder is introduced into the quasiperiodic lattice, the light waves get more extended and the group velocity increases due to less flat bands. This observation is completely contrary to the behavior expected for periodic photonic crystals, where disorder usually disturbs the photon propagation by (partly) destroying their long-range phase coherence, and thus reduces the (former ballistic) transport velocity.

Furthermore, in photonic quasicrystals extended modes also co-exist due to the present long-range order, similarly to photonic crystals. Standing waves coherently spreading throughout the whole quasicrystal can be used to fulfill lasing conditions [189, 244, 360]. In active materials several lasing modes can be found due to the wide variety of available reciprocal lattice vectors, which is in agreement with the dense set of Fourier components associated with a quasiperiodic pattern. The relative intensities of observable diffraction patterns (in the out-of-plane direction of the laser) depends on the local configuration around the lasing mode. Although this has been experimentally verified for two-dimensional systems, demonstration of lasing in three-dimensional structures is absent to date.

7.1.4 Density of states and band structure

The peculiar diffraction patterns give some indications about the expected density of states. For example, the intense diffraction spots, which are associated with strong Bragg diffraction at the quasi-Brillouin zone boundaries in quasiperiodic structures, should be revealed as pseudo-stop bands or pseudogaps (distinct nonzero minima) in the transmittance spectra or density of states; see [409, 463]. As quasicrystals exhibit higher degree rotational symmetries compared with periodic crystals, the constructed quasi-Brillouin zone is more circular or spherical. Accordingly, the transmittance and reflectance spectra along different directions should be rather similar, i.e. more isotropic. Thus, one expects a larger spectral overlap of the stop bands in different spatial directions and to open complete photonic band gaps in photonic quasicrystals, at a lower refractive index contrast than that required in photonic crystals [183, 316, 402, 558, 562, 568]. However, a clear demonstration of a complete band gap in three-dimensional aperiodic structures has not yet been demonstrated.

As the diffraction patterns show a hierarchical character for quasiperiodic structures, this should be reflected in the density of states as a dense set of sharp spikes of narrow width. Each spike will translate into a flat and narrow band of low group velocity. To the best of our knowledge, this fractal character in the density of states has only been confirmed yet for the case of one-dimensional photonic quasicrystals [103, 149, 184].

7.2 Generation of three-dimensional point-sets

In this section we first discuss how to generate a point-set, obeying the higher order symmetries required for quasicrystals, before we turn to describing deterministic aperiodic structures.

While in one and two dimensions, inflation and deflation and tiling or multigrid approaches lead to successful generation of quasiperiodic patterns, these methods fail in most cases for three-dimensional point-sets. Hence, we discuss the cut-and-project method in greater detail, as the preferred method for generating point-sets with almost arbitrary symmetry. This is not restricted to the common icosahedral symmetries found for electronic quasicrystals, as we will show in subsection 7.4.2. Furthermore, the cut-and-project method allows also for the generation of so-called rational approximants, which enable the numerical calculation of the optical properties of finite-sized quasicrystals. Limitations of this approach will be discussed. Being an extremely powerful tool, the cut-and-project method fails for the generation of non-quasiperiodic three-dimensional deterministic aperiodic structures, as here no underlying symmetries can be defined, and which are required for a projection from a higher-dimensional space. Hence, three-dimensional deterministic aperiodic structures have to be generated by combining the underlying series in a clever way.

7.2.1 Quasiperiodic pattern

The patterns of quasicrystals are perfectly deterministic. Their "translational order" is quasiperiodic, i.e. the distribution of the dielectric material can be expressed as a finite sum of periodic functions with incommensurate periods combined with the appropriate unit cell. But unlike just superimposing two periodic lattices with an irrational ratio of their periods, the quasilattice must obey the requirement of "minimal separation" between the quasilattice sites to prevent physically unrealistic short distances.[1] The connections or bond angles between neighboring quasilattice sites must follow a defined orientational order and have long-range correlations.

Several theoretical models have been developed to generate an ideal quasicrystal pattern meeting the above-mentioned conditions, such as e.g. the cut-and-project method or the space-tiling procedure (Fig. 7.4).[2]

---

[1] This can be understood from the fact that quasiperiodic patterns describe the arrangement of real atoms in electronic quasicrystals, see, e.g., Refs. [215, 428].

[2] For further reading see e.g. Refs. [23, 25, 176, 215, 250, 295].

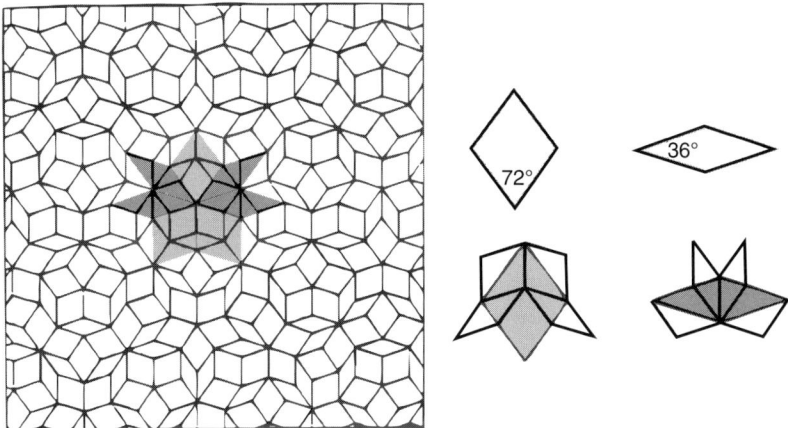

Figure 7.4 Space-tiling procedure: a two-dimensional Penrose tiling is generated by deflation operations, which reflect self-similarity and ensure long-range order. The original two types of rhombic tiles are subdivided into the same types within rescaling. Iteratively deflating an initial tile generates a growing piece of the Penrose tiling. After Ref. [215].

The cut-and-project method

The cut-and-project method [215, 273, 274] is a powerful tool for generating quasiperiodic patterns and related rational approximants. The basic idea of this method is to project points of an N-dimensional periodic lattice into an n-dimensional subspace ($n \leq N/2$), the so-called "physical space." The requested n-dimensional (quasiperiodic) pattern is achieved by using appropriate "selection rules" to specify whether a periodic lattice point is actually projected into physical space or not. These selection rules are deduced from the remaining p-dimensional so-called "internal space" ($p = N - n$, $p \geq n$).

Fundamentals of the cut-and-project method

To illustrate the principle of the cut-and-project method, we generate a one-dimensional quasicrystal; see Fig. 7.5. We assume a two-dimensional simple cubic lattice and introduce a Cartesian coordinate system such that the slope α of the axes with respect to the rows of the lattice is irrational, i.e. each axis passes only through one single lattice point, namely the origin.[3] One axis of the introduced coordinate system represents the one-dimensional physical space r^{physical}, onto which the two-dimensional periodic lattice points are projected to form a quasiperiodic pattern. The second axis, which is orthogonal to the first axis, represents the one-dimensional internal space r^{internal} and provides the selection rules for the

---

[3] The projection of this origin is the single point of exact high-degree rotational symmetry which is inherent in the generated n-dimensional quasiperiodic pattern.

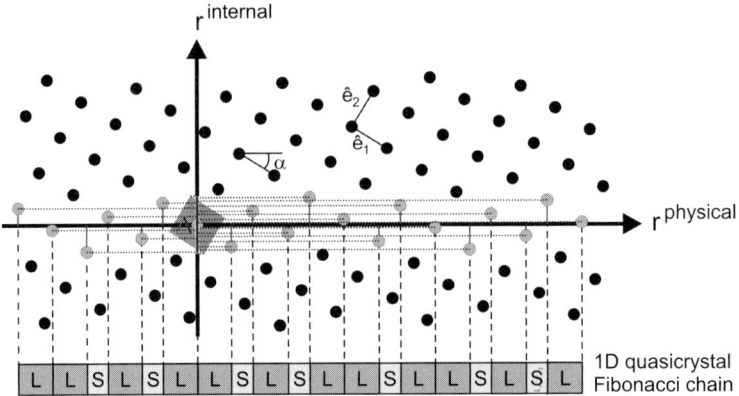

Figure 7.5 Cut-and-project method for generating a one-dimensional quasiperiodic pattern. The Wigner–Seitz cell of the periodic two-dimensional lattice
projected into internal space r^{internal} determines the size of the acceptance domain,
a line of length Δ (depicted in dark gray). Those two-dimensional periodic lattice points, whose coordinates projected into internal space are within Δ, are
highlighted in gray and are projected into physical space r^{physical} to form a
quasiperiodic pattern – if the slope α is irrational. In particular, we choose
$\alpha = 1/\tau = 2/(1 + \sqrt{5})$ and therefore the generated quasicrystal corresponds
to the Fibonacci chain composed of two building blocks L and S with lengths τ
and 1, respectively.

projection into r^{physical}. Without such selection rules, all two-dimensional periodic
lattice points would be projected into physical space and the resulting structure
would consist of an infinite number of points. The selection rules work as follows.
Firstly, one determines the Wigner–Seitz cell of the two-dimensional periodic lattice and projects the edges into internal space r^{internal}. These projections mark the
boundaries of the so-called "acceptance domain" (*here*, a line of length Δ). Now,
each two-dimensional periodic lattice point is projected into internal space first
(compare Fig. 7.5), and only if the projection lies within the acceptance domain,
is the lattice point actually projected into physical space to form the quasiperiodic
pattern.

The choice of the slope α of the introduced coordinate axes with respect to the
rows of the periodic lattice determines the properties of the generated pattern.[4] For
instance, if α is related to the golden mean, $\tau = (1 + \sqrt{5})/2$, the generated one-
dimensional quasiperiodic pattern is the famous Fibonacci chain; compare Fig. 7.5.

At this point, we want to convert the quite illustrative description of the cut-and-
project method into a matrix notation, which can then be easily transferred into
higher dimensions to generate n-dimensional quasiperiodic patterns.

---

[4] Depending on α, the generated pattern is either periodic or quasiperiodic.

To generate a one-dimensional Fibonacci chain, we start from a two-dimensional simple cubic lattice. Each two-dimensional simple cubic lattice point of coordinates (x_1, x_2) transforms into the corresponding coordinates in physical space r^{physical} and internal space r^{internal} via

$$\underbrace{\begin{pmatrix} \tau & 1 \\ 1 & -\tau \end{pmatrix}}_{\mathcal{M}_{2\times 2}} \cdot \begin{pmatrix} x_1 \\ x_2 \end{pmatrix} = \begin{pmatrix} r^{\text{physical}} \\ r^{\text{internal}} \end{pmatrix}. \tag{7.10}$$

The first row of the matrix $\mathcal{M}_{2\times 2}$ implies that a line between two adjacent periodic lattice points is projected to a line of either length 1 or τ, which are the two building blocks L and S of the Fibonacci chain (see Fig. 7.5). The acceptance domain is determined by projecting the vertices of the Wigner–Seitz cell of the two-dimensional simple cubic lattice, a square with vertices at $(\pm 0.5, \pm 0.5)$, into internal space, which is a line of size $\Delta = (\tau + 1)$. The slope α of the coordinate system with respect to the lattice points is $1/\tau$.

In order to generate n-dimensional quasiperiodic patterns, the cut-and-project method can be generally expressed as,

$$\mathcal{M}_{N\times N} \cdot \vec{x}_N = \begin{pmatrix} \vec{x}_n^{\text{ physical}} \\ \vec{x}_p^{\text{ internal}} \end{pmatrix}. \tag{7.11}$$

The N-dimensional periodic lattice points with coordinates $\vec{x}_N$ are transformed by a $N \times N$-dimensional matrix $\mathcal{M}$, the so-called "projection matrix." After this transformation the coordinates of the points are separated into two independent spaces, the physical space of dimension n and the internal space of dimension $p = N - n$. Equivalently, the $N \times N$-dimensional projection matrix $\mathcal{M}$ can be divided into two parts, the $n \times N$-dimensional submatrix $\mathcal{M}^{\text{physical}}$, which determines the physical space, and the $p \times N$-dimensional submatrix $\mathcal{M}^{\text{internal}}$, which is related to the internal space and hence to the selection rules,

$$\mathcal{M}_{N\times N} = \begin{pmatrix} \begin{bmatrix} \mathcal{M}^{\text{physical}}_{n\times N} \end{bmatrix} \\ \begin{bmatrix} \mathcal{M}^{\text{internal}}_{p\times N} \end{bmatrix} \end{pmatrix} = \begin{pmatrix} \vec{u}_1 \\ \vdots \\ \vec{u}_n \\ \vec{v}_1 \\ \vdots \\ \vec{v}_p \end{pmatrix}. \tag{7.12}$$

Here, the N-dimensional vectors $\vec{u}_i$ and $\vec{v}_i$ denote the row vectors of the projection matrix $\mathcal{M}$ or, to be precise, of the individual submatrices $\mathcal{M}^{\text{physical}}$ and $\mathcal{M}^{\text{internal}}$, respectively.

Obviously, determination of the projection matrix $\mathcal{M}$ is crucial for generating the n-dimensional structure on request, as it affects the projected coordinates in physical space as well as the projected coordinates in internal space, and with that the selection rules for the projection.

Projection matrix $\mathcal{M}$

In this subsection, some important properties of the projection matrix $\mathcal{M}$ are specified. A good understanding of this matrix gives us the possibility not only of generating one-, two-, or three-dimensional quasicrystals, but also of creating related structures, e.g., the so-called periodic rational approximants.

The $n \times N$-dimensional submatrix $\mathcal{M}^{\text{physical}}$ consists of N n-dimensional vectors, which represent the N different directions of connecting lines forming the n-dimensional quasicrystal. A line between two adjacent points of the N-dimensional periodic lattice, if they are both projected into the n-dimensional physical space in accordance with the selection rules, corresponds with one of the N n-dimensional vectors of $\mathcal{M}^{\text{physical}}$. This automatically guarantees that these lines have a defined orientational order.[5] The missing matrix elements of $\mathcal{M}$, i.e. the elements of submatrix $\mathcal{M}^{\text{internal}}$, can be completed, as the following properties must be fulfilled:

- condition (1): all row vectors $\vec{u}_i$ and $\vec{v}_j$ of the projection matrix $\mathcal{M}$ must be orthogonal to all other row vectors $\vec{u}_{k \neq i}$ and $\vec{v}_{l \neq j}$
- condition (2): all row vectors $\vec{u}_i$ and $\vec{v}_i$ have the same absolute value
- condition (3): the internal space should be equivalent to the physical space, i.e., physical space and (adequate parts of) the internal space are exchangeable, to account for the fact that the allocation of the physical space is arbitrary.

Following these rules, the submatrix $\mathcal{M}^{\text{internal}}$ can be determined – often with several possible solutions. However, all solutions lead to the same quasicrystal, as the acceptance domains and hence the selection rules are simultaneously affected by the choice of $\mathcal{M}^{\text{internal}}$.

Internal space and rational approximants

This subsection is dedicated to the selection rules of the cut-and-project method, which are derived by projecting the Wigner–Seitz cell of the N-dimensional periodic lattice into the p-dimensional internal space.[6]

To illustrate the influence of the internal space on the generated structure, we turn back to the example of the one-dimensional Fibonacci chain. In Fig. 7.5 the slope

---

[5] Such connecting lines are often required to obtain a mechanically stable three-dimensional photonic quasicrystal, see e.g. Refs. [272, 273, 274].

[6] Please note, if $p > n$, the internal space can be subdivided into x n-dimensional subspaces, which are equivalent to the n-dimensional physical space and one t-dimensional subspace ($t = p - x \cdot n$).

α is chosen irrational, in particular $\alpha = 1/\tau$, to derive the Fibonacci chain. Yet, if the slope α were chosen rational, the generated structure would not be quasiperiodic, but periodic, since the axes pass repeatedly through the two-dimensional periodic lattice points. The distance between such periodically passed lattice points determines the size of the unit cell of the periodic structure.

If the slope α' is set to $1/2$, for instance, by replacing "τ" by "2" in Eq. (7.10), the generated structure is periodic, but consists of two different lines of length 1 and length 2, i.e. compared with the Fibonacci chain, the lengths of the two lines are modified.

Yet, if we replace "τ" by "2" in Eq. (7.10) for the internal space only,

$$\begin{pmatrix} \tau & 1 \\ 1 & -2 \end{pmatrix} \cdot \begin{pmatrix} x_1 \\ x_2 \end{pmatrix} = \begin{pmatrix} r^{\text{physical}} \\ r^{\text{internal}} \end{pmatrix};\qquad(7.13)$$

we end up with a periodic structure which is similar to the Fibonacci chain, as it is also formed by the two lines of length 1 and τ, see Fig. 7.6. Such a periodic structure is called a "rational approximant," since the irrational value of slope α (*here:* $1/\tau \approx 1/1.61803\ldots$) is replaced by a rational value (*here:* $1/2$) in the submatrix $\mathcal{M}^{\text{internal}}$. Depending on the chosen rational value the size of the unit cell

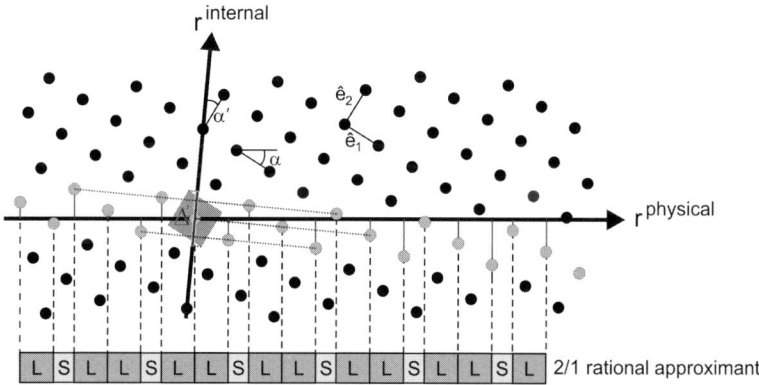

Figure 7.6 Cut-and-project method for generating a one-dimensional periodic rational approximant. The two coordinate axes which represent physical and internal space are no longer orthogonal, as the slopes α and α' are chosen differently, in particular, $\alpha = 1/\tau$ and $\alpha' = 1/2$. The Wigner–Seitz cell of the periodic two-dimensional lattice projected into internal space determines the size of the acceptance domain, a line of length Δ' (depicted in dark gray). Those two-dimensional periodic lattice points, whose coordinates projected into internal space are within Δ', colored in gray, are projected into physical space to form the periodic rational approximant, which is composed of two building blocks L and S with length τ and 1, just as for the Fibonacci chain.

of the rational approximant differs – the better the approximation of the irrational value, the larger is the size of the unit cell.

The concept of periodic rational approximants can be transferred to the n-dimensional case. The p-dimensional submatrix $\mathcal{M}^{\text{internal}}$, which defines the selection rules, is modified, while the n-dimensional submatrix $\mathcal{M}^{\text{physical}}$, which determines the actual coordinates and connecting lines of the requested n-dimensional pattern, is left unchanged. Obviously, as a consequence, the submatrices $\mathcal{M}^{\text{physical}}$ and $\mathcal{M}^{\text{internal}}$ do not define independent and orthogonal subspaces any longer, but within each submatrix the row vectors are still orthogonal to each other.

Three-dimensional icosahedral quasicrystal and its rational approximants

In this subsection, the cut-and-project method is applied to generate three-dimensional icosahedral quasicrystals and related icosahedral rational approximants.

The three-dimensional icosahedral quasicrystal exhibits icosahedral symmetry, which implies the existence of two-fold, three-fold and five-fold rotational symmetry axes. The latter gives clear evidence to quasiperiodicity. Since the overall symmetry of the quasicrystal is icosahedral, it is suggested that we use the icosahedron as a model to determine the vectors defining the submatrix $\mathcal{M}^{\text{physical}}$. The icosahedron has twelve vertices in total, which lie on a sphere around its central point. The icosahedron is characterized by the vectors pointing from the center to the twelve vertices. Typically, the orientation of the coordinate system is chosen such that the z-axis points along a two-fold rotational symmetry axis (see Fig. 7.7 (a)). Correspondingly, the (unit) vectors are defined as,

$$\vec{v}_{1..4}^{\,\text{2fold}} = \text{Norm} \cdot \begin{pmatrix} 0 \\ \pm 1 \\ \pm \tau \end{pmatrix}, \quad \vec{v}_{5..8}^{\,\text{2fold}} = \text{Norm} \cdot \begin{pmatrix} \pm 1 \\ \pm \tau \\ 0 \end{pmatrix}, \quad \vec{v}_{9..12}^{\,\text{2fold}} = \text{Norm} \cdot \begin{pmatrix} \pm \tau \\ 0 \\ \pm 1 \end{pmatrix},$$

$$(7.14)$$

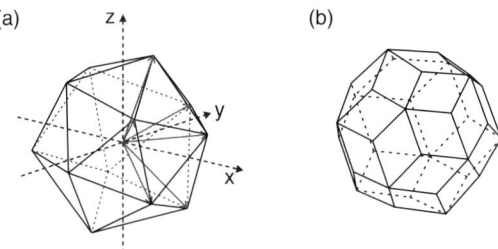

(a)   (b)

Figure 7.7 (a) shows the icosahedron oriented such that the z-axis of the Cartesian coordinate system points along a two-fold rotational symmetry axis. The gray-coloured vectors mark the six vectors, pointing from the central point of the icosahedron to six vertices, which are used for defining the icosahedron (see equations (7.14)). In (b), the triacontahedron is displayed.

with Norm $= 1/\sqrt{\tau^2 + 1}$ and the golden mean $\tau = (1 + \sqrt{5})/2 \approx 1.61803$.

As the icosahedron is symmetrical to its central point, six of these vectors are sufficient to define the icosahedron, and thus to define the respective 3×6-dimensional submatrix $\mathcal{M}^{\text{physical}}$. To complete the projection matrix, $\mathcal{M}^{\text{2f}}$, the likewise 3×6-dimensional submatrix $\mathcal{M}^{\text{internal, 2f}}$ can be determined following conditions (1) to (3) described previously.

The projection matrix of the icosahedral quasicrystal,[7] with the z-axis oriented along a two-fold axis $\mathcal{M}^{\text{2f}}$ is given by (compare Ref. [296]):

$$
\mathcal{M}^{\text{2f}} = \text{Norm} \cdot
\begin{pmatrix}
\tau & \tau & 0 & -1 & 0 & 1 \\
0 & 0 & 1 & \tau & 1 & \tau \\
1 & -1 & -\tau & 0 & \tau & 0 \\
\tau & -\tau & 1 & 0 & -1 & 0 \\
-1 & -1 & 0 & -\tau & 0 & \tau \\
0 & 0 & \tau & -1 & \tau & -1
\end{pmatrix},
\tag{7.15}
$$

with Norm $= 1/\sqrt{\tau^2 + 1}$ and the golden mean $\tau = (1 + \sqrt{5})/2 \approx 1.61803$.

It is obvious that in the projection matrix $\mathcal{M}^{\text{2f}}$, the two submatrices defining physical and internal space are equivalent, as both are formed by six of the twelve vectors of the icosahedron (compare equations (7.14)). The acceptance domain, derived from projecting the six-dimensional Wigner–Seitz cell into internal space, is a three-dimensional polyhedron, the triacontahedron (depicted in Fig. 7.7 (b)) which also exhibits icosahedral symmetry. Besides, this triacontahedron also represents the so-called quasi-Brillouin zone of the icosahedral quasicrystals, the result of projecting the Wigner–Seitz cell of the six-dimensional reciprocal lattice into physical space.

By modifying the submatrix $\mathcal{M}^{\text{internal}}$ of the three-dimensional icosahedral quasicrystal and thus the selection rules, we can generate periodic rational approximants – structures which are identical to the quasicrystal within their specific unit cell and which do qualitatively resemble the quasicrystal appearance outside of their unit cell in a periodic manner.

By substituting the golden mean τ by a rational value a/b in the submatrix $\mathcal{M}^{\text{internal}}$, the icosahedral a/b rational approximant is generated with the modified projection matrix $\mathcal{M}^{\text{2f}}_{\text{approx}}$,

---

[7] The normalization is chosen such that the connecting lines of two adjacent six-dimensional periodic lattice points after projection have a length of 1.

$$
\mathcal{M}^{2\mathrm{f}}_{\mathrm{approx}} = \mathrm{Norm} \cdot
\begin{pmatrix}
\tau & \tau & 0 & -1 & 0 & 1 \\
0 & 0 & 1 & \tau & 1 & \tau \\
1 & -1 & -\tau & 0 & \tau & 0 \\
a/b & -a/b & 1 & 0 & -1 & 0 \\
-1 & -1 & 0 & -a/b & 0 & a/b \\
0 & 0 & a/b & -1 & a/b & -1
\end{pmatrix},
\tag{7.16}
$$

with Norm $= 1/\sqrt{\tau^2 + 1}$ and the golden mean $\tau = (1 + \sqrt{5})/2 \approx 1.61803$.

7.2.2 3D aperiodic point-sets from mathematical sequences

Although the cut-and-project method is a very powerful tool it fails to the best of our knowledge for the creation of non-quasiperiodic point-sets. Hence, a different procedure has to be applied.

A simple way of generating 3D aperiodic point-sets from one-dimensional sequences T_n with elements $(1, -1)$ of generation n is to compute the Cartesian product $(T_n)^3 = T_n \times T_n \times T_n$. One then obtains a tensor of rank 3 with entries t_{ijk} given by

$$
t_{ijk} = T_n(i) \cdot T_n(j) \cdot T_n(k) \quad i, j, k = 1, \ldots 2^n \text{ or } F_{n+2},
\tag{7.17}
$$

where the Fibonacci number F_{n+2} has to be chosen for the case of the Fibonacci sequence while 2^n is chosen for the Thue–Morse and the Rudin–Shapiro sequences. This generalization procedure preserves all characteristics of the specific Fourier-spectrum as pointed out in [31, 32].

As an alternative approach for Thue–Morse and Rudin–Shapiro sequences it is also possible to use their recursive relations in a multi-dimensional form to arrive at the exact same point-set as above. The recursive relation for the Thue–Morse sequence in its three-dimensional form with elements $(1, -1)$ reads, for example,

$$
t(\mathcal{M}\vec{x}) = t(\vec{x}), \; t(\mathcal{M}\vec{x} + \vec{s}) = -t(\vec{x}) \text{ with } t(0) = -1 \text{ and } \vec{x} \in \mathbb{Z}^3.
\tag{7.18}
$$

Choosing the shift vector $\vec{s}$ to be $\vec{s} = (1, 0, 0)^T$ and the expanding Matrix $\mathcal{M}$ as

$$
\mathcal{M}_{\mathrm{Thue-Morse}} =
\begin{pmatrix}
0 & 0 & -2 \\
1 & 0 & 0 \\
0 & 1 & 0
\end{pmatrix},
\tag{7.19}
$$

the recursive equations generate the same three-dimensional Thue–Morse point-set as the direct multiplication. While for Rudin–Shapiro one obtains a point-set with absolutely continuous spectrum irrespective of the particular choice of expanding matrix $\mathcal{M}_{\mathrm{Rudin-Shapiro}}$, it is still under discussion why only special classes

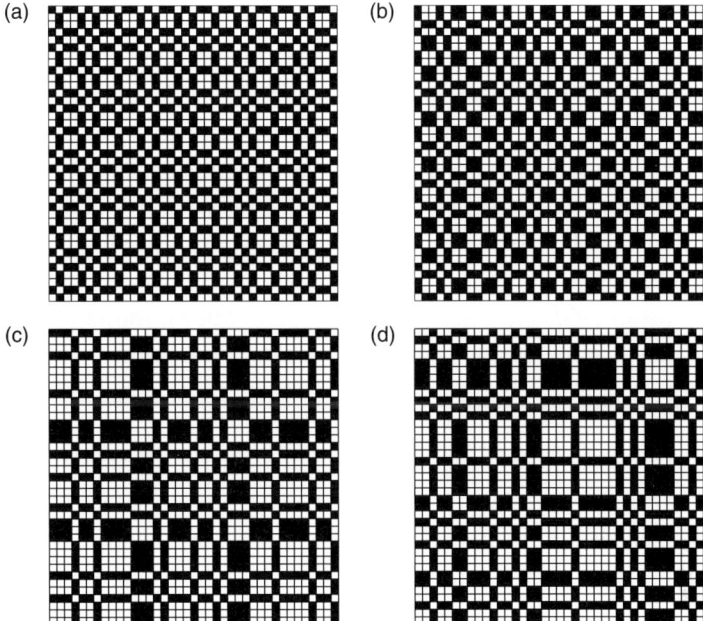

Figure 7.8 Two-dimensional ±1-valued point-sets displayed in the interval $[0, 40]^2$ for (a) Fibonacci, (b) Thue–Morse, (c) Rudin–Shapiro and (d) a random sequence. Black represents the (−1)-element, white the (+1)-element.

of $\mathcal{M}_{\text{Thue–Morse}}$ produce singular-continuous spectra for Thue–Morse recursive equations [32].

Hence, the direct multiplication approach is, up to now, the method of choice for the generation of aperiodic point-sets. An overview of the aperiodic point-sets obtained from direct multiplication is shown in Fig.7.8, displaying a cut through the xy-plane.

In contrast with the quasicrystal case the elements have to be chosen in such a way that the resulting structures are mechanically stable. One possible choice is tripods, as discussed in detail in Section 7.4.

7.3 Fabrication

One of the major challenges for the realization of three-dimensional photonic quasicrystals and deterministic aperiodic structures is the actual fabrication of the desired samples. As already pointed out in the last section, the generation of point-sets with suitable symmetry properties and even connections between the deterministically placed "atoms" is a problem, which has been solved. For these samples to have "photonic" properties, distances between the "atoms" have to be on the scale of the wavelength of light used for spectral characterization, or even below this. For near-infrared light, typical distances are on the scale of a few microns; for visible light, some 100 nanometers.

Since the structural composition of photonic quasicrystals and deterministic aperiodic structures is generally very complex, most scientists have limited themselves to studying one- and two-dimensional photonic quasicrystals operating at infrared frequencies. We only briefly list the most common approaches for 1D and 2D fabrication and discuss their suitability for 3D samples. A recent paper reviews the state-of-the-art in detail [98] and a more detailed view can be found in the preceding chapters. Finally, we review the state-of-the-art for the fabrication of 3D samples.

7.3.1 One- and two-dimensional photonic structures

For one-dimensional structures, layers of two different dielectric materials are stacked in a quasiperiodic fashion, following the famous Fibonacci chain [149, 184, 189, 463] or the Thue–Morse sequence [103, 463], for instance. Such structures can be fabricated, for example, via molecular beam epitaxy or thermal evaporation. In this fashion only one-dimensional variation of materials following a deterministic fashion is possible – even with lithographically pre-structured substrates.

Two-dimensional structures can be realized either by arranging cylindrical dielectric rods vertically on a substrate [83, 218], or by embedding small air rods in a dielectric environment [568]. Also two-dimensional plasmonic representations of deterministic aperiodic structures have been realized along these lines [98]. These two-dimensional structures are mechanically stable, since the dielectric/metallic cylinders are firmly connected to the substrate and the air rods are fixed within the surrounding material. A flexible approach to large-area fabrication of two-dimensional quasicrystal patterns is based on single beam computer-generated holography exposing a polymeric liquid crystal film [567]. This technique offers the possibility of generating two-dimensional quasicrystals of any rotational symmetry, yet has the drawback of fairly low spatial resolution and low refractive index contrast ($\Delta n \approx 0.2$). While holography is in principle well suited to the fabrication of quasicrystalline patterns due to their underlying symmetry, no such realizations are known to date for deterministic aperiodic sequences derived from mathematical series.

7.3.2 Three-dimensional photonic structures

For fabricating three-dimensional photonic structures, one requires a technique with the ability to distribute dielectric or metallic materials arbitrarily in all three spatial directions and, furthermore, to stabilize the generated pattern afterwards.

Several methods have been presented. One method, performed by Roichman *et al.* [410], uses optical tweezers. Colloidal silica microspheres are organized into the required quasiperiodic arrangement by using a holographic optical trapping technique. Computer-generated holograms are projected through a microscope objective of high numerical aperture to create three-dimensional arrays of optical traps. Applying this technique, the generated quasicrystal is quite similar to electronic quasicrystals where the real atoms "float" in a vacuum via their binding potential. Figure 7.9 shows a microscope image of the trapped spheres, as well as a diffraction pattern of the generated quasicrystal. Of course, this technique is limited to the amount of individually controlled optical traps and Brownian motion of the trapped particles might add some statistical fluctuation to the overall structure.

An alternative way of obtaining a stable three-dimensional photonic quasicrystal was shown in 2005 by W. Man *et al.* [316], who fabricated an icosahedral photonic quasicrystal for the microwave regime by stereolithography. The authors generated a mechanically stable quasiperiodic network by connecting the "atoms" at the quasiperiodic lattice sites via plastic rods of 1 cm length in a well-defined manner.

On a smaller lengthscale, three-dimensional icosahedral photonic quasicrystals have been fabricated via laser interference holography [547] and phase-mask lithography [47, 436]. These techniques generate cross-linked mechanically stable quasiperiodic patterns, yet with connections that do not necessarily consist of rods of equal length and do not define a strict orientational order. Furthermore, the three-dimensional nanostructures generated via phase-mask lithography [47, 436] are actually comprised of quasiperiodic planes, i.e. two-dimensional quasiperiodic patterns similar to the phase-mask, periodically stacked in an axial direction. Thus, they are called three-dimensional *axial* photonic quasicrystals.

Although the above-mentioned techniques are principally suited to creating deterministic aperiodic structures, to date no realization of three-dimensional

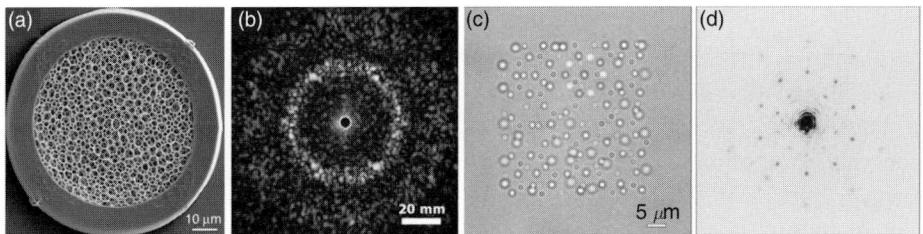

Figure 7.9 (a) Direct laser-written aperiodic sample: hyperuniform point-set. (b) Small angle scattering of a helium neon laser at the sample shown in (a) after infiltration with toluene. (c) Light microscopy image of a quasicrystal generated by trapping dielectric spheres in multiple optical tweezers. (d) Laue diagram of the structure shown in (c). Pictures in (a) and (b) after [180], pictures in (c) and (d) after [410].

deterministic aperiodic structures has been presented. The only realizations coming close to aperiodic structures are self-assembled photonic crystals with a controlled amount of random disorder [147], photonic glasses [35], and hyperuniform structures [180] (see Fig. 7.9 for a hyperuniform structure and its diffraction pattern), which have been created using direct laser writing. This latter technique is very versatile for the creation of almost arbitrary complex three-dimensional structures and will be used for the fabrication of the samples presented in the Examples section at the end of this chapter. Hence, we briefly discuss the mechanisms behind it.

7.3.3 Direct laser writing

For direct laser writing, pulses from an ultrafast laser are focused into a photo-sensitive material using high-numerical-aperture objective lenses. The intensity reached in the very focal volume is sufficiently high to polymerize the material via two- or multi-photon absorption, although the material is completely transparent for the fundamental wavelength of the laser. This allows for polymerization deep in the volume of the material without changing the properties of the surrounding material. Scanning the focus through the volume with a three-axis piezo-stage or with galvanometric mirrors while modulating the laser intensity generates three-dimensional connected structures. Even multiple crossings of already written structures – a necessity during fabrication of quasicrystals – pose no challenge, provided that a photoresist is used that does not significantly change its index of refraction during writing. For instance, the commercially available negative-tone photoresist SU-8 fulfills the above requirements. During subsequent development the unpolymerized material is washed out and the free standing structure remains.

In principle, almost arbitrarily small features can be generated along these lines. In real-world applications two limiting factors appear: (i) The laser sources are not perfectly stable, hence, working directly at the two-photon-absorption threshold is almost impossible, and (ii) common photoresist materials do not possess an ideally sharp threshold and, furthermore, do not allow for resolution better than the average size of the building blocks. Local inhomogeneities of the monomer/photoinitiator mixture further reduce the achievable feature size, which, however, lies on the order of 80 nm for state-of-the-art systems. Because of its high flexibility, this technique has, to the best of our knowledge, produced the greatest variety of different three-dimensional photonic structures [513]: woodpile, slanted-pore, square- and round-spiral photonic crystals – with and without functional defects, chiral photonic crystals, and last but not least, photonic quasicrystals and deterministic aperiodic structures, which are presented as an example in the next section.

7.4 Examples

This section is dedicated to our experimental and theoretical work on three-dimensional photonic quasicrystals and deterministic aperiodic structures. Our studies on quasicrystals have been inspired by the three-dimensional quasiperiodic patterns found in metallic alloys [45, 428], due to their unusual physical properties as well as through their aesthetics.

The icosahedral quasiperiodic pattern obtained from the cut-and-project method (see Section 7.2.1) is realized by two-photon direct laser writing (see Section 7.3.3) in the negative-tone photoresist SU-8. We characterize the high quality of the fabricated samples by scanning electron micrograph (SEM) images and visible-light Laue diffraction experiments. Additionally, we adopt a combination of the rational approximant approach and the scattering matrix formalism [527] which enables us to interpret experimental findings and to distinguish *intrinsic* from *extrinsic* properties (e.g. sample imperfections) [277] of our photonic quasicrystals.

7.4.1 High quality SU-8 icosahedral photonic quasicrystals

Applying the cut-and-project method (see Section 7.2.1), the pattern of the three-dimensional photonic quasicrystal of icosahedral symmetry is calculated. The mechanical connected and stable quasiperiodic network, which is generated by keeping connecting lines between two adjacent six-dimensional periodic lattice points, if both lattice points are projected, consists of rods of equal length l. The thus calculated icosahedral quasiperiodic pattern is realized as an SU-8 microstructure by direct laser writing (see Refs. [112, 273, 320]). To reduce distortion of the fabricated samples, e.g. due to shrinkage of SU-8 during development [325], we surround the photonic quasicrystals with a thick stabilizing wall. The shape of the wall is chosen as cylindrical to not break the symmetry of the icosahedral quasicrystals, especially when studying Laue diffraction patterns. Because of the stabilizing wall, we can access the photonic quasicrystals only from the top (perpendicular to the glass substrate) for optical measurements. To allow an overall optical characterization of the three-dimensional icosahedral photonic quasicrystals, we fabricate the samples such that the surface normal points along one of the principal rotational (real-space) symmetry axes, in particular along a five-fold, a three-fold, or a two-fold symmetry axis.

Corresponding SEM images are depicted in Fig. 7.10 and show evidence of having succeeded in fabricating porous three-dimensional icosahedral photonic quasicrystals with well-aligned smooth and ordered rods. Laue diffraction patterns taken with visible light (532 nm wavelength emitted by a solid-state laser) from the icosahedral photonic quasicrystals oriented along a five-fold, three-fold,

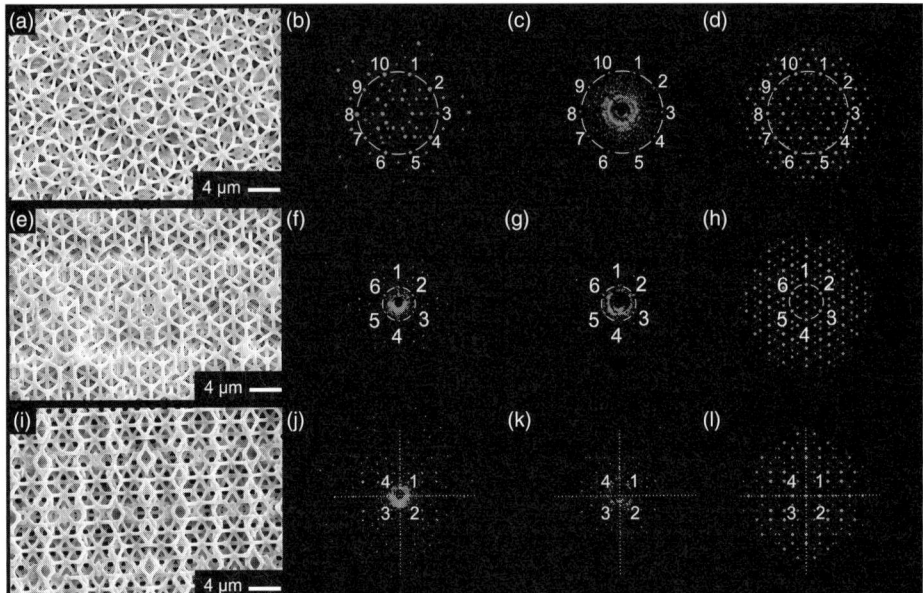

Figure 7.10 The first column (a), (e), (i) depicts SEM images of the icosahedral photonic quasicrystals with $l = 2\,\mu m$ rods and of $8\,\mu m$ thickness oriented along five-fold, three-fold, and two-fold symmetry axes, respectively. Corresponding Laue diffraction patterns measured at $4\,\mu m$ thick icosahedral photonic quasicrystals are shown in the second column (b), (f), (j), while the third column (c), (g), (k) displays the Laue diffraction patterns obtained from $8\,\mu m$ thick samples. The measured Laue diffraction patterns agree very well with those calculated (d), (h), (l) in the fourth (right) column. After Ref. [273].

or two-fold symmetry axis (first column of Fig. 7.10) reveal sharp Laue diffraction spots with the expected rotational symmetries, i.e. 2×5-fold, the 2×3-fold, and the 2×2-fold symmetry, respectively [273]. This is depicted in Fig. 7.10 in the second and third column for $4\,\mu m$ and $8\,\mu m$ thick icosahedral photonic quasicrystals, respectively, which have a diameter of $100\,\mu m$ and a rod length l of $2\,\mu m$. The overly intense non-diffracted zeroth-order spot is blocked so as not to overload the camera. In the fourth (right) column of Fig. 7.10 the calculated Laue diffraction patterns are depicted, which are computed applying the cut-and-project method: The reciprocal lattice of the six-dimensional simple cubic periodic lattice is projected into the three-dimensional physical reciprocal k-space resulting in a dense set of diffraction spots. The intensity of each spot is given by the square of the Fourier transform of the acceptance domain (projected Wigner–Seitz cell). In the case of a spherical acceptance domain of diameter Δ, the intensity is proportional to $sinc^2(\pi k \Delta)$, where k is the modulus of the three-dimensional internal reciprocal k-space vector [215, 272, 273]. The intensity of each spot is visualized by the diameter of the spots. For reasons of clarity, spots below a certain intensity are not shown.

Comparing the measured Laue diffraction patterns (second and third columns of Fig. 7.10) with theory (fourth column) one clearly observes a very nice overall agreement [273]. Furthermore, we can distinguish Laue diffraction spots even in high orders. This demonstrates the high quality of the fabricated icosahedral photonic quasicrystals. As the Laue diffraction patterns are calculated without taking account of the rods, but only considering the vertices of the rods as single scatterers,[8] the good agreement indicates that the rods, which are required to obtain a stable mechanically connected pattern, apparently do not alter the overall rotational symmetry of the icosahedral photonic quasicrystals.

Comparing the Laue diffraction patterns of the second and third columns of Fig. 7.10, one realizes that it is much harder to discern the individual Laue diffraction spots when the sample thickness is increased (from the second to the third column). In thicker samples, the Laue diffraction spots get sharper, probably because the impinging laser beam is scattered multiple times and interacts with an increased number of scattering centers [277]. Additionally, the background intensity is enhanced in the Laue diagrams of thicker samples. Firstly, due to an increased number of scattering events, laser light might be diffracted into some of the many diffraction orders which are of only low intensity in the Laue diagrams of thinner samples (and in the calculated diagrams, where only single scattering is assumed). Having thus gained in intensity, the formerly low-intensity diffraction spots contribute to the background. Secondly, sample imperfections and distortions, e.g. due to shrinkage of SU-8 during development [325], is presumably more pronounced for thicker samples and thus leads to an increase in the diffusive background (speckle).

Optical properties

A systematic microscopic theory of the optical properties of ideal icosahedral photonic quasicrystals accounting for multiple photon-scattering effects [277] is developed on the basis of scattering matrix calculations [527] combined with periodic rational approximants of the icosahedral quasicrystals.[9]

Firstly, we like to confirm the reliability of the rational approximant approach for representing the optical properties of icosahedral photonic quasicrystals. Therefore, icosahedral photonic quasicrystals oriented along a two-fold axis and corresponding rational approximants of different unit cell sizes are fabricated

---

[8] The vertices of the rods denote the positions of the projected six-dimensional periodic lattice sites in physical space and likewise the positions of the real atoms forming electronic quasicrystals [215, 428].

[9] Note that the recently introduced alternative theoretical approach of Ref. [409] has not delivered explicit findings adaptable to our three-dimensional case.

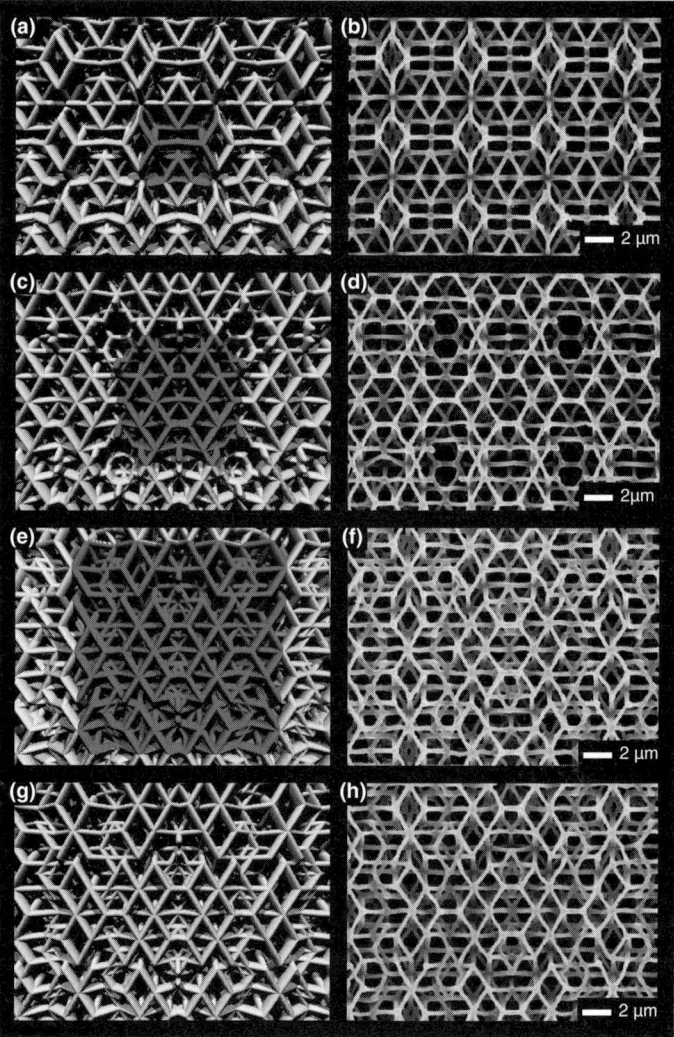

Figure 7.11 From top to bottom, rational approximants (a–f) with increasing size of the respective unit cells (highlighted in dark gray in the left column) and the icosahedral photonic quasicrystal (g), (h) oriented along a two-fold rotational symmetry axis are shown. The left column shows computer-generated images, the right column SEM images of corresponding SU-8 microstructures fabricated via direct laser writing. After Ref. [277].

via direct laser writing. This is illustrated in Fig. 7.11. In the left column, the computer-generated ray-tracing images of a (a) 1/1, (c) 2/1, (e) 3/2 rational approximant, and (g) of the icosahedral quasicrystal are depicted from top to bottom. The respective unit cells of the rational approximants are highlighted in dark gray and successively increase in size. On the right, i.e. (b), (d), (f), (h), the SEM images of correspondingly fabricated SU-8 microstructures with a rod length l of $2\,\mu$m are shown, which agree very well with the blueprint (right column).

As a test for the applicability of the combination of scattering matrix calculations with the rational approximant approach, we compare calculated angle-resolved transmittance spectra for incident linearly polarized light and corresponding measured spectra, the latter obtained experimentally by using a home-built setup. The rational approximants and the icosahedral quasicrystal are fabricated with a rod length $l = 1\,\mu$m, getting significant spectral features within the spectral range of the detecting system of our setup. The thickness of the samples is around $4.5\,\mu$m, the diameter about $50\,\mu$m. The light impinges on the samples within the finite opening angle of about 5 degrees, which we consider in the calculations by appropriate averaging.

In the scattering matrix calculations [527], we account for as many as $g = 8$ orders of the reciprocal lattice vectors to ensure convergence for a given rational approximant. The real-space discretization is 20 nm. The elliptical shape of the voxels with an aspect ratio between axial and lateral extension of about two, due to the fabrication via direct laser writing [275], is also accounted for. This allows reliable calculations for the 2/1 rational approximant, while memory space and CPU times are, however, already excessive for the 3/2 rational approximant. The incident light is linearly s- or p-polarized; see Fig. 7.12.

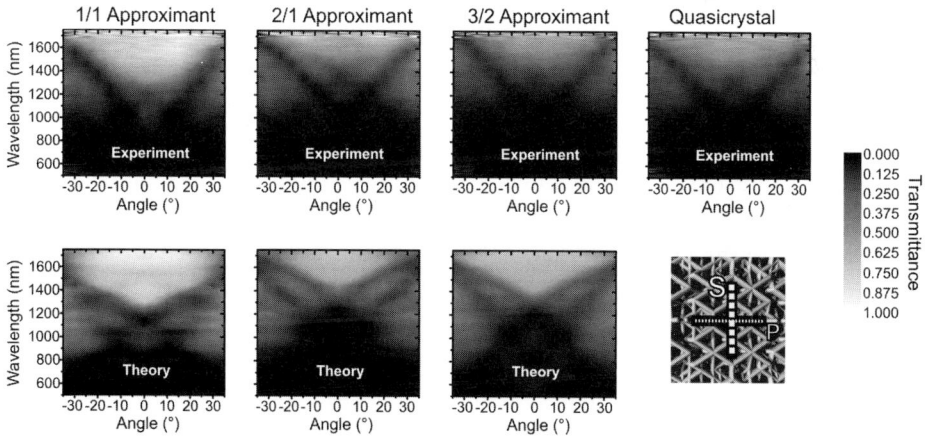

Figure 7.12 Measured (top) and calculated (bottom) angle-resolved transmittance spectra versus wavelength of light and versus angle of incidence with respect to the surface normal of the samples are depicted as gray scale plots. The incident light is s-polarized. The indicated rational approximants are illustrated in Fig. 7.11. After Ref. [277].

The obtained results are summarized in Fig. 7.12 for incident linearly *s*-polarized light. The transmittance properties of the rational approximants rapidly converge to those of the icosahedral photonic quasicrystal with increasing size of the unit cell. Furthermore, the scattering matrix calculations agree well with our experimental data. We also find that the transmittance properties of the icosahedral photonic quasicrystal are already well reproduced by the 2/1 rational approximant, which we have hence used in subsequent calculations [277]. The same observations and conclusions hold for incident *p*-polarized light, which is not shown here. The performed studies on rational approximants of different unit cell sizes give us confidence that the obtained optical properties are not due to the artificially introduced periodicity, but caused by the complex arrangement of the dielectric material within the respective unit cells. Recall that within the respective unit cells the rational approximants are identical to the icosahedral quasicrystal.

7.4.2 Rhombicuboctahedral quasicrystals

Electronic three-dimensional quasicrystals formed by real (metal) atoms have been found only in the icosahedral quasicrystal configuration so far [45, 428, 463]. Accordingly, all man-made, photonic, and phononic three-dimensional quasicrystals following the model of electronic quasicrystals are icosahedral as well. This means that the only example for a local rotational symmetry violating crystalline behavior in three dimensions is the five-fold axis of the icosahedron. Yet, generally, seven- or eight-fold symmetries should also be possible for quasicrystals in three dimensions.

In this subsection, the blueprint of the novel class of three-dimensional *rhombicuboctahedral* quasicrystals [276], calculated via the cut-and-project method is presented and realized as an SU-8 microstructure fabricated by means of two-photon direct laser writing. The anticipated rhombicuboctahedral symmetry, revealing the unusual eight-fold, three-fold, and two-fold rotational symmetry axes, is demonstrated by Laue diffraction diagrams taken with red light.

According to the cut-and-project method, quasicrystals can be constructed by starting with a simple-cubic crystal in a higher dimension that is projected to the physical space with the requested lower dimensionality. Following the procedure shown in Section 7.2.1 we construct the projection matrix for the rhombicuboctahedral quasicrystal by taking the rhombicuboctahedron as the model (see Fig. 7.13 (a)). The vectors $\vec{v}_i$ pointing from the central point to the vertices are given by

$$\vec{v}_{1..8} = \begin{pmatrix} \pm\sigma \\ \pm1 \\ \pm1 \end{pmatrix}, \; \vec{v}_{9..16} = \begin{pmatrix} \pm1 \\ \pm\sigma \\ \pm1 \end{pmatrix}, \; \vec{v}_{17..24} = \begin{pmatrix} \pm1 \\ \pm1 \\ \pm\sigma \end{pmatrix}, \qquad (7.20)$$

with the silver ratio $\sigma = 1 + \sqrt{2}$.

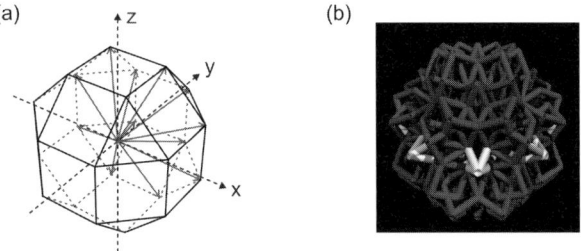

(a)

(b)

Figure 7.13 (a) displays the rhombicuboctahedron. The gray colored vectors mark the twelve vectors, pointing from its central point to twelve vertices, which define the rhombicuboctahedron. In (b), the outer boundaries of the blueprint of the three-dimensional rhombicuboctahedral quasicrystal are truncated according to the rhombicuboctahedron. The rods marked in white emphasize the unusual eight-fold rotational symmetry. After Ref. [276].

As the rhombicuboctahedron is symmetrical to its central point, twelve of these vectors are sufficient to define this polyhedron. Starting from those twelve vectors and following conditions (1) to (3) defined in Section 7.2.1, the following 12×12-dimensional projection matrix $\mathcal{M}_{\text{Rhombicuboctahedral}}$ was finally obtained [276] – after having spent considerable CPU time in our search for possible solutions employing a dedicated home-built C^{++}-programme:

$$\mathcal{M}_{\text{Rhombicuboctahedral}} = \frac{1}{\sqrt{2+\sigma^2}} \begin{pmatrix} 1 & 1 & \sigma & \sigma & -\sigma & -\sigma & 1 & 1 & -1 & -1 & -1 & -1 \\ 1 & 1 & 1 & 1 & 1 & 1 & \sigma & \sigma & \sigma & \sigma & 1 & 1 \\ \sigma & -\sigma & 1 & -1 & 1 & -1 & 1 & -1 & 1 & -1 & \sigma & -\sigma \\ 1 & 1 & -1 & -1 & -1 & -1 & -1 & -1 & \sigma & \sigma & -\sigma & -\sigma \\ 1 & -1 & -\sigma & \sigma & 1 & -1 & -\sigma & \sigma & -1 & 1 & 1 & -1 \\ -\sigma & \sigma & 1 & -1 & \sigma & -\sigma & -1 & 1 & 1 & -1 & 1 & -1 \\ \sigma & \sigma & -\sigma & -\sigma & -1 & -1 & 1 & 1 & -1 & -1 & 1 & 1 \\ 1 & -1 & -1 & 1 & \sigma & -\sigma & 1 & -1 & 1 & -1 & -\sigma & \sigma \\ -1 & 1 & -1 & 1 & 1 & -1 & \sigma & -\sigma & -\sigma & \sigma & 1 & -1 \\ -1 & 1 & -\sigma & \sigma & -1 & 1 & 1 & -1 & \sigma & -\sigma & 1 & -1 \\ -1 & -1 & -1 & -1 & 1 & 1 & \sigma & \sigma & -1 & -1 & -\sigma & -\sigma \\ \sigma & \sigma & 1 & 1 & \sigma & \sigma & -1 & -1 & -1 & -1 & -1 & -1 \end{pmatrix}$$

(7.21)

where $\sigma = 1 + \sqrt{2}$ is the silver ratio.

It is obvious that the total projection matrix can be separated into four different but equivalent three-dimensional subspaces, as each subspace contains twelve vectors that define the rhombicuboctahedron. One subspace is chosen as physical space, the remaining three subspaces form the internal space.

In order to derive the selection rules for the rhombicuboctahedral quasicrystal, the Wigner–Seitz cell of the twelve-dimensional simple cubic lattice is projected into the nine-dimensional internal space. As a result, one gets a complex nine-dimensional polyhedron. This is approximated by a nine-dimensional sphere of radius $r = (2 + 2\sigma)/\sqrt{2 + \sigma^2} \approx 2.4405$.[10] In order to generate a mechanically connected and stable quasiperiodic network, connecting lines between two adjacent twelve-dimensional periodic lattice points, if both lattice points are projected, are kept and realized as rods. Hence, the rods follow a well-defined rotational and orientational order of the generated pattern. The respective blueprint is shown in Fig. 7.13 (b).

In Fig. 7.14, a correspondingly fabricated rhombicuboctahedral quasicrystal oriented along an eight-fold axis is depicted. The rod length is $3\,\mu m$, the structure thickness about $10\,\mu m$. The quasicrystal structure with a diameter of $86\,\mu m$ is surrounded and mechanically supported by a cylindrical wall. These micrographs are directly compared with computer generated ray-tracing images of our blueprint. The real-space images of these highly complex structures seem to conceal the desired eight-fold real-space rotational symmetry, but the white rods in the ray-tracing images illustrate this aspect. To further investigate the rotational symmetry, we consider Laue diffraction diagrams taken with a helium-neon laser (633 nm wavelength). The laser spot size is chosen to be slightly smaller than the inner diameter of the stabilizing cylindrical wall in order to avoid excessive diffraction from the sample edges. The overwhelmingly intense transmitted laser beam is blocked in order not to overload the camera. The measured Laue diagram (left-hand side of Fig. 7.14 (c)) nicely agrees with the calculated one (right-hand side). The latter is computed by applying the cut-and-project method for the twelve-dimensional reciprocal lattice projected into three-dimensional physical k-space and relating the intensity of each spot to the square of the Fourier transform of the acceptance domain, which is approximated by a sphere, see Refs. [215, 272, 273]. This computation is based on the assumption of single scattering events (i.e. on Bragg diffraction) from the fictitious lattice points, i.e. the connecting rods are not accounted for at this point. This approximation is reasonable as it does not affect the quasicrystal symmetry.

By construction, the center of the structure shown in Fig. 7.14 coincides with the projection of the (0,0,0,0,0,0,0,0,0,0,0,0) origin in twelve-dimensional space. This coincidence might introduce artificial additional symmetries. To investigate this aspect and to rule out artifacts, we have fabricated a second sample with a

---

[10] The radius $r \approx 2.4405$ of the spherical acceptance domain is chosen such that a reasonable density of projected points is obtained. The actual choice of the radius does not affect the overall symmetry of the final rhombicuboctahedral quasicrystal.

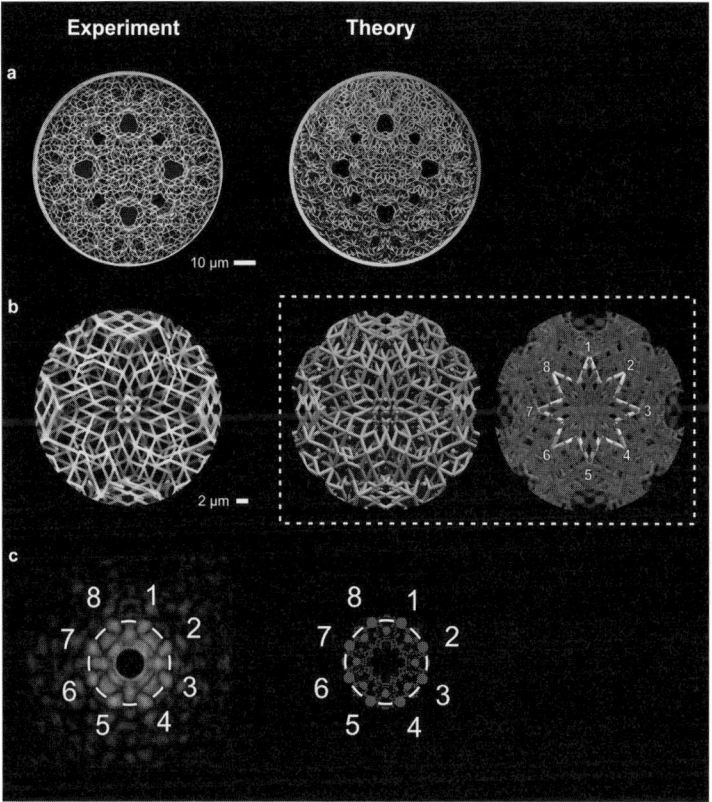

Figure 7.14 A rhombicuboctahedral photonic quasicrystal oriented along the local eight-fold axis is shown. The left column corresponds to experiment, the right column to theory. (a) depicts the electron micrograph and (b) the magnified view thereof. In (c), Laue diagrams taken with red light (central beam blocked) are displayed. The eight-fold symmetry is clearly visible. The ray-tracing simulations in (a) and (b) directly correspond with the blueprint of the fabricated structure. On the right-hand side of (b), the upper part of the identical quasicrystal is artificially made semi-transparent, such that the local eight-fold symmetry becomes apparent from the rods marked in white that are located in a lower layer. In the calculated Laue diagram in (c), the diameter of the spots is a measure of the intensity. For clarity, diffraction spots below a certain threshold intensity are not shown. The central area is blacked out to account for the blocked central beam in the experiment. After Ref. [276].

center that lies far away from the projection of the origin. The result, shown in Fig. 7.15(a), passes this control, as the Laue diagram again shows the expected eight-fold quasicrystal symmetry. To further investigate the symmetry properties of the rhombicuboctahedral quasicrystal, we have fabricated further samples that are oriented differently with respect to the glass substrate. Figure 7.15(b)

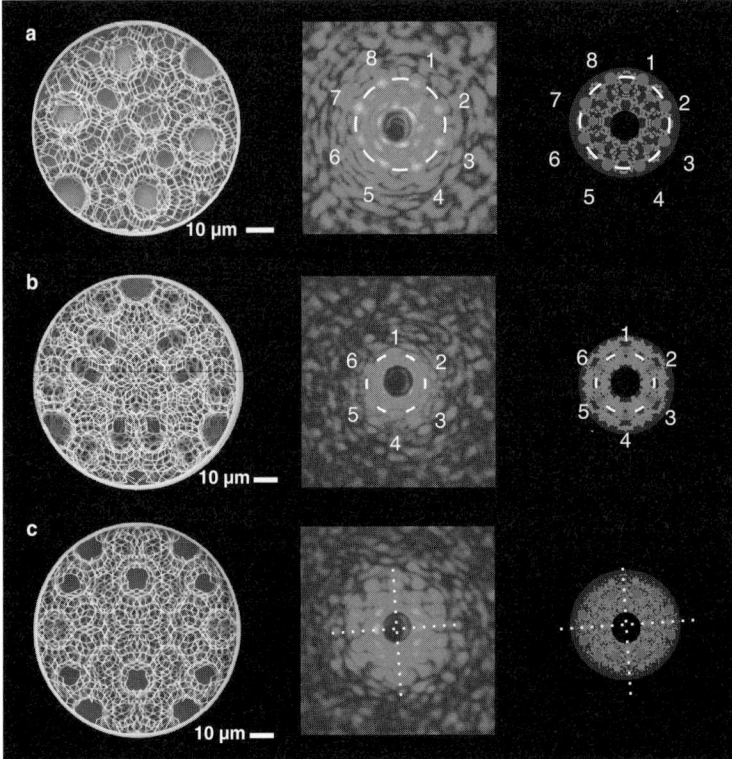

Figure 7.15 Rhombicuboctahedral photonic quasicrystals oriented along the local (a) eight-fold, (b) three-fold, and (c) two-fold symmetry axes are depicted. The left column shows electron micrographs of fabricated structures, the center column Laue diagrams measured with red light, and the right column calculated Laue diagrams. The structure in (a) is constructed such that the projection of the origin in twelve-dimensional space is not in the center of the structure, but rather outside the area depicted. Importantly, the Laue diagram still reveals an eight-fold symmetry, ruling out possible artifacts. The Laue diagrams in (b) and (c) clearly exhibit six- and two-fold rotational symmetry. After Ref. [276].

depicts the result corresponding with the local three-fold axis; Fig. 7.15 (c) that corresponding with the local two-fold axis of the rhombicuboctahedron. Clearly, the real-space electron micrographs and the corresponding measured optical as well as calculated Laue diagrams are consistent with the rhombicuboctahedral symmetry.

While numerous examples are known in one and two dimensions, in three dimensions, rhombicuboctahedral quasicrystals based on the silver ratio are only the second class after icosahedral quasicrystals that has been realized and observed to date in any system.

7.4.3 Deterministic aperiodic structures

There are two different possible ways to define a photonic structure from the three-dimensional aperiodic point-sets introduced in Section 7.2.2. Either one uses two building blocks arranged on a periodic grid or one building block arranged on an aperiodic grid which comprises two different distances.[11] Here, we use the periodic approach. We start with a simple cubic lattice, $\vec{R}(i, j, k) = i\vec{a}_1 + j\vec{a}_2 + k\vec{a}_3$, from which we remove certain lattice points following the rule,

$$\vec{R}_{\text{DAS}}(i, j, k) = \frac{t_{ijk} + 1}{2} \vec{R}(i, j, k) \quad i, j, k = 1, \ldots 2^n \text{ or } F_{n+2}, \tag{7.22}$$

where t_{ijk} is defined in Eq. (7.17).

The nonzero lattice sites $\vec{R}_{\text{DAS}}$ are decorated with tripods formed from the primitive simple cubic lattice vectors. This procedure leads to a connected network of tripods which can either be realized in a negative- or positive-tone resist via direct laser writing.[12] Presence of a tripod represents the element A, absence of a tripod the element B of our two-letter alphabet.

Because of stability considerations we decided to fabricate the structures in a positive-tone resist (AZ 9260, MicroChemicals GmbH). This positive-tone resist features low shrinkage and high structure fidelity although at limited resolution. The lattice constant of the underlying simple cubic lattice is $a = 2\,\mu\text{m}$. Microscopy images of typical resulting structures can be seen in Fig. 7.16. The surface structure is similar to the cross-sections shown in Fig. 7.8. The three-dimensionality of the structures is exemplified by focused ion beam cross-sections shown in Fig. 7.17.

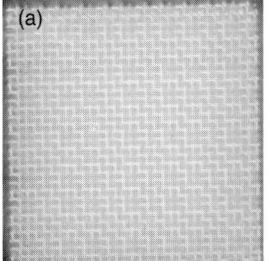 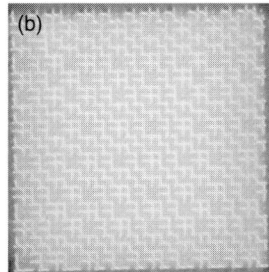 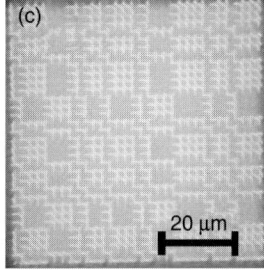

Figure 7.16 Microscopy images of the top side of (a) Fibonacci, (b) Thue–Morse, and (c) Rudin–Shapiro structures.

---

[11] Of course, a third possibility is to combine both approaches.

[12] However, in a positive-tone resist it is necessary to write a trench around the structure to allow the developer to reach all parts of the structure. In a negative-tone resist a wall surrounding the structure is needed to achieve mechanical stability.

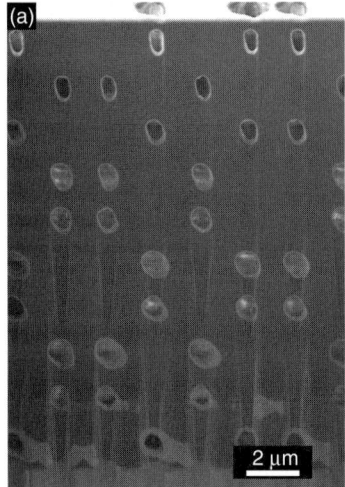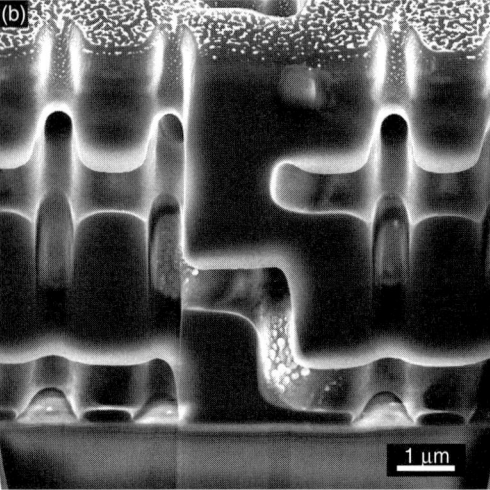

Figure 7.17 Focused ion beam (FIB) cross-sections: (a) Thue–Morse showing only the (air) rods perpendicular to the cutting plane. (b) Rudin–Shapiro structure displaying the in-plane rods.

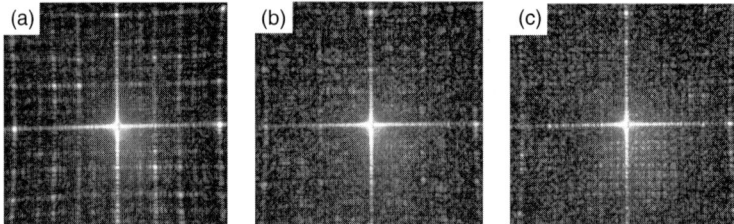

Figure 7.18 Diffraction patterns at 633 nm for (a) Fibonacci, (b) Thue–Morse, and (c) Rudin–Shapiro structures.

Diffraction patterns

Diffraction patterns are recorded in transmission using a helium-neon laser at 633 nm wavelength. Figure 7.18 shows the diffraction patterns of Fibonacci, Thue–Morse, and Rudin–Shapiro structures, all with the same dimensions (footprint $94 \times 94\,\mu m^2$, thickness $22\,\mu m$). Apart from the prominent central cross due to the square-shaped footprint of the structures one can discover the remaining influence of the simple cubic lattice displayed by bright Bragg spots in the corners of the diffraction patterns. This is due to the fact that the lattice decoration with tripods leads to a dielectric distribution which does not completely remove the underlying periodicity. Furthermore, keep in mind that our simple considerations regarding the Fourier spectra of one-dimensional aperiodic structures demonstrated that the characteristic patterns are more pronounced for larger realizations. Our sample dimensions correspond with roughly 50 lattice constants in the xy-plane.

However, the diffraction patterns clearly reveal the properties of the respective Fourier spectra. In the Fibonacci diffraction pattern (Fig. 7.18(a)) one can distinguish bright Bragg spots, as expected for a pure point spectrum, whereas for Rudin–Shapiro (Fig. 7.18(c)) no such features can be identified. Instead, one observes a relatively homogeneous intensity distribution. In the diffraction pattern of the Thue–Morse structure (Fig. 7.18(b)) it is possible to spot regularly arranged brighter parts in the diffraction pattern, in agreement with the expected singular continuous Fourier spectrum. The non-negligible speckle background in all patterns can be attributed to the onset of multiple scattering and hence deviations from the first Born approximation.

Acknowledgments

We acknowledge support through the Nanostructuring Center (NSC) of the University of Kaiserslautern. We further acknowledge support through the Deutsche Forschungsgemeinschaft (DFG) and the State of Baden-Württemberg through the DFG-Center for Functional Nanostructures (CFN) within subproject A 1.4. The research of G.v.F. has further been supported through a DFG Emmy-Noether fellowship (DFG-Fr 1671/4-3).

8

Cavity quantum electrodynamics with three-dimensional photonic band gap crystals

WILLEM L. VOS AND LÉON A. WOLDERING

8.1 Introduction

The propagation of light in periodically ordered photonic crystals bears a strong analogy to the wave propagation of conduction electrons in atomic crystals. The development of Bloch modes and dispersion are determined by the interference of waves that are diffracted by lattice planes [19, 52]. The periodicity commensurate with the wavelength ($a \approx \lambda/2$) gives rise to Bragg diffraction that is associated with frequency windows for which waves are forbidden to propagate in a particular direction. Such *stop gaps* have long been known to arise for electromagnetic waves, notably for X-rays in atomic crystals [214]. At optical frequencies, the stop gaps that occur in one-dimensional periodic structures, known as Bragg stacks, are widely used in optical laboratories as broadband highly reflecting dielectric mirrors [556].

The distinguishing feature of three-dimensional (3D) photonic crystals is that a common stop gap can be achieved for all directions and for all polarizations simultaneously: the widely pursued *3D photonic band gap*. Historically a 3D band gap has never been considered for the propagation of X-rays in periodic media, since a gap requires a high refractive index contrast of order unity, whereas the refractive index varies by less then 10^{-4} in the X-ray range. At frequencies inside the 3D photonic band gap, the density of optical states (DOS) completely vanishes. As the density of states can also be interpreted as the density of vacuum fluctuations [334], a 3D band gap thus serves as an effective shield to these fluctuations. The total absence of optical modes in a photonic band gap has implications beyond classical optics that break the analogy between the behavior of light in photonic crystals and the behavior of electrons in atomic crystals.

Light Localisation and Lasing, ed. M. Ghulinyan and L. Pavesi. Published by Cambridge University Press.
© Cambridge University Press 2015.

3D photonic band gap crystals play an important role in cavity quantum electrodynamics (cQED) [64, 549], where they offer at least five prospects for new physics.[1] Firstly, probably the most eagerly pursued phenomenon is the complete inhibition of spontaneous emission: an excited quantum emitter – such as an atom, molecule, or quantum dot – embedded in a crystal with its transition frequency tuned to within the 3D band gap remains forever excited since it cannot decay to the ground state by emitting a photon. Any interaction mediated by vacuum fluctuations is affected by their suppression in the 3D band gap [182]. Therefore, a crystal with a 3D photonic band gap not only inhibits spontaneous emission – including a shift of the emitter's frequency known as the Lamb shift [226] – it will also modify the spectrum of blackbody radiation [28], it will affect resonant dipole–dipole interactions including the van der Waals and Casimir forces [15, 258], and the well-known Förster resonant energy transfer that is prominent in biology and chemistry [182, 362].

A finite inhibition of spontaneous emission is also feasible in other optical materials such as microcavities or nanowires, if the emitters are carefully positioned in a tiny volume [36, 50]. In contrast, in photonic band gap crystals the extent over which emitters are controlled is only limited by the crystal's volume. Moreover, the complete and radical suppression of electromagnetic modes in the band gap is unique to photonic crystals, and is not found in other optical materials that appear to exclude all light in a particular frequency range. For instance, if we imagine a metal-coated box, then an emitter inside such a box is not seen from the outside, and the emission might be perceived to be inhibited. Nevertheless there are many optical states in the box in which photons can be emitted, before they are ultimately absorbed in the metal walls. In a 3D photonic band gap, however, there are simply *no* electromagnetic states available, hence an excited emitter cannot emit a photon at all and remains forever in the excited state.

Secondly, once a band gap is achieved, the cQED physics becomes even richer by introducing point defects. Such defects locally break the crystal's symmetry, which results in the appearance of isolated electromagnetic resonances in the band gap. At these resonances the field is spatially localized within a tiny nanoscale volume V_{cav} which is typically less than a wavelength cubed and thus less than a cubic micron. In other words, a point defect acts as a tiny cavity that is shielded in all three dimensions from the vacuum by the surrounding crystal [219, 551]. Hence such a photonic band gap cavity is called a "nanobox for light." Since the density of states in a nanobox is proportional to $1/V_{cav}$ and thus greatly enhanced by the tiny volume, an embedded emitter experiences a greatly enhanced emission rate, also known as the Purcell effect [396].

---

[1] Anderson localization of light in a 3D photonic band gap [222] is analogous to Anderson localization of electrons, see Ref. [263].

A third reason why 3D photonic band gaps are relevant to solid-state cQED occurs when a gain medium is introduced in a nanobox. Such a nanobox with gain offers the promise of a thresholdless laser [549]. Since only one resonance exists in a photonic band gap cavity, and the vacuum is shielded, there is no competing spontaneous emission into modes other than the lasing mode, and the laser thus immediately switches on [48]. Moreover, since 3D photonic band gap crystals are typically semiconductor devices, the on-chip integration of such a thresholdless laser is readily foreseeable.

Fourthly, an important research theme in cQED is the breaking of the weak-coupling approximation. There are in essence two ways to break this limit. The most well-known approach consists of embedding a two-level quantum emitter in a high-finesse cavity, and tuning the emitter frequency ω_{eg} to the cavity resonance frequency ω_{cav}. When a quantum of energy is exchanged between the emitter and the cavity at a rate Ω_R – called the vacuum Rabi frequency – that exceeds leakage rates such as the spontaneous emission rate or the cavity escape rate, the emitter and the cavity resonance are in a coherent superposition, called QED strong coupling [182]. While the achievement of this limit has been realized with pillar cavities [405], ring cavities [382], and 2D photonic crystal cavities [559], the realization in nanoboxes is still outstanding. There are several alternative ways of breaking the weak-coupling limit. One approach is to operate close to a van Hove singularity where the density of states has a cusp [19]. A second approach to breaking the weak-coupling limit consists of rapidly modulating the "bath" that surrounds a two-level quantum emitter [262], using ultrafast all-optical switching methods [227].

Finally, in quantum physics there is an active interest in decoherence, that is, the loss of coherence between the components of a system that is in a quantum superposition [569]. A consequence of decoherence is that a quantum system irreversibly reverts to revealing classical behavior, which is undesirable for applications, notably quantum information processing [352]. In the case where the quantum systems are optical emitters, cQED is the relevant field. An important component of decoherence in cQED corresponds with the escape or emission of photons, or by absorption of the photons by the environment [311]. Hence the shielding of vacuum fluctuations by a 3D photonic band gap offers opportunities to make optical quantum systems robust to decoherence.

In this review we summarize recent progress on the five subjects in cQED listed above, with emphasis on experimental work. In addition, we provide an overview of the current status regarding the fabrication of 3D photonic band gap crystals. For the scope of this review, we limit ourselves to (i) classes of photonic crystals with demonstrated 3D photonic band gaps, and (ii) light frequencies in the optical regime, chosen to correspond with wavelengths $\lambda \leq 2500$ nm (or frequencies $\omega/(2\pi) \geq 10^{14}$ s^{-1}) as these are accessible by conventional optics available in many laboratories all over the world. Moreover, at these frequencies stimulated and spontaneous emission rates are highly significant.

8.2 Theory

8.2.1 Spontaneous emission control

In the weak-coupling approximation of quantum electrodynamics (QED), the radiative rate of spontaneous emission γ_{rad} of a two-level dipole quantum emitter is given by Fermi's "golden rule" [131]. It is well known that spontaneous emission is not an immutable property of an emitter – such as an excited atom, molecule, or quantum dot – but it is also influenced by the emitter's nearby environment [236, 396]. The influence of any non-dissipative environment is described by the local density of optical states (LDOS) $N(\omega, \mathbf{r}, \mathbf{e}_d)$ which counts the number of photon modes available for emission weighted with their amplitude squared, and that is interpreted as the density of vacuum fluctuations [454]. Control of QED properties by means of confined light – expressed by a modified LDOS – is generally considered to be the realm of cQED [182]. The radiative rate for the transition from the excited state $|e\rangle$ to ground state $|g\rangle$ is conveniently expressed as [362, 506]

$$\gamma_{rad}(\omega_{eg}, \mathbf{r}, \mathbf{e}_d) = \frac{\pi d^2 \omega_{eg}}{\hbar \epsilon_0} N(\omega = \omega_{eg}, \mathbf{r}, \mathbf{e}_d), \qquad (8.1)$$

with ω_{eg} the emission frequency, $\mathbf{r}$ the position of the emitter, $\mathbf{e}_d$ the dipole orientation, d the modulus of the transition dipole moment matrix element, and $\hbar$ Planck's constant. It is instructive to briefly summarize the approximations and assumptions used to derive the time-evolution of the emitter's excited state and thus Fermi's "golden rule" Eq. (8.1):

(a) The electric-dipole approximation is applied to the emitter to evaluate the electric-field operator.
(b) The perturbation expansion is used to describe the time evolution of the emitter–field system.
(c) The rotating-wave approximation is used to neglect rapidly changing counter-rotating field terms.
(d) The dielectric function $\epsilon(\mathbf{r})$ is taken to be real to properly define modes and hence the LDOS (see Eq. (8.3) below).
(e) The emitter–field interaction is taken to be a *Markovian* process, which assumes that if a photon is emitted, the memory of this event is lost practically instantaneously by the quantum system, thus corresponding with a vanishing coherence time [262, 334].

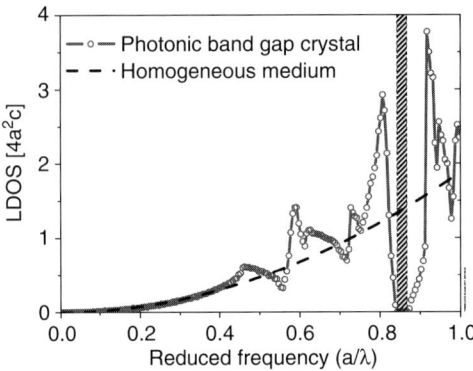

Figure 8.1 Local density of optical states (LDOS) versus frequency in a 3D photonic band gap crystal. The LDOS was calculated for an infinite inverse opal with an $\epsilon = 11.9$ backbone (connected circles), relevant for silicon. The LDOS was averaged over the unit cell such that it represents the total density of states. The hatched bar indicates the range where the LDOS vanishes: the 3D photonic band gap. The dashed curve is the LDOS of a homogeneous medium with an effective refractive index similar to that of the crystal. The frequency is reduced with the lattice constant a. Data from Ref. [354].

From Eq. (8.1) it is apparent that one can control the emission rate by means of the LDOS. The prefactor contains intrinsic emitter properties, namely the transition dipole moment d and transition frequency ω_{eg}. Equation (8.1) reveals the well-known fact that the emission rate depends on the frequency and the position of the emitter. While many efforts are directed at controlling spontaneous emission by various classes of nanophotonic systems – cavities, antennae, plasmonic systems, random media – arguably the most radical change in emission occurs in a 3D photonic band gap where the LDOS vanishes; see Figure 8.1. Therefore, an emitter in a 3D band gap that is in the excited state is radically forbidden to decay by emitting a photon. In other words, while the excited state $|e\rangle$ is usually unstable against decay to the ground state $|g\rangle$ under the emission of a photon, the excited state is stabilized in a 3D photonic band gap!

8.2.2 The local density of optical states

The local density of optical states (LDOS) $N(\omega, \mathbf{r}, \mathbf{e}_d)$ is defined as

$$N(\omega, \mathbf{r}, \mathbf{e}_d) \equiv \frac{6\omega}{\pi c^2}(\mathbf{e}_d^T \cdot \mathrm{Im}(\mathbf{G}(\omega; \mathbf{r}, \mathbf{r})) \cdot \mathbf{e}_d), \qquad (8.2)$$

with $\mathbf{G}(\omega; \mathbf{r}, \mathbf{r})$ the Green dyadic [362] that depends on frequency and position. The LDOS is a classical quantity, which can be appreciated from the absence of Planck's constant from Eq. (8.2). In the case of dissipative optical media with complex ϵ, the Green dyadic is well defined, and the imaginary part of the Green dyadic describes the *total* decay rate, i.e. the sum of the radiative decay rate and the non-radiative quenching rate induced by the dissipative environment. In dielectric photonic crystals, it is more tractable to calculate the LDOS by a summation over all Bloch modes as [454],

$$N(\omega, \mathbf{r}, \mathbf{e}_d) \equiv \frac{1}{(2\pi)^3 \epsilon(\mathbf{r})} \sum_p \int_{-\infty}^{\infty} d\mathbf{k}\, \delta(\omega - \omega_{\mathbf{k},p}) |\mathbf{e}_d \cdot \Lambda_{\mathbf{k},p}(\mathbf{r})|^2, \qquad (8.3)$$

with $\Lambda_{\mathbf{k},p}(\mathbf{r})$ a field mode with wavevector $\mathbf{k}$ and polarization state $p = 1,2$. For cavity QED it is important to note that the field modes are projected on the orientation of the transition dipole moment $\mathbf{e}_d$. Since this expression for the LDOS employs the plane-wave expansion [219], the LDOS in Eq. (8.3) pertains to an infinitely extended crystal $L \to \infty$. In the limit of a homogeneous dielectric medium with a spatially independent refractive index $n = \sqrt{\epsilon}$, the LDOS is equal to

$$N(\omega) \equiv \frac{n\omega^2}{(3\pi)^2 c^3}, \qquad (8.4)$$

which illustrates the well-known dependence on the frequency squared.

Figure 8.1 shows the LDOS calculated for a 3D photonic band gap crystal. At low frequencies the LDOS increases quadratically, which illustrates the feature that if the wavelength is much greater than the lattice parameter ($\lambda \gg a$), the crystal effectively behaves as a homogeneous medium; see Eq. (8.4). With increasing frequency, modulations appear in the LDOS, as well as characteristic cusps called van Hove singularities [19]. The hatched bar highlights the range where the LDOS completely vanishes: the 3D photonic band gap. The complete vanishing of the LDOS over a finite bandwidth, irrespective of position $\mathbf{r}$ or dipole orientation $\mathbf{e}_d$, is a unique feature of 3D photonic crystals. Whereas other systems such as 2D slab photonic crystals [355] or nanowires [50] have deep pseudogaps and concomitant strong emission inhibition, the LDOS never completely vanishes.

At the edge of the photonic band gap the LDOS vanishes as $\sqrt{|\omega - \omega_{edge}|}$, as notably pointed out in Ref. [298]. Hence the edge has a cusp and is thus also a van Hove singularity. For many years, the LDOS was considered to diverge at the edge of the gap, which led to many intricate QED predictions [267]. It appears, however, that such a diverging LDOS is typical for 1D systems [226], but not for 3D crystals.

8.2.3 Quantum efficiency of the emitters and degree of cQED control

In order to choose a suitable QED experiment with emitters in a photonic band gap crystal or to correctly interpret results, it appears to be essential to consider at least one property of the emitters, namely the emission quantum efficiency [238, 239].

The emission quantum efficiency η of one emitter is defined as the ratio of the radiative rate to the sum of the radiative and the non-radiative rates γ_{nrad},

$$\eta(\omega_{eg}) = \frac{\gamma_{rad}(\omega_{eg})}{\gamma_{rad}(\omega_{eg}) + \gamma_{nrad}} = \frac{\gamma_{rad}(\omega_{eg})}{\gamma_{tot}(\omega_{eg})}. \tag{8.5}$$

For the experimentally relevant case of an inhomogeneously broadened ensemble of emitters, the quantum efficiency becomes distributed; this situation is extensively discussed in Ref. [497].

High-efficiency quantum emitters. If the quantum efficiency of the emitters is high ($\eta \to 1$), the experimental method of choice is time-resolved emission. In this method, the decay of the population density of excited emitters $N_{exc}(t)$ is probed by recording a decay curve $g(t)$. Here $g(t)$ is the *total* decay, i.e. the sum of the radiative decay events $f(t)$ and non-radiative decay events $(g - f)(t)$. We are primarily interested in time-resolved emission measurements, which are generally recorded by the well-known time-correlated-single-photon-counting method [266]. The resulting decay curve $f(t)$ is the distribution of arrival times of single photons after exciting the emitters with a short laser pulse at time $t = 0$, averaged over many excitation–detection cycles. Such a histogram is the probability density of emission which is modeled with a probability density function [116]. The emission by the excited emitters at time t is described by a reliability function or cumulative distribution function equal to $\left(1 - \frac{N_{exc}(t)}{N_{exc}(0)}\right)$ [116]. The reliability function tends to 1 in the limit $t \to \infty$ and to zero in the limit $t \to 0$. The relation between the fraction of excited emitters and the decay curve, in other words, between the reliability function and the probability density function is

$$\int_0^t g(t')dt' = 1 - \frac{N_{exc}(t)}{N_{exc}(0)}. \tag{8.6}$$

Physically, Eq. (8.6) means that the decrease of the density of excited emitters at time t is equal to the integral of all prior decay events [497]. Equivalently, the total decay $g(t)$ is proportional to the time derivative of the fraction of excited emitters. In many reports the distinction between the reliability function and the probability density function is neglected; the intensity of the decay curve $g(t)$ is assumed to be directly proportional to the density of excited emitters $N_{exc}(t)$. This proportionality only holds for single-exponential decay but not for the general case of non-single-exponential decay; this distinction has important consequences for the interpretation of non-single-exponential decay which often occurs in photonic crystals [497].

From the slope of the photon arrival time distribution, we obtain the total decay rate which is equal to $\gamma_{tot} = \gamma_{rad} + \gamma_{nrad}$. Since an efficient emitter has $\gamma_{rad} \gg \gamma_{nrad}$, a time-resolved experiment yields, to a good approximation, the radiative rate γ_{rad} that one is interested in. Since the radiative rate depends on the density of states, see Eq. (8.1), it is generally independent of the direction of emission. While there are special situations where the rate may depend on detection angle (see Ref. [514]), one should beware of possible artifacts in this respect.

The average arrival time of the emitted photons or the average excited-state lifetime is given by the first moment of the time-resolved emission curve $f(t)$ [497]. This result confirms that the dynamics of the population of the excited state $N_{exc}(t)$ is controlled by embedding emitters in a photonic band gap crystal.

Low-efficiency quantum emitters. If the quantum efficiency of the emitters is low ($\eta \ll 1$), the experimental method of choice is the continuous-wave (cw) observation of the emitted intensity [238, 239]. From rate equations one can derive that the emitted intensity $I(\omega_{eg})$ is equal to

$$I(\omega_{eg}) = \gamma_{tot}(\omega_{eg})N_{exc}\eta(\omega_{eg}) = P\eta(\omega_{eg}) = P\frac{\gamma_{rad}(\omega_{eg})}{\gamma_{tot}(\omega_{eg})}, \qquad (8.7)$$

where P is the rate of excitation of the emitters [238]. If the quantum efficiency is low, then $\gamma_{rad} \ll \gamma_{nrad}$, hence $\gamma_{tot} \to \gamma_{nrad}$ so that γ_{tot} is independent of the LDOS. Consequently the cw intensity $I(\omega_{eg})$ of the emitters becomes proportional to γ_{rad}, hence $I(\omega_{eg})$ is then an excellent probe of the LDOS. If one studies the cw intensity $I(\omega_{eg})$ to obtain information on the crystal's LDOS, care must be taken to distinguish spectral features from several other effects, such as angular effects related to photonic bandstructures, as well as the angle-dependent excitation and the angle-dependent collection efficiency [203, 238]. To interpret results in terms of the LDOS, the observed intensity spectrum $I(\omega_{eg})$ is normalized to a reference that employs the same light sources in the same chemical environment. Hence the same transition dipole moment and the same quantum efficiency pertain, so that Eqs. (8.1, 8.5, 8.7) are properly interpreted. Secondly, a reference must have a well-known LDOS. Thirdly, a reference is required with an environment where the emitters experience the same refractive index as in the photonic band gap crystal, to avoid complications arising from Lorentz local field factors [421]. Fourthly, since cw intensity is very sensitive to the collected solid angle, it is important that both the crystal and the reference have similar escape functions for internal light escaping to the detection system, or at least very well-known escape characteristics. It is our experience that the best suited reference is a similar photonic crystal in the long wavelength limit, for several reasons. The LDOS is well known and proportional to ω^2, the average index is the same, hence Lorentz local field factors cancel, and the escape function is well known and matches a Lambertian distribution [238]. Finally, this choice ensures a reference made by the same fabrication process, hence the same chemistry and thus the same non-radiative rate. Other considered reference systems such as a random medium, or a bulk material do not share these strengths and are therefore less reliable reference systems for the interpretation of cw intensity.

In the case of low-efficiency quantum emitters, the total decay rate measured in a time-resolved experiment hardly depends on the LDOS as it is dominated by non-radiative decay events $((g - f)(t))$ [239]. In this situation the dynamics of the excited-state population is hardly controlled by the photonic environment, in contrast with the case of efficient emitters whose population $N_{exc}(t)$ is truly controlled by the photonic band gap environment.

8.2.4 Beyond weak coupling

In the Markov approximation we assume that the quasi-continuum of the field modes – the *bath* – equilibrates on timescales much faster than the oscillation period of the dipole τ_d. Since the LDOS $N(\omega, \mathbf{r}, \mathbf{e}_d)$ has a dimension of *per frequency*, it denotes a *timescale* that is usually considered to be the correlation time of the bath τ_b [262]. From Eq. (8.4) we calculate the correlation time τ_b for the LDOS at an optical frequency of 10^{15} s^{-1} in a quantization volume $V = (2\pi c/\omega)^3$ to be $\tau_b = 10^{-18}$ s, which is much shorter than $\tau_d \simeq 10^{-15}$ s. The interaction with the fast bath destroys the emitter's memory of the past, which leads to the exponential decay of the excited state. The situation $\tau_b \ll \tau_d$ is also known as the weak-coupling limit of cQED [182, 334], and pertains to the majority of spontaneous emission studies in nanophotonics with photonic crystals.

Highly interesting quantum physics arises when the weak-coupling limit is violated. When the LDOS is strongly increased, the relaxation of the bath becomes slower. Once the correlation time τ_b becomes of the order of, or even longer than the dipole oscillation period τ_d, the quantum system retains a memory of earlier times, hence approximation *e* of the weak-coupling limit is violated.[2] This means that the quantum emitter and the bath are becoming strongly coupled, and the evolution of the emitter's excited state $|e\rangle$ does not show exponential decay any more. In this situation – typically studied with a high-Q cavity [382, 405, 559] – the emitter resonance and the cavity resonance hybridize, such that the emitters's excited state population oscillates in time, as illustrated in Figure 8.2.

In addition to raising the LDOS by a cavity resonance, there are other ways to violate the weak-coupling limit. One approach is to operate close to a van Hove singularity [19]. Such a singularity manifests as a cusp in the density of states, causing the density of states to become non-analytical. The non-analytic behavior means that approximations *b* and *e* of the weak-coupling approximations are

---

[2] The approximations *a* through *e* referred to here are the approximations listed in Section 8.2.1 above.

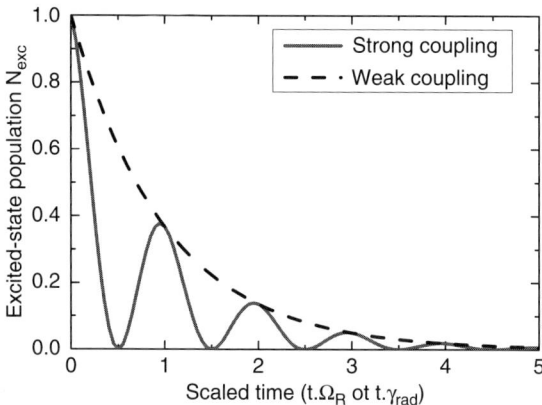

Figure 8.2 Excited-state population N_{exc} of a two-level emitter with time t. In the strong-coupling limit, damped vacuum Rabi oscillations appear (full curve). In the weak-coupling limit, the population of the excited state $|e\rangle$ decays exponentially with the spontaneous emission rate γ_{rad} (dashed curve). Time is scaled with the vacuum Rabi frequency Ω_R, or the spontaneous emission rate γ_{rad} in the case of weak coupling. In the strong-coupling case, γ_{rad} is taken to be equal to Ω_R.

violated. As a result, a single emitter tuned to a van Hove singularity is predicted to exhibit non-exponential dynamics, including intricate time-dependent oscillations of the excited-state population that are called "fractional decay" [224, 252, 506].

Another approach to violating the weak-coupling limit is to rapidly modulate the bath. Hence approximations *b* and *c* of the weak-coupling limit are violated. Rapid modulation of the bath can be achieved by ultrafast switching of the optical properties of a 3D photonic band gap crystal on ultrafast timescales, using all-optical methods [227].

8.3 Ultimate tools for 3D photonic band gap cavity QED

8.3.1 Requirements for 3D photonic band gap crystals

There are four main requirements for periodically ordered nanostructures in order to function successfully as 3D photonic band gap crystals. Firstly, the photonic interaction strength S between the light and the nanostructures should be elevated. We have seen that S is defined as the polarizability per volume. The photonic interaction strength equals, to a very good approximation, the relevant frequency bandwidth $\Delta\omega/\omega$ of a stop gap: $\Delta\omega/\omega = S$. For photonic crystals composed of spherical scatterers, S can be expressed analytically as [237, 517]

$$S = \frac{4\pi\alpha}{V} = 3\phi\frac{\epsilon - 1}{\epsilon + 2}g(K.R), \tag{8.8}$$

where ϵ is the relative dielectric function of the spheres, equal to $\epsilon = m^2$ with m the ratio of the refractive indices of the spheres and of the surrounding medium: $m = n_1/n_2$. Thus a high refractive index contrast m is highly desired, which dictates the choice of the constituent materials. Air with index $n = 1.0$ is a convenient low-index material, and semiconductors such as silicon or GaAs are often the high-index material of choice, with an index $n \simeq 3.5$.[3] In Eq. (8.8), $g(K.R)$ is the form factor of the scatterers (here, spheres in the Rayleigh–Gans limit) as a function of scatterer radius R and modulus of the diffraction vector $K = |\mathbf{K}| = |\mathbf{k_{out}} - \mathbf{k_{in}}|$. The form factor dictates both the shape of the scatterers, and the optimal filling fraction of the scatterers in the crystal [517].

The second main requirement is that the optical absorption of the constituent materials should be as small as possible. This is borne out by the fact that a photonic band gap is a multiple scattering phenomenon, hence at every chance that photons are absorbed, it's "game over." The role of absorption has been studied via the imaginary part of the frequency ω'' [253]. Near the edge of the gap the square root singularity disappears for the weakest absorption considered. Within the band gap the density of states becomes nonzero and proportional to ω''/ω_{gap}. Hence for cQED in a 3D band gap, absorption of light should be reduced as much as possible. This requirement limits the use of semiconductors to frequencies below the electronic band gap. Thus photonic crystals made from silicon or GaAs are limited to wavelengths longer than 1100 nm or 870 nm. Indeed, in the early days of semiconductor research, an empirical relation between refractive index and electronic band gap E_g was established called Moss' rule: $n^4 E_g = 77$ [346]. In the visible range, TiO_2 is a versatile high-index material with $n \simeq 2.7$, depending on its atomic crystal structure. Thus the choice for a high-index material limits the band gap width (via the maximum index) and the operating frequency (via the electronic gap).

The third main requirement is the photonic crystal topology. In Ref. [120] it was found that for electromagnetic waves the network topology is more favorable for the appearance of gaps than the Cermet topology with isolated scatterers. In the network topology, both low- and high-refractive index materials form a continuous network. This requirement strongly limits the classes of relevant crystal structures, as for instance the important class of isolated scatterers in a medium – typical for suspension of colloidal nanoparticles – do not meet this requirement.

The fourth main requirement to photonic crystal structures is that the unavoidable statistical variations of the crystal dimensions – in other words, random disorder – must be as small as possible. It is convenient to express random disorder

---

[3] A very high photonic strength may also be realized with atoms trapped on lattice sites [490]. At frequencies near their resonance, such as alkalis near the D-transition, atoms have a very high polarizability, concomitant with photonic strengths near $S = 1$ [181].

Table 8.1 *Overview of three main classes of 3D photonic crystals. The maximum calculated width of the band gap is given with the relevant references.*

Type of photonic crystal	Calculated max. width of the band gap	Proposed in reference
Inverse air-sphere opals (*fcc*)	12% for $n_{Si} = 3.45$	[237]
Inverse woodpiles (diamond-like)	25% for $n_{Si} = 3.6$	[193]
Woodpiles (diamond-like)	18% for $n_{Si} = 3.6$	[193]

in terms of relative variations of the lattice parameter $\delta a/a$, and of size variations of the constituent building blocks, such as the relative variation of sphere radii $\delta R/R$. Random structural variations cause light to be scattered, leading to the extinction of coherent beams [240]. In the case of a high photonic strength typical of band gap formation, the extinction length is limited to about 200 lattice spacings in the case of small structural variations of 1%, typical of a high-quality structure. We will see below that significant photonic band gap effects have recently been observed on excited-state populations, from which it is concluded that state-of-the-art structures have a sufficiently high order for cQED. Therefore, we note that a 3D photonic band gap seems to be more robust to disorder than slow-light phenomena, which are also considered in the context of strong light–matter interaction [201].

A large variety of methods have been proposed to obtain many different types of 3D photonic crystals. Examples of the fabrication of these structures are described in a multitude of reviews. Refs. [63, 219] offer excellent introductions to the field of photonic crystals where the theory of photonic crystals, fabrication methods, and embedding of cavities are extensively discussed. Ref. [142] discusses self-assembly as a tool to obtain both ordered 3D photonic crystals and random structures. Ref. [138] reviews photonic crystal fabrication by direct laser writing.

This section focuses on fabrication methods for 3D photonic crystals with demonstrated band gaps for light with wavelengths up to 2500 nm. A few notable classes of photonic band gap crystals are identified, including inverse opals, and diamond-like woodpiles and inverse woodpiles; see Table 8.1. In addition 3D photonic crystals containing optical cavities are discussed.

8.3.2 Optical signature of a 3D photonic band gap

An important aspect of 3D photonic crystal fabrication is their characterization. Scanning electron micrographs (SEM) are often used to investigate the outside of the resulting structure and obtain a first impression of whether the fabrication method has been successful; see the SEM images in almost any photonic crystal paper in a materials science journal. To investigate the interior quality of the crystals it is possible to "open" the structures after their fabrication. For instance, focused ion beam milling was employed to (destructively) observe the inner structure in Refs. [420, 492]. A non-destructive characterization method for 3D nanostructures is for instance small-angle X-ray scattering [537]. Experimental demonstration that 3D crystals exhibit a photonic band gap, however, is more challenging and requires advanced optical methods and a careful analysis. Structures with a 3D photonic band gap have, by definition, the following properties: (i) stopbands with overlapping frequencies for all wavevectors and polarizations simultaneously, and (ii) a vanishing density of states. These properties allow the presence of a band gap to be probed experimentally.

Experimentally the widths of peaks in reflectivity or of troughs in transmission provide a good estimate of the width of a stopband. To assess the photonic strength of the crystal, the width of the stopbands in experiments is often compared with the width of stop gaps from calculated bandstructures; see Refs. [459, 469]. Care must be exerted, however, as stopbands in reflectivity or transmission may also occur due to so-called silent modes. These modes occur when incident plane waves cannot couple to a field mode inside the crystal with a peculiar spatial symmetry [219, 408, 412]. Furthermore, deriving a conclusion based on comparing the measured stopbands with calculated band structures may be impeded by the fact that the calculated geometries potentially differ from real crystals due to inevitable unknowns in the fabrication or the characterization. Moreover, a complication can arise whereby field modes only couple to a specific external polarization [192]. Thus polarization-resolved experiments are necessary to demonstrate that stopbands occur for all polarizations [460]. It is also important to exclude spurious boundary effects by verifying that measured stopbands reproduce at different locations on the fabricated photonic crystal. Consequently, for reflectivity or transmission experiments to serve as a reliable indicator for the presence of a band gap they must be performed (i) such that all incident angles are probed (2π sr solid angle), (ii) with resolved polarizations, (iii) on different exposed surfaces of the structure, and (iv) shown to be position independent [202].

The main property of photonic band gap crystals, namely a vanishing density of states, allows for a robust investigation of the band gap. However, it is challenging to experimentally probe the density of states. The main approach is to embed suitable emitters in the crystals and measure their emission rates, as originally done for inverse opals with a pseudogap but no 3D band gap in the LDOS [239, 304]. In a photonic band gap the spontaneous emission of the emitters is inhibited, thus resulting in longer lifetimes of the excited state, as described by Fermi's golden rule Eq. (8.1). Therefore the lifetime is an observable that relates directly to the density of states and one can argue that measuring the emission rate of embedded emitters is the most suitable experimental method to demonstrate a 3D photonic band gap. Relevant experiments are discussed in the next section on cQED.

When describing how photonic crystals were obtained, in many papers the fabrication data are complemented with reflectivity and transmission spectra from one or a few directions only. In light of the discussion above, in a strict sense such results are not experimental proof of a band gap. To give an overview of a wide spectrum of available fabrication procedures, we take a pragmatic approach and include results in which a band gap was inferred from such measurements in this review. Table 8.2 provides an overview of the fabrication methods and different types of photonic band gap crystals discussed, including relevant references.

8.3.3 *Inverse opals*

In inverse opals, spherical voids are stacked in a face-centered-cubic (fcc) structure. These air spheres are embedded in a backbone material with a high index of refraction. Typical fabrication procedures for these photonic crystals employ template-assisted assembly based on templates of close-packed, fcc-ordered colloidal spheres, infiltrating the templates with a high refractive index material, and subsequently removing the template material to obtain air-sphere crystals [196, 538]. For completely infiltrated inverse opals with a high-index volume fraction of $\phi = 26\%$ it was established that the refractive index contrast must exceed $m = 2.8$ in order to open a band gap [452]. Subsequent calculations revealed that intricate incomplete filling of high-index material in the form of shells surrounding the spherical voids and connecting windows between the spheres yields a maximum possible band gap width of $\Delta\omega/\omega = 12\%$ for fcc silicon inverse opals [62, 237].

Silicon is an excellent backbone material due to its high dielectric constant of 11.9 (refractive index $n = 3.45$) at $\lambda = 1550$ nm and the availability of routine deposition methods. Silicon inverse opals were first demonstrated in Refs. [49, 511]. In both cases, the template was infiltrated with silicon using chemical vapor deposition. The calculated expected widths of the band gaps were 5 and 7%, respectively. Reflectivity was used to demonstrate the photonic behavior of the crystals. Silicon inverse opals have been reviewed in Ref. [473]. Inverse opals of other high-index semiconductors such as GaAs have also been pursued; see Ref. [394].

As an alternative to opals, templates have also been fabricated by holographic lithography [65], or by interference lithography [401] where silicon inverse opals were obtained. Large defect channels are written in these structures by an additional two-photon polymerization process step. Two-photon polymerization was also used to obtain large waveguide-like structures in silicon inverse opals obtained by self-assembly [407].

Table 8.2 *Overview of various types of 3D photonic band gap crystals that have been realized. This table provides information on the fabrication methods employed and the high-index backbone material of the crystals. Relevant references are provided in which detailed information on these structures and their fabrication can be found. For some of these references, values for calculated or otherwise expected widths of the band gaps are added between brackets. The last column provides additional remarks. The following abbreviations are used: 2PhP, two photon polymerization, CVD, chemical vapor deposition, DLW, direct-laser-writing, FIB, focused ion beam milling, Inv. woodpile, inverse woodpile, QD, quantum dots, RIE, reactive ion etching, SEM, scanning electron microscopy.*

Type	Fabrication method	Material	References (indication of band gap width)	cQED feature
Inverse opal	Inversion of templates	Si	[49](5%),[511](7%),[401],[407]	Demonstrated ultrafast switching [37, 127]
Diamond-like				
-Yablonovite	RIE in three directions	GaAs	[82](19%)	
-Spiral crystal	Glancing angle deposition	Si	[557](10 to 14%)	
-Biotemplated	Double inversion of beetle scales	Titania	[143](5%)	Time-resolved emission in crystal with pseudogap [228]
-Inv. woodpile	Macroporous etching + FIB	Si	[420](17%)	
-Inv. woodpile	DLW + inversion	Si	[190](14%)	
-Inv. woodpile	2PhP of a template + inversion	Si	[437](15%)	
-Inv. woodpile	Two-directional etching	Si	[468](19%),[469](>14%)	Embedded quantum well layer
-Inv. woodpile	Two-directional etching	GaAs	[472](18%)	
-Inv. woodpile	Two-directional etching with holder	Si	[492](24%)	Observed inhibited and modified emission of QDs in 3D band gap [282]

Woodpile	Combination of many fabrication methods	Si	[135]	Demonstrated ultrafast switching [126]
Woodpile	Wafer fusion of layers	III–V, Si	[356](16%),[367],[368],[231],[211]	Incorporated cavities and quantum wells
Woodpile	E-beam lithography + CVD	Si	[397](21%)	Incorporated point defects
Woodpile	DLW + double inversion	Si	[474](8.6%),[459](6.9%),[460](15%),[458]	Incorporated waveguides
Woodpile	Micromanipulation of layers	GaAs, Si	[16],[470](17%),[471],[66],[185](19%)	Cavity lasing observed

While inverse opals are very popular on account of the relatively easy fabrication routes [142], the maximum width of the band gap is relatively narrow, and requires intricate optimization. Since the band gap is of a higher order in the band structures (second-order Bragg) [516], the gap is narrower and more sensitive to unavoidable disorder and fabrication deviations than for structures with a lower-order band gap, which is a main disadvantage for inverse opals. From calculations in Ref. [299] it is concluded that variations of air-sphere radii and lattice positions of less than 2% of the lattice parameter are already sufficient to completely close the band gap. These results indicate that in order to display a photonic band gap, inverse opals must be made with an extremely high precision, which may be beyond the present state-of-the-art in nanofabrication.

8.3.4 Diamond-like photonic crystals

In the early 1990s the possibility of using photonic crystals with diamond-like structures was described in a number of papers; see Refs. [192, 193]. These photonic crystals are extremely interesting because they have significant potential for wide photonic band gaps. Therefore, materials with relatively low refractive index contrasts of $m > 1.9$ suffice in order to obtain a photonic band gap. A simultaneous advantageous feature of 3D diamond-like photonic crystals is that the band gap is robust to unavoidable fabrication deviations and random disorder; see Refs. [191, 439, 540] for calculations on this aspect.

In one of the earliest studies calculations showed that air-spheres arranged in a diamond structure with a refractive index contrast of 3.6 would give wide band gaps of up to 28% bandwidth [192]. Unfortunately such diamond structures remain elusive to date. In a 2004 review diamond-like photonic crystals and the efforts to obtain them have been reviewed [315]. Here, several main results from that review are highlighted, as well as subsequent results.

8.3.5 Woodpiles

3D woodpile photonic crystals were originally proposed by the Iowa State group as a practical way to realize powerful diamond-structured photonic crystals [193]. The expected maximum width of the band gap was predicted to be a sizable 18% [193]. The crystal structure resembles a pile of logs of wood, hence their name. One may argue that the analogy also pertains to the way that they are often fabricated: by sequential stacking of layers of semiconductor rods; see Fig. 8.3. This strategy has a distinct advantage: since layers are stacked in sequential fashion it is possible to alter the layout of individual layers. This freedom of design has been successfully used to incorporate high-quality optical cavities and waveguides into 3D photonic

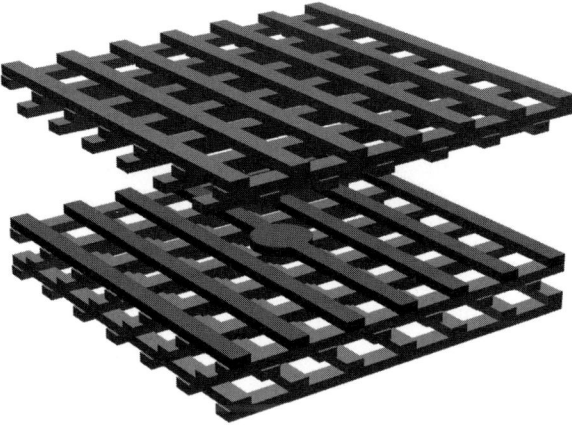

Figure 8.3 Schematic illustration of a 3D woodpile photonic crystal that is fabricated using a layer-by-layer approach. The top part of the structure is lifted in order to provide a view on the central layer in which a point-defect with excess high-index material is visible. This point-defect acts as a 3D optical cavity, or a nanobox for light.

band gap crystals, which are promising tools for cQED. In this section a few prominent examples of woodpiles and of woodpiles with embedded optical cavities are discussed. Alternative methods for obtaining woodpiles by direct-writing and two-photon polymerization are highlighted.

In a pioneering paper by the Sandia team, an intricate combination of several thin film deposition and (silicon) fabrication processes was employed to obtain woodpile crystals [135]. Structures consisting of up to four layers of poly-crystalline rods were obtained. The geometrical properties of these woodpiles were chosen such that a band gap is expected near $\lambda = 1500$ nm. Measured spectra indicate stop gaps with a transmission reduced to 15% at that wavelength. While no convincing evidence for a band gap was presented and the structures seem to have a significant misalignment between consecutive layers, kudos are in order for this pioneering work, considering that a broad stopgap with a high reflectivity has been observed [126].

The Kyoto group, led by Noda, has been extremely successful in fabricating woodpile photonic crystals from III–V semiconductors and from silicon using a wafer fusion technique [356, 367]. In this fabrication method, photonic crystals are fabricated by stacking semiconductor stripes, see Fig. 8.3. Initially, patterns of stripes – i.e. the rods in the woodpile – are formed by electron-beam lithography and reactive ion etching in a single crystalline layer. In the case of GaAs crystals, the layer is GaAs that sits on an (Al)GaAs etch-stop layer, on a GaAs substrate. A pair of patterned wafers is stacked and bonded in a crossed configuration with precise control over the alignment, and one of the substrates is removed. These process steps are repeated. At each fusing step, the number of layers doubles, until the desired structure is obtained. One limitation of the fabrication method is that it is so advanced that it has only as yet been successfully realized by the Kyoto group. While initial crystals were only four to eight layers thick, much thicker crystals were later realized. Notably, impressive GaAs woodpile structures have been reported with a thickness of 17 layers of rods [368]. To investigate the presence of the 3D photonic band gap, normal incidence transmission and reflection spectra were collected, together with transmission spectra as a function of angle of incidence. In the latter experiments the stopband shifts to lower frequencies (longer wavelengths) with increasing incident angles, and the shift saturates at 40° to 50°. If the range of the 3D band gap is defined as the range where the attenuation exceeds 80%, it covers the range from 1300 to 1550 nm.

In an elegant experiment the Kyoto group investigated the surface modes on a 3D band gap crystal [212]. It was demonstrated experimentally that photons can be confined and manipulated at the surface of 3D photonic crystals. GaAs woodpile crystals were studied with a thickness of eight layers and a 3D band gap at wavelengths between 1300 and 1600 nm. Since the surface states appear in the gap at frequencies below the light line, an evanescent coupling method with a prism was used to excite the surface states; see Fig. 8.4(A). Light is sent under total internal reflection conditions into the prism that is placed just above the photonic crystals. Once the evanescent waves excite a surface mode of the photonic crystal, the reflected optical power shows a distinct minimum. By measuring reflection spectra as a function of angle of incidence and tracking the reflectivity minimum versus wavelength, the dispersion relations of the surface states were obtained for various crystal directions. Figures 8.4(B–E) show images of the photonic-crystal surface observed from the top with a camera. Figure 8.4(D) clearly shows that light with a wavelength $\lambda_0 = 1430$ nm propagates along the surface of the 3D photonic crystal, at an angle of incidence $\theta = 45.7°$. The light propagates to the end of the crystal surface where it is scattered. In contrast, at $\lambda_0 = 1400$ nm and $\lambda_0 = 1460$ nm light does not propagate on the surface as there are no surface modes at these wavelengths. The wavelength of the surface modes could be tuned by scanning θ. This study is a creative demonstration of band gap behavior, since the observations reveal that the light is strictly confined to the surface of the crystal; this is beautiful evidence that light is forbidden to propagate into the bulk of the crystal.

Reference [212] has also presented even more advanced manipulation of "flat light": by controlling the surface termination, the Kyoto team was able to demonstrate gaps for the surface modes. Moreover, point-defects were fabricated that behave as cavities for the surface-trapped light. Impressive cavity quality factors up to $Q = 9000$ were reported. This work takes an important step in opening a new

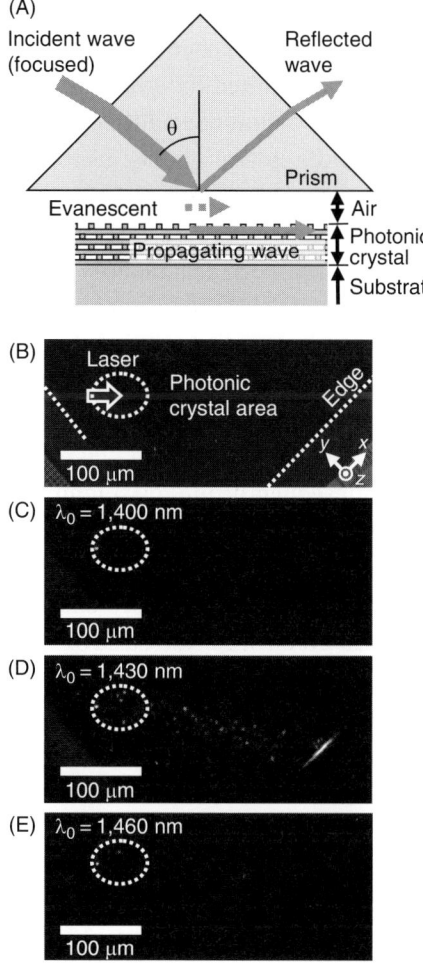

Figure 8.4 Experimental demonstration of surface states of a 3D photonic crystal with a 3D band gap. (A) Experimental setup used to couple light into the surface modes. (B) Optical microscope image of the surface of the 3D photonic crystal, showing the excitation point of the incident light and the edge of the photonic crystal. (C–E) Experimental results for light propagation through the surface mode, in which the irradiating angle θ was set to 45.7°, and the wavelength to $\lambda_0 = 1400$ nm, 1430 nm, or 1460 nm. (D) Light with $\lambda_0 = 1430$ nm propagates through the surface modes of the 3D photonic band gap. From Ref. [212] with kind permission.

route for the manipulation of light by 3D photonic crystals, as well as pioneering the surface science of 3D photonic crystals. Moreover, the results bear intriguing analogies to surface plasmon–polaritons physics [33]; the absorption-free nature of a 3D photonic crystal surface is expected to lead to new sensing applications, as well as to novel light–matter interactions that are at the heart of cQED.

In Refs. [367, 368] optical cavities were realized in woodpile crystals. To this end, a central layer containing point-defects of different sizes was incorporated in the woodpile stack. The intensity emitted by embedded InGaAsP quantum wells served to probe the cavity quality factors. Several peaks were observed that could be identified as cavity resonances, and a quality factor of up to $Q = 350$ was measured for a thick 17-layer woodpile. Subsequently, GaAs and silicon wood-piles were fabricated with intricate waveguide structures [211, 231]. The guiding of light along these waveguides has been successfully demonstrated. These examples clearly demonstrate that woodpiles show great potential for control over propagation and emission of light, including advanced functionality.

Silicon woodpiles have also been fabricated by direct laser writing of a template and subsequent double inversion to amorphous silicon [474]. A woodpile was first fabricated from photoresist, which was fully infiltrated with silicon dioxide through chemical vapor deposition. Subsequently, the resist template was removed to yield a silicon dioxide inverse woodpile. The filling fraction of the inverse woodpile can be tuned by additional deposition of silicon dioxide. In the last step, the inverse woodpile is infiltrated with amorphous silicon by chemical vapor deposition to obtain the woodpile crystal. Based on scanning electron micrographs, the expected band gap for this structure was predicted to have a width of nearly 9%. Optical transmission measurements displayed a strong stop gap in the range of the expected band gap. Innovations were reported wherein the amorphous silicon was converted to polycrystalline silicon [459]. Moreover, inverse woodpiles were fabricated with embedded waveguides [458, 460].

A team from Tokyo University has successfully made woodpiles by an impressive stacking of pre-fabricated layers of semiconductor rods using a micromanipulation technique [16]. Up to 17 layers were stacked with an accuracy of 50 nm. From bottom to top, the crystals consist of eight GaAs layers, an active layer containing InAsSb quantum dots, and up to eight GaAs layers. In some cases the rods in the active layer were designed such that point defects could be incorporated; see the illustration in Figure 8.3. For these cavity structures, quality factors up to $Q = 2300$ were reported. It was observed that the quality factor increases by adding more top layers to the crystal. The quality factors were improved to an impressive $Q = 38\,500$ by stacking a total of 25 layers and fine-tuning the size of the optical cavity [470, 471].

A team from MIT has made 3D silicon photonic crystals that are in essence hybrids of woodpile and air-sphere crystals [397]. The structures were made with an electron beam lithographic approach, where in each fabricated layer a hole section and rod section are vertically combined. E-beam lithography allows one to align each subsequent layer to the previous one with high accuracy. Each layer is fabricated from deposited amorphous silicon and subsequent layers are deposited on top of the underlying ones. Since the lithographic pattern is defined for each layer separately, incorporation of point-defects is possible. The calculated maximum width of the band gap for these crystals is 21%. A high reflectivity up to 90% was observed at wavelengths at around 1300 nm. Troughs typical of cavity resonances were observed in the reflectivity peaks. Unfortunately, the scanning electron micrographs reveal significant fabrication-induced structural variations in the crystals. Nevertheless, this work is a beautiful example of layer-by-layer fabrication.

It is noted that many of the woodpile photonic crystals discussed above appear not to be cubic and thus not truly diamond-like. This is caused by the fact that the individual stacked layers do not always possess an optimal thickness compared with the periodicity of the rods in the layers. Hence the crystals have band gap widths that differ from those calculated for true diamond-like structures. Furthermore, layer-by-layer fabrication methods typically introduce relatively large alignment errors in the structures, which is expected to have an adverse effect on the photonic band gap [90].

8.3.6 Inverse woodpiles

Broad band gaps with widths exceeding 25% have been predicted for 3D photonic crystals, known as "inverse woodpiles" [193]. These crystals consist of pores that run in two perpendicular directions in a high-index backbone. Thus, the structure is the inverse of the woodpile structure. Compared with woodpiles, inverse woodpiles have several advantageous features. Firstly, the layout and pore alignment are defined such that it is straightforward to obtain a cubic diamond-like structure with a broad gap; the cubic structure is not distorted by imperfect stacking. Secondly, the pore diameters may be varied to optimize the volume fraction of the high-index material, which is an essential tuning parameter for broad band gaps. In contrast, in woodpile structures optimization of the volume fraction would entail a change of the nanorod dimensions which leads to structure distortions.

By combining macroporous etching and focused ion beam milling, beautiful inverse woodpiles were first fabricated in silicon [420]. Impressive large structures were obtained although unintended misalignment reduced the width of the expected band gap to 17%. These crystals were shown to have photonic stopbands by means of reflectivity measurements along a high-symmetry crystal axis.

Inverse woodpiles have also been fabricated by means of a sequence of (i) direct laser writing to obtain a polymer template, (ii) inversion through deposition of silicon, and (iii) removal of the template [190]. The deposition of silicon was assisted by an intermediate silicon dioxide layer. The resulting structures have an expected band gap width of 14% near a wavelength of 2500 nm. The occurrence of a stop gap was confirmed by means of optical reflectivity and transmission measurements.

Two-photon polymerization was used to obtain a template for a silicon inverse woodpile [437]. Inversion of the polymer template is achieved through a sequence of conformal coating of a layer of Al_2O_3, chemical vapor deposition of silicon, and removal of both Al_2O_3 and template. While the demonstrated inverse woodpiles are not truly cubic and thus not diamond-like, a band gap with a maximum width of 15% was predicted for the optimal structure. Measured reflectivity revealed a stop gap near $\lambda = 1100$ nm, with maxima up to 60%. Features in the scanning electron micrographs and absence of interference fringes in the spectra suggest that the structures made by this method suffer from imperfections such as a nonuniform surface on large length scales and roughness on smaller scales, which adversely affect band gap formation.

In 2006, a two-directional etching method was introduced by the Kyoto group as a means of fabricating 3D inverse woodpile crystals [468]. It was shown that cubic diamond-like structures are obtained by reactive ion etching through a suitable mask of two perpendicular sets of pores under angles of 45° with respect to the wafer surface. While the structures are thin, they show strong 60% reflectivity peaks near 1600 nm. In 2009, a more extended study resulted in structures displaying around 97% reflectivity [469]. In these crystals a band gap with a width up to 14% is expected. In addition, clear changes in cw intensity were observed with embedded light emitters (see Section 8.3.7). These results illustrate that these inverse woodpiles are very powerful optical structures.

A similar two-directional etching method was used at Duke University to obtain woodpiles in GaAs [472]. The maximum expected width of the band gap of optimal structures is about 18%. Reflectivity experiments reveal a strong 92% peak near a wavelength of 1300 nm, indicative of a high photonic strength. An outlook was presented on how to obtain microcavities at the intersection of unit cell modulations and line defects.

A new two-directional etching approach using masks in inclined planes was demonstrated at the University of Twente to realize high quality diamond-like inverse woodpiles from silicon with expected band gap widths up to 24% [492]. The method was developed to be CMOS-compatible, in close collaboration with high-tech industrial partners. In this method, high-purity single crystalline wafers are first etched in one direction by reactive ion etching to obtain large arrays of deep pores [541]. Secondly, the sample is cleaved, rotated by 90°, and placed in a dedicated holder wafer that was developed to carefully align the samples. Thirdly, by using the holder, a second etch mask is defined in an inclined plane to the first pattern with a high translational alignment accuracy, better than 30 nm, and a high rotational accuracy, better than 0.71° [481]. Finally, a second set of pores is etched perpendicular to the first set by deep reactive ion etching. The overlap region of the two perpendicular sets of pores form the 3D inverse woodpile

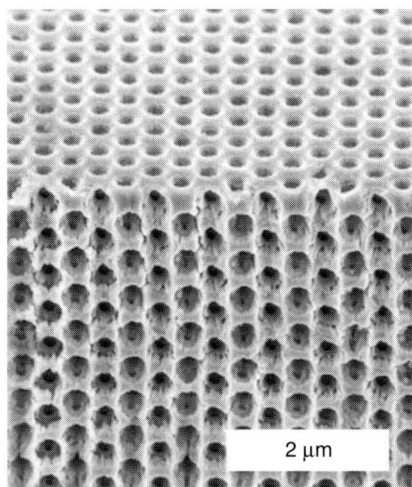

Figure 8.5 Scanning electron micrograph of a 3D inverse woodpile photonic band gap crystal. Such a 3D nanostructure is fabricated from monocrystalline silicon using a CMOS-compatible two-directional etching technique combined with advanced alignment, see Refs. [481, 492]. This crystal has a diamond-like structure.

crystals; see Figure 8.5. The signature of the band gap was demonstrated by extensive polarization-resolved optical reflectivity measurements on different crystal faces, which revealed stop bands that occur over large solid angles spanning 1.76π sr [202]. The maximum observed reflectivity of 65% was found to be limited by the finite crystal thickness and by surface roughness. From the experimental observations, the band gap was determined to have a broad relative bandwidth of 16%. We will see in the next section that these crystals are optically very powerful as they reveal prominent cQED effects on excited-state lifetimes [282].

8.3.7 Other diamond-like structures

In 1991, a type of structure with the potential for wide band gaps was proposed. These structures are obtained by etching or milling pores of air sequentially in three different directions in a backbone with a high index of refraction [550]. Such crystals were later fabricated from GaAs [82]. Transmission measurements were used to study the optical properties of these structures and an attenuation up to 80% was reported. A width of the band gap of around 19% was reported, inferred from the width of the troughs in transmittance. In addition, this paper emphasizes the importance of avoiding making these structures by etching or milling of tapered pores. This fabrication deviation was shown to have a significant effect on the band gap.

Diamond-like silicon spiral photonic crystals have been obtained by glancing angle deposition [557]. In this method, silicon is grown by electron beam evaporation on substrates that contain tungsten seeds. These seeds are arranged in a suitable square lattice with a lattice constant near 1000 nm. Spirals were grown by carefully rotating the substrate during deposition. An optimal crystal has an expected band gap width of nearly 15%. For the realized crystals a narrower band gap width is expected due to a mismatch of the obtained geometry with the ideal geometry. The photonic properties were analyzed by optical reflectivity, where peaks up to 80% were observed, centered near a wavelength $\lambda = 2000$ nm.

An intriguing method for fabricating diamond-like photonic crystals is biotemplating [143]. In this method, the diamond-like scale of a beetle is used as a template. By a double, sol–gel based, inversion method, the template is replicated in titania. The periodicity of these diamond-like structures is in the order of the wavelength of visible light. The expected width of the band gap of these structures is calculated to be around 5%, but the gap is probably reduced in width or even closed due to the significant structural distortions that are apparent in the scanning electron micrographs of the structures.

In summary, we have seen that 3D photonic crystals have been fabricated by many different fabrication methods. It appears that the potential and promise of wide band gaps has challenged many groups all over the world to expand the boundaries of materials science and nanotechnology, resulting in novel ways to sculpt and fashion 3D nanostructures. It is exciting to foresee how this know-how will increasingly be used to apply photonic crystals in innovations where complete control of light, including single photons, is essential. For many of these strategies it remains a challenge to embed optical cavities. In particular for strategies employing two-directional etching, the realization of embedded cavities must be demonstrated. Adapting the above – or new – manufacturing strategies to this end remains an inspiring goal for future materials science research.

8.4 Cavity quantum electrodynamics

8.4.1 Inhibited spontaneous emission in a 3D photonic band gap

Time-resolved emission. Recently a first experiment was reported to study the control of the excited-state lifetime of emitters in 3D photonic band gap crystals [282]. To this end, 3D inverse-woodpile crystals with a cubic diamond-like structure were made from silicon with fixed lattice parameters and a range of pore radii, so as to tune the band gap relative to the spectrum of the emitters. As emitters PbS colloidal quantum dots were studied at room temperature. The dots were immersed into the crystals as a dilute suspension, hence the transition dipole orientations $\mathbf{e}_d$ sampled all directions. The dots were kept in a toluene suspension to minimize non-radiative decay γ_{nrad}. Although toluene as a low-index medium reduces the refractive index contrast with silicon to $m = 2.3$, inverse woodpile crystals have from the outset such a broad band gap that the gap retained a relative width of 5%.

Time-correlated single photon counting was used to precisely measure the distribution of arrival times $f(t)$ of the emitted photons [266]. Figure 8.6 shows time-resolved spontaneous emission for quantum dots in two different photonic crystals, compared with a reference [282]. Emission in a crystal outside the band gap decays faster than the reference, confirming that the excited-state lifetime of the quantum dots is controlled by the photonic crystals. A limitation of this study is that the signal is not only emission from within the crystal, but also a background of quantum dots outside the crystal. Thus at frequencies in the band gap the emission is at short arrival times ($t < 500$ ns) dominated by the fast decaying background signal. Beyond 500 ns a slow decay is apparent which corresponds to a strongly inhibited emission. Figure 8.6 reveals that in the band gap the distribution of photon arrival times decays monotonically in time, as predicted by Ref. [298]; no fractional decay or oscillations were detected, as predicted elsewhere [252, 506, 522]. Possible reasons for this discrepancy could be that the predicted features are not robust to ensemble averaging in the experiment. Conversely, theory studies consider excited-state population dynamics instead of photon arrival times, while these two phenomena can strongly differ [497], or sometimes assume excessively large transition dipole moments.

To interpret the time-resolved spontaneous emission, the dynamics of the quantum dots in the crystal was modeled with single exponential decay, where

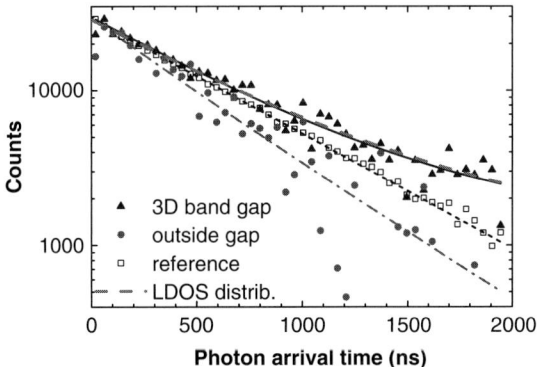

Figure 8.6 Distributions of photon arrival times emitted by PbS quantum dots in silicon photonic crystals with a 3D photonic band gap (at 0.893 eV and $r = 170$ nm, triangles), detuned from the band gap (at 0.850 eV and $r = 136$ nm, circles), and in a reference (suspension, squares). The curves are exponential models. The long dashed curve is calculated from the spatial distribution of the LDOS. The cubic crystals have lattice parameters $a = 693$ nm and $c = 488$ nm, with $a = c\sqrt{2}$, and pore radii $136 < r < 186$ nm. Data from Ref. [282].

care was taken to properly account for signal and background statistics [282]. Outside the band gap the emission rate was found to be up to two-fold enhanced, in agreement with calculated density of states (see Eq. (8.3)). Within the photonic band gap the emission rates are strongly inhibited by a factor $10\times$ compared with emission rates outside the band gap. From additional observations it was concluded that the quantum dots have a very high quantum efficiency ($> 90\%$), thus the low emission rates correspond with 10-fold enhanced excited state lifetimes, which is a clear step towards the stabilization of the excited state in a photonic band gap. The resulting lifetime of $T_1 = 5.5\,\mu s$ is very long for quantum dots and good news for applications in quantum information processing.

The experimental observations within the band gap cannot be interpreted with current theories as these generally use the plane wave expansion typical of infinite crystals. Therefore, a heuristic model was made for the spatial dependency of the LDOS in a finite photonic crystal, where it is postulated that the unit-cell averaged LDOS decreases exponentially with distance z into the crystal, with a new decay length called ℓ_{LDOS} [282]. This model leads to a distribution of emission rates with a minimum rate corresponding to $160\times$ inhibition. The corresponding time-resolved emission curve is shown in Fig. 8.6, where it is apparent that the curve is non-exponential due to the broad distribution, and the model agrees very well with the experiments. The estimated LDOS decay length $\ell_{LDOS} = 1.03a$ is $6\times$ smaller than the Bragg length that can be derived from the photonic strength S (see Section 8.1), which confirms that this is a new length scale typical for 3D band gaps. Finally, the good agreement of model and experiment suggests that this approach is fruitful for future ab-initio cQED theory in 3D photonic band gaps.

Continuous-wave intensity. In 2004, the Kyoto group reported the first study of emission in 3D photonic band gap crystals [367]. To this end, InGaAsP multiple quantum wells were incorporated in the central layer of 3D woodpile crystals with thicknesses between five and nine layers. Cw emission intensity spectra $I(\omega_{eg})$ were collected at room temperature as a function of the angle of observation. Broad stop bands were observed at constant angle, in agreement with broad stopgaps in the dispersion relation. A strongly suppressed emission was observed in the wavelength range between 1450 and 1600 nm, independent of the angle of observation, which agrees with the expected inhibition in a 3D photonic band gap. Moreover, the observed range agrees well with the band gap range identified by transmission measurements and by FDTD simulations.

Subsequently, the Kyoto group has studied emission intensity in inverse woodpile crystals [469]. A layer of InGaAsP quantum wells was embedded in the crystal to serve as internal emitters. By employing advanced additional wafer bonding procedures, the layer emitters were sandwiched between two halves of the photonic crystal, and carefully aligned. Emission spectra were measured at low temperatures

(< 90 K). The cw intensity $I(\omega_{eg})$ was found to be reduced compared with emitters embedded in bulk silicon. While this observation represents a directional stop band, a logical next step is the demonstration of angle-independent emission inhibition.

In all studies above, the reference sample was a bulk layer of semiconductor. Here, emitters experience a different environment and refractive index (near 3.5) from that in the photonic band gap crystal (effective index probably ≤ 2). Therefore it is not possible to interpret the level of inhibition, since relatively large Lorentz local field effects might come into play [421]. Since all studies concern cw intensity $I(\omega_{eg})$ whose features likely correlate with the LDOS, it is concluded that the emitters must have a relatively low quantum efficiency. This conclusion is borne out from the analysis of similar observations of inhibited cw emission in photonic crystals without a band gap [238, 239].

Intensity studies have also been performed on various types of emitters embedded in GaAs or Si woodpile crystals, made by micromanipulation by the Tokyo group. Remarkably, no inhibited emission has been reported, see, e.g. Ref. [16]. It is surmised that the layer-by-layer stacking results in variations of the vertical layer spacings, which cause leaking-in of vacuum fluctuations to the crystal and thus adversely affect the inhibition. We conclude that the role of fabrication imperfections on spontaneous emission control is currently not very well known and therefore merits further attention.

8.4.2 Emission in a nanobox for light

An important driving force for the pursuit of 3D photonic crystals is the study of emission from a cavity in a 3D photonic band gap that confines light in all three dimensions: a nanobox for light. Pioneering steps in this direction were taken by the Kyoto group [367, 368]. A central layer containing point-defects of different sizes was incorporated in the woodpile stack. The cw intensity emitted by embedded InGaAsP quantum wells was collected to probe the optical properties of the embedded cavity. With increasing size of the point-defect, an increasing number of resonances were identified, whose intensity has an increased baseline compared with inhibited emission in the pristine crystal. Apparently, increasing point-defect size causes an increased leakage into the photonic crystal and thus increasingly defies the shielding of the band gap. For the smallest point-defects, a single optical resonance was observed, whose intensity had a baseline equal to the signal in the band gap range. The small cavities thus appear to be close realizations of a nanobox with a true 3D shielding by the band gap.

Emission in a 3D photonic band gap cavity has also been studied by the Tokyo group [16, 185, 470]. In the first study, a layer of InAsSb quantum dots was embedded in the central layer where the cavity was located [16]. A strong polarization-dependent variation of the cw intensity was reported, which could be assigned to a cavity resonance with a quality factor in the region of 2000. The main features were reproduced in a second study at low temperatures, where the quality factor was boosted to more than 8000 [470].

In a third study, woodpile photonic crystals with cavities were fabricated from silicon [185]. Ge islands were embedded, whose cw emission intensity was recorded at low temperature (25 K). Two cavity resonances with narrow linewidths were observed. The intensity of these resonances was reported to be 30 to 60 times greater compared with the reference. Since the reference consisted of emitters in bulk semiconductor, it is very challenging to extract cavity enhancement factors (Purcell factors) from the measurements. Firstly, it is likely that the collection efficiency substantially differs between the photonic crystal and the bulk reference. In bulk semiconductor, the solid angle in which emission is collected is limited by total internal reflection to a small fraction of the total solid angle. In contrast, light in a photonic crystal is scattered over all 4π solid angles, thus the collection efficiency is greater so that the intensity extracted from a photonic crystal may be overestimated. Secondly, a different reference environment corresponds with a different Lorentz local field factor [421]. If the reference is a semiconductor, its refractive index is much greater, leading to a much higher emission rate. Hence, the intensity in the reference is overestimated compared with the photonic crystal. Such effects require care to be properly accounted for.

To date, studies of emission in a nanobox concern the cw intensity $I(\omega_{eg})$ whose sharply peaked features likely correlate with the LDOS. Therefore, the emitters have a relatively low quantum efficiency [239]. This conclusion is confirmed by Ref. [185], where it was reported that the Ge islands have a substantial non-radiative recombination rate γ_{nrad}.

We conclude that at this time, the challenge is open to demonstrate Purcell-enhanced emission in a 3D nanobox. Such a demonstration requires the use of time-resolved emission, and thus the embedding of quantum emitters with an elevated quantum efficiency into a nanobox for light.

8.4.3 *Laser action in 3D photonic crystal nanocavities*

The promise of miniature "thresholdless" laser action [48] in 3D photonic band gap crystals has long been a strong motivation for the field of photonic crystals. Important progress in this aspect has recently been reported by the Tokyo group. GaAs woodpile crystals were fabricated with a layer of point-defects that act as cavities [471]. The cavity layer was covered with an increasing number of crystal layers to increase the cavity confinement to the point of achieving a quality factor up to

38 000. The cavity layer contained highly efficient InAs quantum dots that were excited with ns laser pulses to demonstrate lasing oscillation at low temperatures. It was observed that the peak pump power required to reach the laser threshold decreased with increasing cavity quality factors, as controlled by increasing the number of top GaAs layers. From measurements of output power versus input power, the spontaneous emission coupling factor was determined to be $\beta = 0.54$, 0.67, and 0.92 for the structures with 6, 8, and 12 upper layers, respectively. Here β is the fraction of spontaneous emitted light that contributes to lasing; it is defined as the ratio of the power emitted into the laser mode to the total emitted power over all modes. For the best confined structure, the value of β was impressively close to the theoretical limit of unity for a miniature thresholdless laser. In future, introducing a single quantum dot into the nanobox would establish an ideal solid-state system for the study of interactions between 3D confined photons and electrons enclosed in a completely controlled optical environment.

The Tokyo group has also fabricated woodpile photonic crystals with cavities [66]. The crystals contained a GaAs active layer with embedded InAs quantum dots that were excited with ns laser pulses to investigate lasing at low temperatures (11 K). From output power versus input power measurements, the spontaneous emission coupling factor was found to be as high as $\beta = 0.78$, which is probably a record for silicon microlasers. The coupling factor is a little lower than in the GaAs woodpiles mentioned above, which makes sense since the quality factor is also a little lower ($Q = 22\,000$), hence the confinement is slightly lower. The promise for lasing in these photonic crystals has been clearly demonstrated in these impressive experiments. It is a thrilling prospect to see thresholdless lasing approaching room temperature, thereby opening prospects for the application of such intricate miniature lasers.

8.4.4 *Ultrafast all-optical switching of 3D photonic band gap crystals*

It is an exciting prospect in cQED to rapidly modulate the "bath" that surrounds a quantum emitter, and thereby enter new physical regimes [262, 310]. The notion of "switching" the density of states on ultrafast timescales was first considered for 3D photonic band gap crystals [227], where it was theoretically proposed to quickly modify the refractive index of the semiconductor backbone by exciting free carriers with short laser pulses. As a result the 3D photonic band gap will exhibit a large shift in frequency and a change in width. At frequencies near the gap, the LDOS may be switched from a high value to zero, or from zero to a high value, or from a high to zero to a high value on 100-fs timescales, independent of material relaxation times. Such fast changes are expected to yield rich cQED behavior of excited quantum emitters. To this end, rate equations were subsequently derived for the excited state population of two-level emitters in a time-dependent environment [479]. The weak-coupling approximations were used,[4] and the LDOS was modified on timescales faster than the excited-state lifetime. It was found that a short increase of the radiative decay rate depletes the excited state and drastically increases the emission intensity $f(t)$ during the switch event. The time-dependent spontaneous emission revealed a distribution of photon arrival times that strongly deviated from ubiquitous exponential decay; a deterministic burst of photons will be spontaneously emitted during the ultrashort switch event.

Several experiments have been performed to study ultrafast all-optical switching of 3D photonic band gap crystals. In a pioneering study reflectivity changes were reported on silica opaline matrices infiltrated with Si [322]. Unfortunately, this work suffered from several limitations: firstly, the low refractive index contrast was insufficient for a band gap. Secondly, the experiments were performed at frequencies above the electronic band gap of Si, so that light is being absorbed. Finally, the maximum feasible refractive index change was limited by the large induced absorption. Soon after, the Karlsruhe group reported transmission changes on Si inverse opals [37]. The induced absorption in their crystal was strongly reduced by annealing the Si backbone, resulting in drastically decreased Drude damping. The frequency range in this study was limited to the range of first-order Bragg diffraction where a pseudogap occurs, but no photonic band gap [62, 452, 516]. The authors of Ref. [127] performed experiments on 3D Si inverse opals in the range of the 3D photonic band gap, and similar results were obtained at Minnesota [526]. Induced absorption was limited by a judicious choice of pump conditions, allowing the demonstration of a large shift of the photonic band gap [127]. Fast dynamics was observed – 500 fs up and 21 ps down – implying that switching could potentially be repeated at GHz rates. Ultrafast switching has also been performed on Si woodpile crystals that were probed by reflectivity over an octave in frequency, including the telecom range [126]. Only 300 fs after the switching pulse, the complete band gap shifted to higher frequencies before quickly relaxing within 18 ps. The switched spectra were successfully analyzed with a theory for finite photonic crystals.

While ultrafast switching of the optical properties of 3D photonic crystals has been clearly demonstrated, the demonstration of spontaneous emission switching of emitters that experience a fast change of the LDOS is currently open. It will be an ultimate challenge to observe the breaking of the weak-coupling approximation by fast time-dependent modulation. Relevant questions are: what is the relevant timescale for the frequency of a photon in a cavity to adjust to the shifting band gap,

---

[4] See Section 8.2.1 above.

and to what extent do such considerations apply to excited emitter states near a gap? It will be truly exciting to see the first light shine on this subject.

8.5 Applications and prospects

Quantum decoherence. There is currently a fast growing interest in quantum information science, where the goal is to store, process, and transmit information encoded in inherently quantum mechanical systems [352]. While many types of physical systems are being pursued, cQED systems involving quantum light interacting with quantum matter receive much attention. Solid-state cQED offers many advantageous prospects for qubits such as system scalability and on-chip architecture, miniaturization and high speeds, and the spatial localization of quantum emitters [365]. For the manipulation of quantum states, it is paramount to prevent decoherence, otherwise the system will behave classically [569]. It is thus desirable to increase the dephasing time T_2 of the system, which depends on the excited state lifetime T_1 and the pure dephasing T_{deph} within one state: $T_2^{-1} = T_1^{-1} + T_{deph}^{-1}$ [262]. We note in passing that the inverse of the dephasing time equals the linewidth that is measured in an absorption experiment. Earlier on in this review, we have seen that the T_1 of quantum emitters is controlled by the LDOS, leading already to 10-fold enhanced lifetimes observed in a 3D photonic band gap.

Pure dephasing depends on both radiative and non-radiative effects. Examples of non-radiative effects are collisions in gas phase, vibrations in liquids, or phonons in solid state. These dephasing effects are typically controlled by cooling systems to low temperature, or by mechanically decoupling the quantum system as much as possible from the environment using nifty spacers. The main source of radiative dephasing is spontaneous emission from the excited state, which depends on the density of vacuum fluctuations, hence the LDOS. Yet again, this effect is amenable to control by the nanophotonic environment. In the extreme case of a photonic band gap, one can therefore also enhance the radiative dephasing time. Therefore, we conclude – in agreement with elaborate theoretical calculations [39, 542] – that a 3D photonic band gap crystal offers a favorable environment to shield qubits operating at optical frequencies from noise and fluctuations.

Resonant energy transfer. In cQED, it is a central question as to how multiple interacting emitters are controlled [182, 362]. A well-known optical emitter–emitter interaction is resonant energy transfer between pairs of dipoles, where an energy quantum is transferred from one emitter, called donor, to a second emitter, called acceptor, see e.g. [266]. The involved dipole–dipole interactions are crucial to quantum information science, and (Förster) energy transfer plays a central role in photosynthesis, as well as in photovoltaics, lighting, and molecular sensing. Recently, the question was addressed of whether energy transfer can be controlled by the photonic environment, namely the LDOS. Dye molecules were separated by a short strand of double-stranded DNA, and the LDOS was controlled by positioning the FRET pairs near a mirror [53]. Contrary to early predictions, it was found that the energy transfer efficiency does change with LDOS, whereas the energy transfer rate is independent of the LDOS, in agreement with theoretical considerations.

It was predicted that in a 3D photonic band gap the efficiency of resonance energy transfer is maximal when the band gap is tuned to the donor emission frequency. In the case of a vanishing non-radiative decay, the efficiency will reach a perfect 100%. The observation of changing energy transfer efficiency implies a change in the characteristic Förster distance, in contrast with common lore that this distance is fixed for a given pair of dipoles. In the case of a band gap, this interaction distance will undergo the greatest changes. Thus, future control of resonant energy transfer in a 3D photonic band gap promises favorable vistas that are relevant to applications ranging from quantum information science to physical chemistry and even biophysics.

Lighting. Photonic crystals offer the opportunity for spontaneous emission to be strongly controlled both in spatial terms (direction of emission) or in absolute terms (rate of emission). Therefore, it is timely to discuss the usage as practical light-emitting sources, including light emitting diodes (LED), for everyday lighting applications. In a recent lucid review [108], it has been asserted that photonic crystals offer favorable strategies to efficiently couple out the light; in comparison with usually employed random surface roughening, some curious common physical limitations were noted. The Purcell enhancement of the spontaneous emission rate was also considered as a means to enhance the internal efficiency of the light sources. It was found that such an approach is effective only for sources with specific properties, such as a small spatial extent and a narrow bandwidth so that it fits with the necessary resonant cavity. In the review, it is also justifiably concluded that a 3D photonic band gap is not a desired feature for lighting, as a band gap results in the inevitable decrease in the emission quantum efficiency (see Eq. (8.5)) [108]. Interestingly, however, it is not widely appreciated that 3D photonic crystals reveal extended frequency ranges with enhanced density of states, which could serve to boost the emission rate and the (internal) quantum efficiency. Calculations have revealed that silicon inverse opals have $3\times$ enhanced emission rates over a broad 10% bandwidth [354]. And provided that one controls the orientation of the transition dipole moment, a $4\times$ enhancement is feasible over a huge octave-broad bandwidth ($\geq$ 100% relative bandwidth). Therefore, we propose that the study of 3D photonic crystals at frequencies outside their band gap may yield fruitful applications.

In the context of lighting, it is also important to consider the important question of how to efficiently couple out of a 3D photonic band gap crystal, or how to couple light in, e.g. to excite emitters, or to address waveguides and cavities. In comparison with 2D slab crystals, it is not very obvious how to use in-coupling gratings in 3D [108]. We anticipate that the powerful new method of optical wavefront shaping – that is, the spatial addressing of incident waves with a spatial light modulator [345] – will open new avenues to provide access to photonic crystals. In particular, based on studies where one light emitting particle in an optically thick scattering medium was selectively addressed, we are optimistic that these new techniques will, for instance, allow effective coupling of light into and out of a nanobox buried deep inside a photonic crystal.

Novel 3D nanofabrication strategies. In high-speed computing, power and especially heat dissipation issues are becoming increasingly important. Moreover, it is being realized that continued miniaturization will meet boundaries set by fundamental physical limits [30, 96]. Therefore in state-of-the-art circuitry in the CMOS industry, an increasing number of researchers are currently contemplating the array of circuits in 3D grids and networks. A second approach to alleviate heat loads is to perform data communication on even shorter length scales by optical instead of electrical signals. It is our firm belief that 3D nanofabrication approaches that have been stimulated by 3D photonic crystals fabrication [468, 481] will serve as an inspiration for novel 3D CMOS fabrication strategies that are relevant for future high-speed computing [96].

8.6 Summary

In summary, we have seen that three-dimensional (3D) photonic crystals with a 3D photonic band gap play a fundamental role in cavity quantum electrodynamics (cQED), especially in phenomena where the local density of optical states plays a central role. One can say that photonic band gap crystals offer a knob to dial the density of states for broad frequency bandwidths over a wide range, from near zero to several times the value in vacuum. We have given an overview of the current status of the fabrication of 3D photonic crystals with a band gap at optical frequencies. Many different methods yield powerful 3D photonic crystals. At this time, the widely pursued woodpile crystals offer the widest versatility, as embedded high-Q cavities and waveguides have been demonstrated. The optical experiments have been discussed that provide signatures of 3D band gap behavior, such as broadband and wide-angle reflectivity or the observation of intricate surface modes. We have discussed the main implications of 3D band gaps for cQED, in particular spontaneous emission inhibition of emitters embedded in a 3D band gap crystal. Here, important progress has occurred in the last decade, bringing inhibition from what was a theoretical prediction to experimental reality. We have discussed the progress in spontaneous emission and laser action of emitters placed in a photonic band gap cavity. Near thresholdless laser action has been observed, and its realization is now approaching room temperature operation, thereby opening avenues for applications. The steps towards the breaking of the weak-coupling limit of cQED have been outlined, in particular by ultrafast modulation, where experimental tools are steadily ripening. In the final section, we have reviewed several exciting applications of 3D photonic band gaps, namely the shielding of decoherence for quantum optical systems, the manipulation of multiple coupled emitters, including resonant energy transfer, lighting, and possible spin-off to 3D nanofabrication for future high-end computing. While inhibited spontaneous emission in a 3D photonic band gap has tested our perseverance as it has taken twenty-five to forty years since the predictions by Bykov, Yablonovitch, and John [64, 226, 549], the many recent efforts on 3D band gaps with favorable outcomes bode well for exciting contributions to nanophotonics and beyond.

Acknowledgments

We are grateful to all colleagues in our group for many years of pleasant and fruitful collaboration, including Ad Lagendijk, Allard Mosk, Pepijn Pinkse, Lydia Bechger, Hannie van den Broek, Tijmen Euser, Philip Harding, Alex Hartsuiker, Cock Harteveld, Simon Huisman, Bart Husken, Arie Irman, Femius Koenderink, Merel Leistikow, Peter Lodahl, Juan Galisteo Lopez, Mischa Megens, Ivan Nikolaev, Karin Overgaag, Willem Tjerkstra, Floris van Driel, Judith Wijnhoven, Elahe Yeganegi, and many, many others. We also thank Irwan Setija (ASML), Fred Roozeboom (TNO, TUE), John Kelly and Daniël Vanmaekelbergh (Utrecht University), Ruud Balkenende (Philips), and many others for successful collaborations. We are grateful to Kenji Ishizaki and Susumu Noda for their kind permission to reproduce their results. This work is part of the research program of the "Stichting voor Technische Wetenschappen (STW)," and the "Stichting voor Fundamenteel Onderzoek der Materie (FOM)," which are supported financially by the "Nederlandse Organisatie voor Wetenschappelijk Onderzoek (NWO)."

References

[1] Abrahams, E. (eds). 2010. *50 Years of Anderson Localization*. World Scientific Publishing Co. Pte. Ltd.

[2] Abrahams, E., Anderson, P. W., Licciardello, D. C., and Ramakrishnan, T. V. 1979. Scaling theory of localization: Absence of quantum diffusion in two dimensions. *Phys. Rev. Lett.*, **42**, 673–676.

[3] Abrikosov, A. A. and Ryzhkin, I. A. 1978. Conductivity of quasi-one-dimensional metal systems. *Adv. Phys.*, **27**, 147–230.

[4] Adar, R., Henry, C. H., Milbrodt, M. A., and Kistler, R. C. 1994. Phase coherence of optical waveguides. *J. Lightwave Technol.*, **12**(4), 603–606.

[5] Agarwal, V., Soto-Urueta, J. A., Becerra, D., and Mora-Ramos, M. E. 2005. Light propagation in polytype Thue–Morse structures made of porous silicon. *Photonics and Nanostructures – Fundamentals and Applications*, **3**(2-3), 155–161. The Sixth International Symposium on Photonic and Electromagnetic Crystal Structures (PECS-VI) – PECS-VI.

[6] Akkermans, E. and Montambaux, G. 2007. *Mesoscopic Physics of Electrons and Photons*. Cambridge University Press, Cambridge.

[7] Akkermans, E., Wolf, P. E., and Maynard, R. 1986. Coherent backscattering of light by disordered media: Analysis of the peak lineshape. *Phys. Rev. Lett.*, **56**, 1471–1474.

[8] Altshuler, B. L. 1985. *Pis'ma Zh. Eksp. Teor. Fiz.*, **41**, 530.

[9] Altshuler, B. L., Lee, P. A., and Webb, R. A. (eds). 1991. *Mesoscopic Phenomena in Solids*. Elsevier, Amsterdam.

[10] Ambartsumyan, R. V., Basov, N. G., Kryukov, P. G., and Letokhov, V. S. 1966. Laser with a nonresonant feedback. *JETP Lett.*, **3**, 167–169.

[11] Anderson, P. W. 1958. Absence of diffusion in certain random lattices. *Phys. Rev.*, **109**, 1492–1505.

[12] Anderson, P. W., Thouless, D. J., Abrahams, E., and Fisher, D. S. 1980. New method for a scaling theory of localization. *Phys. Rev. B*, **22**, 3519–3526.

[13] Andreasen, J., Asatryan, A. A., Botten, L. C., *et al.* 2011. Modes of random lasers. *Adv. Opt. Photonics*, **3**(1), 88–127.

[14] Angelani, L., Conti, C., Ruocco, G., and Zamponi, F. 2006. Glassy behavior of light. *Phys. Rev. Lett.*, **96**(6), 065702.

[15] Antonoyiannakis, M. I. and Pendry, J. B. 1999. Electromagnetic forces in photonic crystals. *Phys. Rev. B*, **60**, 2363–2374.

[16] Aoki, K., Guimard, D., Nishioka, M., *et al.* 2008. Coupling of quantum-dot light emission with a three-dimensional photonic-crystal nanocavity. *Nature Photon.*, **2**, 688–692.

[17] Apalkov, V. M., Raikh, M. E., and Shapiro, B. 2004. Anomalously localized states in the Anderson model. *Phys. Rev. Lett.*, **92**, 066601.

[18] Armitage, A., Skolnick, M. S., Kavokin, A. V., *et al.* 1998. Polariton-induced optical asymmetry in semiconductor microcavities. *Phys. Rev. B*, **58**, 15367–15370.

[19] Ashcroft, N. W. and Mermin, N. D. 1976. *Solid State Physics.* Holt, Rinehart, and Winston, USA.

[20] Astratov, V. N., Franchak, J. P., and Ashili, S. P. 2004. Optical coupling and transport phenomena in chains of spherical dielectric microresonators with size disorder. *Appl. Phys. Lett.*, **85**, 5508–5510.

[21] Aulbach, J., Gjonaj, B., Johnson, P. M., Mosk, A. P., and Lagendijk, A. 2011. Control of light transmission through opaque scattering media in space and time. *Phys. Rev. Lett.*, **106**, 103901.

[22] Azbel, M. Ya. 1983. Eigenstates and properties of random systems in one dimension at zero temperature. *Phys. Rev. B*, **28**, 4106–4125.

[23] Baake, M. 1999. A Guide to Mathematical Quasicrystals. *arXiv:math-ph*, 9901014 v1.

[24] Baake, M. and Grimm, U. 2009. Kinematic diffraction is insufficient to distinguish order from disorder. *Phys. Rev. B*, **79**(2), 20203.

[25] Baake, M., Grimm, U., and Moody, R. V. 2002. What is Aperiodic Order? *arXiv:math.HO*, 0203252v1.

[26] Baake, M., and Grimm, U. 2010. Surprises in aperiodic diffraction. *J. Phys. - Conf. Series*, **226**(Apr.), 012023.

[27] Baba, T. 2008. Slow light in photonic crystals. *Nature Photon.*, **2**, 465–473.

[28] Babuty, A., Joulain, K., Chapuis, P.-O., Greffet, J.-J., and De Wilde, Y. 2013. Blackbody spectrum revisited in the near field. *Phys. Rev. Lett.*, **110**, 146103.

[29] Bachelard, N., Andreasen, J., Gigan, S., and Sebbah, P. 2012. Taming random lasers through active spatial control of the pump. *Phys. Rev. Lett.*, **109**, 033903.

[30] Ball, P. 2012. Feeling the heat. *Nature*, **492**, 175–176.

[31] Barbé, A. and von Haeseler, F. 2005. Correlation and spectral properties of higher-dimensional paperfolding and Rudin–Shapiro sequences. *J. Phys. A - Math. Gen.*, **38**(12), 2599–2622.

[32] Barbé, A. and Von Haeseler, F. 2007. Correlation and spectral properties of multi-dimensional Thue–Morse sequences. *Int. J. Bifurcat. Chaos*, **17**(04), 1265–1303.

[33] Barnes, W. L., Dereux, A., and Ebbesen, T. W. 2003. Surface plasmon subwavelength optics. *Nature*, **424**, 824–830.

[34] Barthelemy, P., Ghulinyan, M., Gaburro, Z., *et al.* 2007. Optical switching by capillary condensation. *Nature Photon.*, **1**, 172–175.

[35] Barthelemy, P., Bertolotti, J., and Wiersma, D. S. 2008. A Lévy flight for light. *Nature*, **453**(7194), 495–498.

[36] Bayer, M., Reinecke, T. L., Weidner, F., *et al.* 2001. Inhibition and enhancement of the spontaneous emission of quantum dots in structured microresonators. *Phys. Rev. Lett.*, **86**, 3168–3171.

[37] Becker, C., Linden, S., von Freymann, G., *et al.* 2005. Two-color pump–probe experiments on silicon inverse opals. *Appl. Phys. Lett.*, **87**, 091111.

[38] Beenakker, C. W. J. 1997. Random-matrix theory of quantum transport. *Rev. Mod. Phys.*, **69**, 731–808.

[39] Bellomo, B., Lo Franco, R., Maniscalco, S., and Compagno, G. 2008. Entanglement trapping in structured environments. *Phys. Rev. A*, **78**, 060302.

[40] Bendickson, J. M., Dowling, J. P., and Scalora, M. 1996. Analytic expressions for the electromagnetic mode density in finite, one-dimensional, photonic band-gap structures. *Phys. Rev. E*, **50**, 4107–4121.

[41] Berger, G. A., Kempe, M., and Genack, A. Z. 1997. Dynamics of stimulated emission from random media. *Phys. Rev. E*, **56**, 6118–6122.

[42] Bertolotti, J., Galli, M., Sapienza, R., *et al.* 2006. Wave transport in random systems: Multiple resonance character of necklace modes and their statistical behavior. *Phys. Rev. E*, **74**, 035602(R).

[43] Bertolotti, J., Gottardo, S., Wiersma, D. S., Ghulinyan, M., and Pavesi, L. 2005. Optical necklace states in Anderson localized 1D systems. *Phys. Rev. Lett.*, **94**, 113903.

[44] Bertolotti, J., van Putten, E. G., Blum, C., *et al.* 2012. Non-invasive imaging through opaque scattering layers. *Nature*, **491**, 232–234.

[45] Bindi, L., Steinhardt, P. J., Yao, N., and Lu, P. J. 2009. Natural quasicrystals. *Science*, **324**(5932), 1306.

[46] Birowosuto, M. D., Skipetrov, S. E., Vos, W. L., and Mosk, A. P. 2010. Observation of spatial fluctuations of the local density of states in random photonic media. *Phys. Rev. Lett.*, **105**, 013904.

[47] Bita, I., Choi, T., Walsh, M. E., Smith, H. I., and Thomas, E. L. 2007. Large-area 3D nanostructures with octagonal quasicrystalline symmetry via phase-mask lithography. *Adv. Mater.*, **19**(10), 1403.

[48] Björk, G., Karlsson, A., and Yamamoto, Y. 1994. Definition of a laser threshold. *Phys. Rev. A*, **50**, 1675–1680.

[49] Blanco, A., Chomski, E., Grabtchak, S., *et al.* 2000. Large-scale synthesis of a silicon photonic crystal with a complete three-dimensional bandgap near 1.5 micrometres. *Nature*, **405**, 437–439.

[50] Bleuse, J., Claudon, J., Creasey, M., *et al.* 2011. Inhibition, enhancement, and control of spontaneous emission in photonic nanowires. *Phys. Rev. Lett.*, **106**, 103601.

[51] Bliokh, K. Yu., Bliokh, Yu. P., *et al.* 2008. Coupling and level repulsion in the localized regime: From isolated to quasiextended modes. *Phys. Rev. Lett.*, **101**, 133901.

[52] Bloch, F. 1929. Über die Quantenmechanik der Elektronen in Kristallgittern. *Z. Physik*, **52**, 555–600.

[53] Blum, C., Zijlstra, N., Lagendijk, A., *et al.* 2012. Nanophotonic control of the Förster resonance energy transfer efficiency. *Phys. Rev. Lett.*, **109**, 203601.

[54] Boer, J. F. de., van Rossum, M. C. W., van Albada, M. P., Nieuwenhuizen, T. M., and Lagendijk, A. 1994. Probability distribution of multiple scattered light measured in total transmission. *Phys. Rev. Lett.*, **73**, 2567–2570.

[55] Bohren, C. F. and Huffmann, D. R. 1983. *Absorption and Scattering of Light by Small Particles*. Wiley, New York.

[56] Boriskina, S. V., Gopinath, A., and Dal Negro, L. 2008. Optical gap formation and localization properties of optical modes in deterministic aperiodic photonic structures. *Opt. Express*, **16**(23), 18813–18826.

[57] Boriskina, S. V., Gopinath, A., and Dal Negro, L. 2009. Optical gaps, mode patterns and dipole radiation in two-dimensional aperiodic photonic structures. *Physica E: Low-dimensional Systems and Nanostructures*, **41**(6), 1102–1106.

[58] Brenner, N. and Fishman, S. 1999. Pseudo-randomness and localization. *Nonlinearity*, **5**(1), 211–235.

[59] Brouwer, P. W. 1998. Transmission through a many-channel random waveguide with absorption. *Phys. Rev. B*, **57**(17), 10526–10536.

[60] Bruggeman, D. A. G. 1935. Berechnung verschiedener physikalischer Konstanten von heterogenen Substanzen. *Ann. Phys. (Leipzig)*, **24**, 636–679.

[61] Burresi, M., Radhalakshmi, V., Savo, R., *et al.* 2012. Weak localization of light in superdiffusive random systems. *Phys. Rev. Lett.*, **108**(11), 110604.

[62] Busch, K. and John, S. 1998. Photonic band gap formation in certain self-organizing systems. *Phys. Rev. E*, **58**, 3896–3908.

[63] Busch, K., von Freymann, G., Linden, S., *et al.* 2007. Periodic nanostructures for photonics. *Phys. Rep.*, **444**, 101–202.

[64] Bykov, V. P. 1972. Spontaneous emission in a periodic structure. *Sov. Phys. JETP*, **35**, 269–273.

[65] Campbell, M., Sharp, D. N., Harrison, M. T., Denning, R. G., and Turberfield, A. J. 2000. Fabrication of photonic crystals for the visible spectrum by holographic lithography. *Nature*, **404**, 53–56.

[66] Cao, D., Tandaechanurat, A., Nakayama, S., *et al.* 2012. Silicon-based three-dimensional photonic crystal nanocavity laser with InAs quantum-dot gain. *Appl. Phys. Lett.*, **101**, 191107.

[67] Cao, H. 2003. Lasing in disordered media. Chap. 6, pages 317–370 in: Wolf, E. (ed.), *Progress in Optics*, vol. 45. Elsevier.

[68] Cao, H. 2005. Review on latest developments in random lasers with coherent feedback. *J. Phys. A*, **38**, 10497.

[69] Cao, H., Ling, Y., Xu, J. Y., Cao, C. Q., and Kumar, P. 2001. Photon statistics of random lasers with resonant feedback. *Phys. Rev. Lett.*, **86**(20), 4524–4527.

[70] Cao, H., Xu, J. Y., Zhang, D. Z., *et al.* 2000. Spatial confinement of laser light in active random media. *Phys. Rev. Lett.*, **84**(24), 5584–5587.

[71] Cao H., Zhao, Y. G., Ho, S. T., *et al.* 1999. Random laser action in semiconductor powder. *Phys. Rev. Lett.*, **82**, 2278–2281.

[72] Capaz, R. B., Koiller, B., and de Queiroz, S. L. A. 1990. Gap states and localization properties of one-dimensional Fibonacci quasicrystals. *Phys. Rev. B*, **42**, 6402–6407.

[73] Castellanos-Beltran, M. A., Ngo, D. Q., Sharks, W. E., Jayich, A. B., and Harris, J. G. E. 2013. Measurement of the full distribution of persistent current in normal-metal rings. *Phys. Rev. Lett.*, **110**, 156801.

[74] Cerdán, L., Enciso, E., Martín, V., *et al.* 2012. FRET-assisted laser emission in colloidal suspensions of dye-doped latex nanoparticles. *Nature Photon.*, **6**, 621–626.

[75] Chabanov, A. A. and Genack, A. Z. 2001. Statistics of dynamics of localized waves. *Phys. Rev. Lett.*, **87**(23), 233903.

[76] Chabanov, A. A. and Genack, A. Z. 2005. Statistics of the mesoscopic field. *Phys. Rev. E*, **72**, 055602.

[77] Chabanov, A. A., Hu, B., and Genack, A. Z. 2004. Dynamic correlation in wave propagation in random media. *Phys. Rev. Lett.*, **93**, 123901.

[78] Chabanov, A. A., Stoytchev, M., and Genack, A. Z. 2000. Statistical signatures of photon localization. *Nature*, **404**, 850–853.

[79] Chabanov, A. A., Zhang, Z. Q., and Genack, A. Z. 2003. Breakdown of diffusion in dynamics of extended waves in mesoscopic media. *Phys. Rev. Lett.*, **90**, 203903.

[80] Chaikin, P. M. and Lubensky, T. C. 2000. *Principles of Condensed Matter Physics*. Cambridge University Press, Cambridge.

[81] Chandrasekhar, S. 1950. *Radiative Transfer*. Oxford University Press, Oxford.

[82] Cheng, C. C., Arbet-Engels, V., Scherer, A., and Yablonovitch, E. 1996. Nanofabricated three dimensional photonic crystals operating at optical wavelengths. *Phys. Scr.*, **T68**, 17–20.

[83] Cheng, S. S. M., Li, L.-M., Chan, C. T., and Zhang, Z. Q. 1999. Defect and transmission properties of two-dimensional quasiperiodic photonic band-gap systems. *Phys. Rev. B*, **59**(6), 4091.

[84] Cheng, Z., Savit, R., and Merlin, R. 1988. Structure and electronic properties of Thue–Morse lattices. *Phys. Rev. B*, **37**(9), 4375–4382.

[85] Cherroret, N., Peña, A., Chabanov, A. A., and Skipetrov, S. E. 2009. Nonuniversal dynamic conductance fluctuations in disordered systems. *Phys. Rev. B*, **80**, 045118.

[86] Cheung, S. K., Zhang, X., Zhang, Z. Q., Chabanov, A. A., and Genack, A. Z. 2004. Impact of weak localization in the time domain. *Phys. Rev. Lett.*, **92**, 173902.

[87] Ching, E. S. C., Leung, P. T., Suen, W. M., Tong, S. S., and Young, K. 1998. Waves in open systems: Eigenfunction expansions. *Rev. Mod. Phys.*, **70**, 1545–1554.

[88] Choi, Y., Yoon, C., Kim, M., *et al.* 2012. Scanner-free and wide-field endoscopic imaging by using a single multimode optical fiber. *Phys. Rev. Lett.*, **109**, 203901.

[89] Chu, S. T., Little, B. E., Pan, W., Kaneko, T., and Kokubun, Y. 1999. Cascaded microring resonators for crosstalk reduction and spectrum cleanup in add-drop filters. *IEEE Photon. Technol. Lett.*, **11**(11), 1423–1425.

[90] Chutinan, A. and Noda, S. 1999. Effects of structural fluctuations on the photonic bandgap during fabrication of a photonic crystal. *J. Opt. Soc. Am. B*, **16**, 240–244.

[91] Conti, C., Leonetti, M., Fratalocchi, A., Angelani, L., and Ruocco, G. 2008. Condensation in disordered lasers: Theory, $3D + 1$ simulations, and experiments. *Phys. Rev. Lett.*, **101**(14), 143901.

[92] Cooper, M. L., Gupta, G., Green, W. M. J., *et al.* 2010. 235-Ring coupled-resonator optical waveguides. CLEO 2010 Proceedings of the Conference on Lasers and Electro-Optics CTUHH3.

[93] Cooper, M. L., Gupta, G., Ong, J. R., *et al.* 2011. Correlations between light at spectrally distant wavelengths in coupled microring resonator waveguides. Proceedings of the Conference on Lasers and Electro-Optics CWMH.

[94] Cooper, M. L., Gupta, G., Schneider, M. A., *et al.* 2010. Statistics of light transport in 235-ring silicon coupled-resonator optical waveguides. *Opt. Express*, **18**, 26505–26516.

[95] Cooper, M. L. and Mookherjea, S. 2011. Modeling of multiband transmission in long silicon coupled-resonator optical waveguides. *IEEE Photon. Technol. Lett.*, **23**(13), 872–874.

[96] Coteus, P. W., Knickerbocker, J. U., Lam, C. H., and Vlasov, Y. A. 2011. Technologies for exascale systems. *IBM J. Res. Dev.*, **55**, paper nr. 14.

[97] Cullis, A. G., Canham, L. T., and Calcott, P. D. J. 1997. The structural and luminescence properties of porous silicon. *J. Appl. Phys.*, **82**, 909–966.

[98] Dal Negro, L. and Boriskina, S. V. 2012. Deterministic aperiodic nanostructures for photonics and plasmonics applications. *Laser Photon. Rev.*, **6**, 178–218.

[99] Dal Negro, L. and Feng, N.-N. 2007. Spectral gaps and mode localization in Fibonacci chains of metal nanoparticles. *Opt. Express*, **15**(22), 14396–14403.

[100] Dal Negro, L., Feng, N.-N., and Gopinath, A. 2008. Electromagnetic coupling and plasmon localization in deterministic aperiodic arrays. *J. Opt. A - Pure Appl. Op.*, **10**(6), 064013.

[101] Dal Negro, L., Lawrence, N., and Trevino, J. 2012. Analytical light scattering and orbital angular momentum spectra of arbitrary Vogel spirals. *Opt. Express*, **20**(16), 18209.

[102] Dal Negro, L., Oton, C. J., Gaburro, Z., *et al.* 2003. Light transport through band edge states of Fibonacci quasicrystals. *Phys. Rev. Lett.*, **90**, 055501.

[103] Dal Negro, L., Stolfi, M., Yi, Y., *et al.* 2004. Photon band gap properties and omnidirectional reflectance in Si/SiO_2 Thue–Morse quasicrystals. *Appl. Phys. Lett.*, **84**(25), 5186–5188.

[104] Dal Negro, L., Yi, J. H., Nguyen, V., Yi, Y., Michel, J., and Kimerling, L. C. 2005. Spectrally enhanced light emission from aperiodic photonic structures. *Appl. Phys. Lett.*, **86**(26), 261905.

[105] Dalfovo, F., Giorgini, S., Pitaevskii, L. P., and Stringari, S. 1999. Theory of Bose–Einstein condensation in trapped gases. *Rev. Mod. Phys.*, **71**(Apr.), 463–512.

[106] Dalichaouch, R., Armstrong, J. P., Schultz, S., Platzman, P. M., and McCall, S. L. 1991. Microwave localization by two-dimensional random scattering. *Nature*, **354**, 53–55.

[107] Davanco, M., Ong, J., Rong, S., *et al.* 2012. Telecommunications-band heralded single photons from a silicon nanophotonic chip. *Appl. Phys. Lett.*, **100**(26), 261104.

[108] David, A., Benisty, H., and Weisbuch, C. 2012. Photonic crystal light-emitting sources. *Rep. Prog. Phys.*, **75**, 126501.

[109] Davy, M., Shi, Z., and Genack, A. Z. 2012. Focusing through random media: Eigenchannel participation number and intensity correlation. *Phys. Rev. B*, **85**, 035105.

[110] Davy, M., Shi, Z., Wang, J., and Genack, A. Z. 2013. Transmission statistics and focusing in single disordered samples. *Opt. Express*, **21**, 10367–10375.

[111] Davy, M., Shi, Z., Wang, J., and Genack, A. Z. 2014. Transmission eigenchannels and the densities of states of random media. arxiv.org/abs/1403.3811.

[112] Deubel, M., von Freymann, G., Wegener, M., *et al.* 2004. Direct laser writing of three-dimensional photonic-crystal templates for telecommunications. *Nature Mater.*, **3**(7), 444.

[113] Dorokhov, O. N. 1982. Transmission coefficient and the localization length of an electron in N bond disorder chains. *Pis'ma Zh. Eksp. Teor. Fiz.*, **36**, 259.

[114] Dorokhov, O. N. 1984. On the coexistence of localized and extended electronic states in the metallic phase. *Solid State Commun.*, **51**, 381–384.

[115] Douady, S. and Couder, Y. 1996. Phyllotaxis as a dynamical self organizing process. *J. Theor. Biology*, **178**, 255–274.

[116] Dougherty, E. R. 1990. *Probability and Statistics for the Engineering, Computing and Physical Sciences*. Prentice-Hall International, Inc., Englewood, New Jersey.

[117] Dowling, J. P., Scalora, M., and Bloemer, M. J. 1994. The photonic band edge laser: A new approach to gain enhancement. *J. Appl. Phys.*, **75**, 1896–1899.

[118] Dyson, F. J. and Mehta, M. L. 1962. Statistical theory of the energy levels of complex systems. I-V. *J. Math. Phys.*, **3**, 140.

[119] Economou, E. N. 2006. *Green's Functions in Quantum Physics, Third Edn.* Springer, Berlin.

[120] Economou, E. N. and Sigalas, M. M. 1993. Classical wave propagation in periodic structures: Cermet versus network topology. *Phys. Rev. B*, **48**, 13434–13438.

[121] Eiselt, M. H., Clausen, C. B., and Tkach, R. W. 2003. Performance characterization of components with group delay fluctuations. *IEEE Photon. Technol. Lett.*, **15**(8), 1076–1078.

[122] El-Dardiry, R. G. S., and Lagendijk, A. 2011. Tuning random lasers by engineered absorption. *Appl. Phys. Lett.*, **98**(16), 161106–161106.

[123] El-Dardiry, R. G. S., Mosk, A. P., *et al.* 2010. Experimental studies on the mode structure of random lasers. *Phys. Rev. A*, **81**(4), 043830.

[124] Engelen, R. J. P., Mori, D., Baba, T., and Kuipers, L. 2008. Two regimes of slow-light losses revealed by adiabatic reduction of group velocity. *Phys. Rev. Lett.*, **101**(10), 03901.

[125] Esaki, L. and Tsu, R. 1970. Superlattice and negative differential conductivity in semiconductors. *IBM J. Res. Dev.*, **14**, 61–65.

[126] Euser, T. G., Molenaar, A. J., Fleming, J. G., *et al.* 2008. All-optical octave-broad ultrafast switching of Si woodpile photonic band gap crystals. *Phys. Rev. B*, **77**, 115214.

[127] Euser, T. G., Wei, H., Kalkman, J., *et al.* 2007. All-optical octave-broad ultrafast switching of Si woodpile photonic band gap crystals. *J. Appl. Phys.*, **102**, 053111.

[128] Faez, S., Strybulevych, A., Page, J. H., Lagendijk, A., and van Tiggelen, B. A. 2009. Observation of multifractality in Anderson localization of ultrasound. *Phys. Rev. Lett.*, **103**, 155703.

[129] Fallert, J., Dietz, R. J. B., Sartor, J., *et al.* 2009. Co-existence of strongly and weakly localized random laser modes. *Nature Photon.*, **3**(5), 279–282.

[130] Feng, S., Kane, C., Lee, P. A., and Stone, A. D. 1988. Correlations and fluctuations of coherent wave transmission through disordered media. *Phys. Rev. Lett.*, **61**, 834–837.

[131] Fermi, E. 1932. Quantum theory of radiation. *Rev. Mod. Phys.*, **4**, 87–132.

[132] Ferry, D. K., Alkis, R., and Gilbert, M. J. 2007. Semiconductor device scaling: The role of ballistic transport. *J. Comput. Theor. Nanos.*, **4**(6), 1149–1152.

[133] Fink, M. 1992. Time reversal of ultrasonic fields-Part I: Basic principles. *IEEE T. Ultrason. Ferr.*, **39**, 555–566.

[134] Fischer, J. and Wegener, M. 2011. Three-dimensional direct laser writing inspired by stimulated-emission-depletion microscopy. *Opt. Mater. Express*, **1**(4), 614–624.

[135] Fleming, J. G. and Lin, S. Y. 1999. Three-dimensional photonic crystal with a stop band from 1.35 to 1.95 μm. *Opt. Lett.*, **24**, 49–51.

[136] Florescu, L. and John, S. 2004. Photon statistics and coherence in light emission from a random laser. *Phys. Rev. Lett.*, **93**(1), 13602.

[137] Folli, V., Puglisi, A., Leuzzi, L., and Conti, C. 2012. Shaken granular lasers. *Phys. Rev. Lett.*, **108**(24), 248002.

[138] Freymann, G., Ledermann, A., Thiel, M., *et al.* 2010. Three dimensional nanostructures for photonics. *Adv. Func. Mater.*, **20**, 1038–1052.

[139] Frolov, S. V., Vardeny, Z. V., and Yoshino, K. 1999. Cooperative and stimulated emission in poly(p-phenylene-vinylene) thin films and solutions. *Phys. Rev. B*, **57**, 9141–9147.

[140] Fujiwara, T., Kohmoto, M., and Tokihiro, T. 1989. Multifractal wave functions on a Fibonacci lattice. *Phys. Rev. B*, **40**, 7413–7416.

[141] Fussell, D. P., Hughes, S., and Dignam, M. M. 2008. Influence of fabrication disorder on the optical properties of coupled-cavity photonic crystal waveguides. *Phys. Rev. B*, **78**(14), 144201.

[142] Galisteo-López, J. F., Ibisate, M., Sapienza, R., *et al.* 2011. Self-assembled photonic structures. *Adv. Mater.*, **23**, 30–69.

[143] Galusha, J. W., Jorgensen, M. R., and Bartl, M. H. 2010. Diamond-structured titania photonic-bandgap crystals from biological templates. *Adv. Mater.*, **22**, 107–110.

[144] Garcia, N. and Genack, A. Z. 1989. Crossover to strong intensity correlation for microwave radiation in random media. *Phys. Rev. Lett.*, **63**, 1678–1681.

[145] Garcia, N. and Genack, A. Z. 1991. Anomalous photon diffusion at the threshold of the Anderson localization transition. *Phys. Rev. Lett.*, **66**, 1850–1853.

[146] Garcia, N., Genack, A. Z., and Lisyansky, A. A. 1992. Measurement of the transport mean free path of diffusing photons. *Phys. Rev. B*, **46**, 14475–14479.

[147] Garcia, P. D., Sapienza, R., Toninelli, C., Lopez, C., and Wiersma, D. S. 2011. Photonic crystals with controlled disorder. *Phys. Rev. A*, **84**, 023813.

[148] García-Martín, A. and Sáenz, J. J. 2001. Universal conductance distributions in the crossover between diffusive and localization regimes. *Phys. Rev. Lett.*, **87**, 116603.

[149] Gellermann, W., Kohmoto, M., Sutherland, B., and Taylor, P. C. 1994. Localization of light waves in Fibonacci dielectric multilayers. *Phys. Rev. Lett.*, **72**, 633–636.

[150] Genack, A. Z. 1987. Optical transmission in disordered media. *Phys. Rev. Lett.*, **58**, 2043–2046.

[151] Genack, A. Z. and Drake, J. M. 1990. Relationship between optical intensity, fluctuations and pulse propagation in random media. *Europhys. Lett.*, **11**, 331.

[152] Genack, A. Z. and Drake, J. M. 1994. Scattering for super-radiation. *Nature*, **368**, 400–401.

[153] Genack, A. Z. and Garcia, N. 1991. Observation of photon localization in a three-dimensional disordered system. *Phys. Rev. Lett.*, **66**, 2064–2067.

[154] Genack, A. Z., Garcia, N., and Polkosnik, W. 1990. Long-range intensity correlation in random media. *Phys. Rev. Lett.*, **65**, 2129–2132.

[155] Genack, A. Z., Sebbah, P., Stoytchev, M., and van Tiggelen, B. A. 1999. Statistics of wave dynamics in random media. *Phys. Rev. Lett.*, **82**(4), 715.

[156] Gertsenshtein, M. E. and Vasil'ev, V. B. 1959. Waveguides with random inhomogeneities and Brownian motion in the Lobachevsky plane. *Theor. of Probab. its Appl.*, **4**, 391–398.

[157] Ghulinyan, M. 2007. Formation of optimal-order necklace modes in one-dimensional random photonic superlattices. *Phys. Rev. A*, **76**, 013822.

[158] Ghulinyan, M. 2007. Periodic oscillations in transmission decay of Anderson localized one-dimensional dielectric systems. *Phys. Rev. Lett.*, **99**, 063905.

[159] Ghulinyan, M., Gaburro, Z., Wiersma, D. S., and Pavesi, L. 2006. Tuning of resonant Zener tunneling by vapor diffusion and condensation in porous optical superlattices. *Phys. Rev. B*, **74**, 045118.

[160] Ghulinyan, M., Galli, M., Toninelli, C., *et al.* 2006. Wide-band transmission of nondistorted slow waves in one-dimensional optical superlattices. *Appl. Phys. Lett.*, **88**, 241103.

[161] Ghulinyan, M., Oton, C. J., Bonetti, G., Gaburro, Z., and Pavesi, L. 2003. Free-standing porous silicon single and multiple optical cavities. *J. Appl. Phys.*, **93**, 9724–9729.

[162] Ghulinyan, M., Oton, C. J., Dal Negro, L., *et al.* 2005. Light-pulse propagation in Fibonacci quasicrystals. *Phys. Rev. B*, **71**, 094204.

[163] Ghulinyan, M., Oton, C. J., Gaburro, Z., Bettotti, P., and Pavesi, L. 2003. Porous silicon free-standing coupled microcavities. *Appl. Phys. Lett.*, **82**, 1550–1552.

[164] Ghulinyan, M., Oton, C. J., Gaburro, Z., *et al.* 2005. Zener tunneling of light waves in an optical superlattice. *Phys. Rev. Lett.*, **94**, 127401.

[165] Gifford, D. K., Soller, B. J., Wolfe, M. S., and Froggatt, M. E. 2005. Optical vector network analyzer for single-scan measurements of loss, group delay, and polarization mode dispersion. *Appl. Opt.*, **44**(34), 7282–7286.

[166] Gjonaj, B., Aulbach, J., Johnson, P. M., *et al.* 2011. Active spatial control of plasmonic fields. *Phys. Rev. E*, **5**, 360–363.

[167] Goetschy, A. and Stone, A. D. 2013. Filtering random matrices: the effect of incomplete channel control in multiple scattering. *arXiv:*, 1304.5562.

[168] Goh, T., Suzuki S., and Sugita, A. 1997. Estimation of waveguide phase error in silica-based waveguides. *J. Lightwave Technol.*, **15**(11), 2107–2113.

[169] Gopar, V. A., Muttalib, K. A., and Wölfle, P. 2002. Conductance distribution in disordered quantum wires: Crossover between the metallic and insulating regimes. *Phys. Rev. B*, **66**, 174204.

[170] Gopinath, A., Boriskina, S. V., Feng, N.-N., Reinhard, B. M., and Dal Negro, L. 2008. Photonic-plasmonic scattering resonances in deterministic aperiodic structures. *Nano Lett.*, **8**(8), 2423–2431.

[171] Gopinath, A., Boriskina, S. V., Reinhard, B. M., and Dal Negro, L. 2009. Deterministic aperiodic arrays of metal nanoparticles for surface-enhanced Raman scattering (SERS). *Opt. Express*, **17**(5), 3741–3753.

[172] Gottardo, S., Cavalieri, S., Yaroshchuk, O., and Wiersma, D. S. 2004. Quasi-two-dimensional diffusive random laser action. *Phys. Rev. Lett.*, **93**(26), 263901.

[173] Gottardo, S., Sapienza, R., García, P. D., *et al.* 2008. Resonance-driven random lasing. *Nature Photon.*, **2**(7), 429–432.

[174] Gouedard, C., Husson, D., Sauteret, C., Auzel, F., and Migus, A. 1993. Generation of spatially incoherent short pulses in laser-pumped neodymium stoichiometric crystals and powders. *J. Opt. Soc. Am. B*, **10**(12), 2358–2363.

[175] Grésillon, S., Aigouy, L., Boccara, A. C., *et al.* 1999. Experimental observation of localized optical excitations in random metal-dielectric films. *Phys. Rev. Lett.*, **82**, 4520–4523.

[176] Grimm, U. and Schreiber, M. 1999. Aperiodic tilings on the computer. *arXiv:cond-mat*, 9903010v1.

[177] Griniasty, M. and Fishman, S. 1988. Localization by pseudorandom potentials in one dimension. *Phys. Rev. Lett.*, **60**(13), 1334–1337.

[178] Grünbaum, B. and Shephard, G. C. 1987. *Tilings and Patterns*. W. H. Freeman, New York.

[179] Gumbs, G., Dubey, G. S., Salman, A., Mahmoud, B. S., and Huang, D. 1995. Statistical and transport properties of quasiperiodic layered structures: Thue–Morse and Fibonacci. *Phys. Rev. B*, **52**(July), 210–219.

[180] Haberko, J. and Scheffold, F. 2013. Fabrication of mesoscale polymeric templates for three-dimensional disordered photonic materials. *Opt. Express*, **21**(1), 1057.

[181] Harding, P. J. 2008. *Photonic crystals modified by optically resonant systems*. Ph.D. thesis, (University of Twente) available from: www.photonicbandgaps.com.

[182] Haroche, S. 1992. Cavity quantum electrodynamics. Pages 767–940 in: *Fundamental Systems in Quantum Optics*. North Holland, Amsterdam.

[183] Hase, M., Miyazaki, H., Egashira, M., *et al.* 2002. Isotropic photonic band gap and anisotropic structures in transmission spectra of two-dimensional fivefold and eightfold symmetric quasiperiodic photonic crystals. *Phys. Rev. B*, **66**(21), 214205.

[184] Hattori, T., Tsurumachi, N., Kawato, S., and Nakatsuka,, H. 1994. Photonic dispersion relation in a one-dimensional quasicrystal. *Phys. Rev. B*, **50**, 4220–4223.

[185] Hauke, N., Tandaechanurat, A., Zabel, T., *et al.* 2012. A three-dimensional silicon photonic crystal nanocavity with enhanced emission from embedded germanium islands. *New J. Phys.*, **14**, 083035.

[186] Haus, H. A. 2000. Mode-locking of lasers. *IEEE J. Sel. Top. Quant.*, **6**(6), 1173–1185.

[187] He, S. and Maynard, J. D. 1986. Detailed measurements of inelastic scattering in Anderson localization. *Phys. Rev. Lett.*, **57**, 3171–3174.

[188] Heebner, J. E., Chak, P., Pereira, S., Sipe, J. E., and Boyd, R. W. 2004. Distributed and localized feedback in microresonator sequences for linear and nonlinear optics. *J. Opt. Soc. Am. B*, **21**(10), 1818–1832.

[189] Hendrickson, J., Richards, B. C., Sweet, J., *et al.* 2008. Excitonic polaritons in Fibonacci quasicrystals. *Opt. Express*, **16**(20), 15382.

[190] Hermatschweiler, M., Ledermann, A., Ozin, G. A., Wegener, M., and von Freymann, G. 2007. Fabrication of silicon inverse woodpile photonic crystals. *Adv. Func. Mater.*, **17**, 2273–2277.

[191] Hillebrand, R. and Hergert, W. 2004. Scaling properties of a tetragonal photonic crystal design having a large complete bandgap. *Photonics Nanostruct.*, **2**, 33–39.

[192] Ho, K. M., Chan, C. T., and Soukoulis, C. M. 1990. Existence of a photonic gap in periodic dielectric structures. *Phys. Rev. Lett.*, **65**, 3152–3155.

[193] Ho, K, M., Chan, C. T., Soukoulis, C. M., Biswas, R., and Sigalas, M. 1994. Photonic band gaps in three dimensions: New layer-by-layer periodic structures. *Solid State Commun.*, **89**, 413–416.

[194] Hoeffe, M. and Baake, M. 2000. Surprises in diffuse scattering. *Z. Kristallogr*, **215**, 441–444.

[195] Hohenberg, P. C. 1967. Existence of long-range order in one and two dimensions. *Phys. Rev.*, **158**, 383–386.

[196] Holland, B. T., Blanford, C. F., and Stein, A. 1998. Synthesis of macroporous minerals with highly ordered three-dimensional arrays of spheroidal voids. *Science*, **281**, 538–540.

[197] Hsieh, I., Pu, Y., Grange, R., and Psaltis, D. 2010. Digital phase conjugation of second harmonic radiation emitted by nanoparticles in turbid media. *Opt. Express*, **18**, 12283–12290.

[198] Hu, H., Strybulevych, A., Page, J. H., Skipetrov, S. E., and van Tiggelen, B. 2008. Localization of ultrasound in a three-dimensional elastic network. *Nature Phys.*, **4**, 945–948.

[199] Huang, X. and Gong, Ch. 1998. Property of Fibonacci numbers and the periodic-like perfectly transparent electronic states in Fibonacci chains. *Phys. Rev. B*, **58**, 739–744.

[200] Hughes, S., Ramunno, L., Young, J. F., and Sipe, J. E. 2005. Extrinsic optical scattering loss in photonic crystal waveguides: Role of fabrication disorder and photon group velocity. *Phys. Rev. Lett.*, **94**(3), 033903.

[201] Huisman, S. R., Ctistis, G., Stobbe, S., *et al.* 2012. Measurement of a band-edge tail in the density of states of a photonic-crystal waveguide. *Phys. Rev. B*, **86**, 155154.

[202] Huisman, S. R., Nair, R. V., Woldering, L. A., *et al.* 2011. Signature of a three-dimensional photonic band gap observed on silicon inverse woodpile photonic crystals. *Phys. Rev. B*, **83**, 205313.

[203] Husken, B. H., Koenderink, A. F., and Vos, W. L. 2013. Angular redistribution of near-infrared emission from quantum dots in three-dimensional photonic crystals. *J. Phys. Chem. C*, **117**, 3431–3439.

[204] Iglói, F., Turban, L., and Rieger, H. 1999. Anomalous diffusion in aperiodic environments. *Phys. Rev. E*, **59**(2), 1465–1474.

[205] Il'chishin, I. P. and Vakhnin, A. Yu. 1995. Detecting of the structure distortion of cholesteric liquid crystal using the generation characteristics of the distributed feedback laser based on it. *Mol. Cryst. Liq. Cryst.*, **265**, 687–697.

[206] Imagawa, S., Edagawa, K., Morita, K., *et al.* 2010. Photonic band-gap formation, light diffusion, and localization in photonic amorphous diamond structures. *Phys. Rev. B.*, **82**, 155116.

[207] Imry, Y. 1986. Active transmission channels and universal conductance fluctuations. *Europhys. Lett.*, **1**, 249–256.

[208] Imry, Y. and Landauer, R. 1999. Conductance viewed as transmission. *Rev. Mod. Phys.*, **71**, S306–S312.

[209] Ioffe, A. F. and Regel, A. R. 1960. Noncrystalline, amorphous and liquid electronic semiconductors. *Prog. Semicond.*, **4**, 237.

[210] Ishimaru, A. 1978. *Wave Propagation and Scattering in Random Media*. Academic Press, New York.

[211] Ishizaki, K., Koumura, M., Suzuki, K., Gondaira, K., and Noda, S. 2013. Realization of three-dimensional guiding of photons in photonic crystals. *Nature Photon.*, **7**, 133–137.

[212] Ishizaki, K. and Noda, S. 2009. Manipulation of photons at the surface of three-dimensional photonic crystals. *Nature*, **460**, 367–371.

[213] Jahnke, F. and Koch, S. 1995. Many-body theory for semiconductor microcavity lasers. *Phys. Rev. A*, **52**(2), 1712–1727.

[214] James, R. W. 1954. *The Optical Principles of the Diffraction of X-rays*. Bell & Hyman, London.

[215] Janot, C. 1992. *Quasicrystals: A Primer*. Clarendon Press, Oxford.

[216] Janot, C. 1994. Hierarchical phase transitions and vibrational modes localisation in quasicrystals. *Int. J. Mod. Phys. B*, **08**(17), 2245–2281.

[217] Jiang, X., Zhang, Y., Feng, S., *et al.* 2005. Photonic band gaps and localization in the Thue–Morse structures. *Appl. Phys. Lett.*, **86**(20), 201110.

[218] Jin, C., Cheng, B., Man, B., *et al.* 1999. Band gap and wave guiding effect in a quasiperiodic photonic crystal. *Appl. Phys. Lett.*, **75**(13), 1848.

[219] Joannopoulos, J. D., Johnson, S. G., Winn, J. N., and Meade, R. D. 2008. *Photonic Crystals – Molding the Flow of Light, Second Edition*. Princeton University Press.

[220] Joannopoulos, J. D., Johnson, S. G., Winn, J. N., and Meade, R. D. 2011. *Photonic Crystals: Molding the Flow of Light*. Princeton University Press.

[221] John, S. 1984. Electromagnetic absorption in a disordered medium near a photon mobility edge. *Phys. Rev. Lett.*, **53**, 2169–2172.

[222] John, S. 1987. Strong localization of photons in certain disordered dielectric superlattices. *Phys. Rev. Lett.*, **58**, 2486–2489.

[223] John, S. 1991. Localization of light. *Phys. Today*, **44**, 32–40.

[224] John, S. and Quang, T. 1994. Spontaneous emission near the edge of a photonic bandgap. *Phys. Rev. A*, **50**, 1764–1769.

[225] John, S., Sompolinsky, H., and Stephen, M. J. 1983. Localization in a disordered elastic medium near two dimensions. *Phys. Rev. B*, **27**, 5592–5603.

[226] John, S. and Wang, J. 1990. Quantum electrodynamics near a photonic band gap: Photon bound states and dressed atoms. *Phys. Rev. Lett.*, **64**, 2418–2421.

[227] Johnson, P. M., Koenderink, A. F., and Vos, W. L. 2002. Ultrafast switching of photonic density of states in photonic crystals. *Phys. Rev. B*, **66**, 081102.

[228] Jorgensen, M. R., Galusha, J. W., and Bartl, M. H. 2011. Strongly modified spontaneous emission rates in diamond-structured photonic crystals. *Phys. Rev. Lett.*, **107**, 143902.

[229] Kao, T. S., Jenkins, S. D., Ruostekoski, J., and Zheludev, N. I. 2011. Coherent control of nanoscale light localization in metamaterial: Creating and positioning isolated subwavelength energy hot spots. *Phys. Rev. Lett.*, **106**, 085501.

[230] Katz, O., Small, E., Bromberg, Y., and Silberberg, Y. 2011. Focusing and compression of ultrashort pulses through scattering media. *Nature Photon.*, **5**, 372–377.

[231] Kawashima, S., Ishizaki, K., and Noda, S. 2009. Light propagation in three-dimensional photonic crystals. *Opt. Express*, **18**, 386–392.

[232] Kempe, M., Berger, G. A., and Genack, A. Z. 1997. Stimulated emission from amplifying random media. Pages 301–330 in: Hummel, R. E., and Wissmann, P. (eds) *Handbook of Optical Properties*. CRC Press, Boca Raton, FL.

[233] Khurgin, J. B. and Tucker, R. S. 2009. *Slow Light: Science and Applications*. CRC Press, Boca Raton, Florida.

[234] Kim, M., Choi, Y., Yoon, C., *et al.* 2012. Maximal energy transport through disordered media with the implementation of transmission eigenchannels. *Nature Photon.*, **6**, 581–585.

[235] Kim, S.-K., Lee, J.-H., Kim, S.-H., *et al.* 2005. Photonic quasicrystal single-cell cavity mode. *Appl. Phys. Lett.*, **86**(3), 031101.

[236] Kleppner, D. 1981. Inhibited spontaneous emission. *Phys. Rev. Lett.*, **47**, 233–236.

[237] Koenderink, A. F. 2003. *Emission and transport of light in photonic crystals*. Ph.D. thesis, (University of Amsterdam) available at: www.photonicbandgaps.com.

[238] Koenderink, A. F., Bechger, L., Lagendijk, A., and Vos, W. L. 2003. An experimental study of strongly modified emission in inverse opal photonic crystals. *Phys. Stat. Sol. B*, **197**, 648–661.

[239] Koenderink, A. F., Bechger, L., Schriemer, H. P., Lagendijk, A., and Vos, W. L. 2002. Broadband fivefold reduction of vacuum fluctuations probed by dyes in photonic crystals. *Phys. Rev. Lett.*, **88**, 143903.

[240] Koenderink, A. F., Lagendijk, A., and Vos, W. L. 2005. Optical extinction due to intrinsic structural variations of photonic crystals. *Phys. Rev. B*, **72**, 153102.

[241] Kogan, E. and Kaveh, M. 1995. Random-matrix-theory approach to the intensity distributions of waves propagating in a random medium. *Phys. Rev. B*, **52**, R3813–R3815.

[242] Kohmoto, M., Sutherland, B., and Iguchi, K. 1987. Localization in optics: Quasiperiodic media. *Phys. Rev. Lett.*, **58**, 2436–2438.

[243] Kohmoto, M., Sutherland, B., and Tang, C. 1987. Critical wave functions and a Cantor-set spectrum of a one-dimensional quasicrystal model. *Phys. Rev. B*, **35**, 1020–1033.

[244] Kok, M. H., Lu, W., Tam, W. Y., and Wong, G. K. L. 2009. Lasing from dye-doped icosahedral quasicrystals in dichromate gelatin emulsions. *Opt. Express*, **17**(9), 7275.

[245] Kolář, M., Ali, M., and Nori, F. 1991. Generalized Thue–Morse chains and their physical properties. *Phys. Rev. B*, **43**(1), 1034–1047.

[246] Kopp, V. I., Fan, B., Vithana, H. K. M., and Genack, A. Z. 1998. Low-threshold lasing at the edge of a photonic stop band in cholesteric liquid crystals. *Opt. Lett.*, **23**, 1707–1709.

[247] Kopp, V. I., Zhang, Z.-Q., and Genack, A. Z. 2003. Lasing in chiral photonic structures. *Prog. Quant. Electron.*, **27**, 369–416.

[248] Kottos, T. 2005. Statistics of resonances and delay times in random media: Beyond random matrix theory. *J. Phys. A*, **38**, 10761.

[249] Krachmalnicoff, V., Castanié, E., De Wilde, Y., and Carminati, R. 2010. Fluctuations of the local density of states probe localized surface plasmons on disordered metal films. *Phys. Rev. Lett.*, **105**, 183901.

[250] Kramer, P. and Papadopolos, Z. (eds). 2002. *Coverings of Discrete Quasiperiodic Sets: Theory and Applications to Quasicrystals.* Springer Tracts in Modern Physics, vol. 180. Berlin: Springer-Verlag.

[251] Krauss, T. F. 2007. Slow light in photonic crystal waveguides. *J. Phys. D*, **40**, 2666–2670.

[252] Kristensen, P. T., Koenderink, A. F., Lodahl, P., Tromborg, B., and Mork, J. 2008. Fractional decay of quantum dots in real photonic crystals. *Opt. Lett.*, **33**, 1557–1559.

[253] Krokhin, A. A. and Halevi, P. 1996. Influence of weak dissipation on the photonic band structure of periodic composites. *Phys. Rev. B*, **53**, 1206–1214.

[254] Kroon, L., Lennholm, E., and Riklund, R. 2002. Localization-delocalization in aperiodic systems. *Phys. Rev. B*, **66**(9).

[255] Kroon, L. and Riklund, R. 2004. Absence of localization in a model with correlation measure as a random lattice. *Phys. Rev. B*, **69**(9).

[256] Kuga, Y. and Ishimaru, A. 1984. Retroreflectance from a dense distribution of spherical particles. *J. Opt. Soc. Am. A*, **1**, 831–835.

[257] Kuhl, U. and Stockmann, H. J. 1998. Microwave realization of the Hofstadter butterfly. *Phys. Rev. Lett.*, **80**(15), 3232.

[258] Kurizki, G. and Genak, A. 1988. Suppression of molecular interactions in periodic dielectric structures. *Phys. Rev. Lett.*, **61**, 2269–2271.

[259] Labonté, L., Vanneste, C., and Sebbah, P. 2012. Localized mode hybridization by fine tuning of two-dimensional random media. *Opt. Lett.*, **37**, 1946–1948.

[260] Ladouceur, F. 1997. Roughness, inhomogeneity, and integrated optics. *J. Lightwave Technol.*, **15**(6), 1020–1025.

[261] Ladouceur, F. and Love, J. D. 1995. Effect of roughness and inhomogeneity on evanescent single-mode optical couplers. IEE *P. Optoelectron.*, **142**(6), 288–292.

[262] Lagendijk, A. 1993. Vibrational relaxation studied with light. Pages 197–238 in: *Ultrashort Processes in Condensed Matter.* Plenum, New York.

[263] Lagendijk, A., van Tiggelen, B., and Wiersma, D. S. 2009. Fifty years of Anderson localization. *Phys. Today*, **62**, 24–29.

[264] Lagendijk, A., Vreeker, R., and de Vries, P. 1989. Influence of internal reflection on diffusive transport in strongly scattering media. *Phys. Lett. A*, **136**, 81–88.

[265] Lahini, Y., Avidan, A., Pozzi, F., *et al.* 2008. Anderson localization and nonlinearity in one-dimensional disordered photonic lattices. *Phys. Rev. Lett.*, **100**, 013906.

[266] Lakowicz, J. R. 2008. *Principles of Fluorescence Spectroscopy, Third Edition.* Springer, Berlin.

[267] Lambropoulos, P., Nikolopoulos, G. M., Nielsen, T. R., and Bay, S. 2000. Fundamental quantum optics in structured reservoirs. *Rep. Prog. Phys.*, **63**, 455–503.

[268] Landauer, R. 1970. Electrical resistance of disordered one-dimensional lattices. *Philos. Mag.*, **21**, 863–867.

[269] Lawandy, N. M., Balachandran, R. M., Gomes, A. S. L., and Sauvain, E. 1994. Laser action in strongly scattering media. *Nature*, **368**, 436–438.

[270] Lawrence, N., Trevino, J., and Dal Negro, L. 2012. Control of optical orbital angular momentum by Vogel spiral arrays of metallic nanoparticles. *Opt. Lett.*, **37**(24), 5076–5078.

[271] Lawrence, N., Trevino, J., and Dal Negro, L. 2012. Aperiodic arrays of active nanopillars for radiation engineering. *J. Appl. Phys.*, **111**(11), 113101.

[272] Ledermann, A. 2006 (August). *Three-dimensional icosahedral photonic quasicrystals: fabrication via direct laser writing and optical characterization.* M.Phil. thesis, Universität Karlsruhe (TH).

[273] Ledermann, A., Cademartiri, L., Hermatschweiler, M., *et al.* 2006. Three-dimensional silicon inverse photonic quasicrystals for infrared wavelengths. *Nature Mater.*, **5**, 942–945.

[274] Ledermann, A., von Freymann, G., and Wegener, M. 2007. Laue-Beugung auf dem Schreibtisch. Photonische Quasikristalle. *Physik in unserer Zeit*, **38**(6), 300.

[275] Ledermann, A., von Freymann, G., and Wegener, M. 2009. Optical arrangement and its use. *European patent*, **No. DE 102007032181A1 / WO 002009006976A1**.

[276] Ledermann, A., Wegener, M., and von Freymann, G. 2010. Rhombicuboctahedral three-dimensional photonic quasicrystals. *Adv. Mater.*, **22**, 2363.

[277] Ledermann, A., Wiersma, D. S., Wegener, M., and von Freymann, G. 2009. Multiple scattering of light in three-dimensional photonic quasicrystals. *Opt. Express*, **17**(3), 1844.

[278] Lee, P. A. and Stone, A. D. 1985. Universal conductance fluctuations in metals. *Phys. Rev. Lett.*, **55**, 1622–1625.

[279] Lee, P. T., Lu, T. W., and Tsai, F. M. 2007. Octagonal quasi-photonic crystal single-defect microcavity with whispering gallery mode and condensed device size. *IEEE Photon. Technol. Lett.*, **19**(9), 710–712.

[280] Lee, P. T., Lu, T. W., Tsai, F. M., Lu, T. C., and Kuo, H. C. 2006. Whispering gallery mode of modified octagonal quasiperiodic photonic crystal single-defect microcavity and its side-mode reduction. *Appl. Phys. Lett.*, **88**(20), 201104.

[281] Lee, S. D., Shin, S. J., Choi, S. J., *et al.* 2006. Si-based Coulomb blockade device for spin qubit logic gate. *Appl. Phys. Lett.*, **89**(2), 023111.

[282] Leistikow, M. D., Mosk, A. P., Yeganegi, E., *et al.* 2011. Inhibited spontaneous emission of quantum dots observed in a 3D photonic band gap. *Phys. Rev. Lett.*, **107**, 193903.

[283] Lemoult, F., Lerosey, G., de Rosny, J., and Fink, M. 2010. Resonant metalenses for breaking the diffraction barrier. *Phys. Rev. Lett.*, **104**, 203901.

[284] Leonetti, M., Conti, C., and Lopez, C. 2011. The mode-locking transition of random lasers. *Nature Photon.*, **5**(10), 615–617.

[285] Leonetti, M., Conti, C., and Lopez, C. 2012. Tunable degree of localization in random lasers with controlled interaction. *Appl. Phys. Lett.*, **101**(5), 051104.

[286] Leonetti, M., Conti, C., and Lopez, C. 2012. Random laser tailored by directional stimulated emission. *Phys. Rev. A*, **85**(Apr), 043841.

[287] Leonetti, M., Conti, C., and Lopez, C. 2013. Nonlocality and collective emission in disordered lasing resonators. *Light: Science and Applications*, in press.

[288] Leonetti, M. and Lopez, C. 2012. Random lasing in structures with multi-scale transport properties. *Appl. Phys. Lett.*, **101**(25), 251120.

[289] Leonetti, M. and Lopez, C. 2013. Active subnanometer spectral control of a random laser. *Appl. Phys. Lett.*, **102**(7), 071105.

[290] Leonetti, M., Sapienza, R., Ibisate, M., Conti, C., and López, C. 2009. Optical gain in DNA-DCM for lasing in photonic materials. *Opt. Lett.*, **34**(24), 3764–3766.

[291] Lepri, S., Cavalieri, S., Oppo, G.-L., and Wiersma, D. S. 2007. Statistical regimes of random laser fluctuations. *Phys. Rev. A*, **75**(Jun), 063820.

[292] Letokhov, V. V. 1968. Generation of light by a scattering medium with negative resonance. *Sov. Phys. JETP*, **26**, 835–840.

[293] Leung, P. T., Liu, S. Y., and Young, K. 1994. Completeness and orthogonality of quasinormal modes in leaky cavities. *Phys. Rev. A*, **49**, 3057–3067.

[294] Leuzzi, L., Conti, C., Folli, V., Angelani, L., and Ruocco, G. 2009. Phase diagram and complexity of mode-locked lasers: From order to disorder. *Phys. Rev. Lett.*, **102**(8), 83901.

[295] Levine, D. and Steinhardt, P. J. 1986. Quasicrystals. I. Definition and structure. *Phys. Rev. B*, **34**(2), 596.

[296] Li, F. H. and Wang, L. C. 1988. Analytical formulation of icosahedral quasi-crystal structures. *J. Phys. C*, **21**(3), 495.

[297] Li, J. H., Lisyansky, A. A., Cheung, T. D., Livdan, D., and Genack, A. Z. 1993. Transmission and surface intensity profiles in random media. *Europhys. Lett.*, **22**, 675.

[298] Li, Z. Y. and Xia, Y. N. 2001. Full vectorial model for quantum optics in three-dimensional photonic crystals. *Phys. Rev. A*, **63**, 043817.

[299] Li, Z.-Y. and Zhang, Z.-Q. 2000. Fragility of photonic band gaps in inverse-opal photonic crystals. *Phys. Rev. B*, **62**, 1516–1519.

[300] Liew, S. F., Noh, H., Trevino, J., Dal Negro, L., and Cao, H. 2011. Localized photonic band edge modes and orbital angular momenta of light in a golden-angle spiral. *Opt. Express*, **19**(24), 23631–23642.

[301] Liew, S. F., Yang, J. K., Noh, H., *et al.* 2011. Photonic band gaps in three-dimensional network structures with short-range order. *Phys. Rev. A*, **84**, 063818.

[302] Lifshitz, R. 2002. The square Fibonacci tiling. *J. Alloy. Compd.*, **342**(1–2), 186–190.

[303] Liu, N. H. 1997. Propagation of light waves in Thue–Morse dielectric multilayers. *Phys. Rev. B*, **55**(Feb.), 3543–3547.

[304] Lodahl, P., van Driel, A. F., Nikolaev, I. S., *et al.* 2004. Controlling the dynamics of spontaneous emission from quantum dots by photonic crystals. *Nature*, **430**, 654–657.

[305] Lončar, M., Nedeljković, D., Doll, T., *et al.* 2000. Waveguiding in planar photonic crystals. *Appl. Phys. Lett.*, **77**, 1937–1939.

[306] Lourtioz, J.-M., Benisty, H., Berger, V., *et al.* 2008. *Photonic Crystals: Towards Nanoscale Photonic Devices*. Springer, Heidelberg.

[307] Lubatsch, A., Kroha, J., and Busch, K. 2005. Theory of light diffusion in disordered media with linear absorption or gain. *Phys. Rev. B*, **71**(18), 184201.

[308] Luck, J. 1989. Cantor spectra and scaling of gap widths in deterministic aperiodic systems. *Phys. Rev. B*, **39**(9), 5834–5849.

[309] Luck, J. M., Godreche, C., Janner, A., and Janssen, T. 1993. The nature of the atomic surfaces of quasiperiodic self-similar structures. *J. Phys. A - Math. Gen.*, **26**, 1951–1999.

[310] Ma, X. and John, S. 2009. Ultrafast population switching of quantum dots in a structured vacuum. *Phys. Rev. Lett.*, **103**, 233601.

[311] Mabuchi, H. and Doherty, A. C. 2002. Cavity quantum electrodynamics: Coherence in context. *Science*, **298**, 1372–1377.

[312] Maciá, E. 2006. The role of aperiodic order in science and technology. *Rep. Prog. Phys.*, **69**(2), 397.

[313] Maciá, E. and Domínguez-Adame, F. 1996. Physical nature of critical wave functions in Fibonacci systems. *Phys. Rev. Lett.*, **76**, 2957–2960.

[314] Mahler, L., Tredicucci, A., Beltram, F., *et al.* 2010. Quasi-periodic distributed feedback laser. *Nature Photon.*, **4**(3), 165–169.

[315] Maldovan, M. and Thomas, E. L. 2004. Diamond-structured photonic crystals. *Nature Mater.*, **3**, 593–600.

[316] Man, W., Megens, M., Steinhardt, P. J., and Chaikin, P. M. 2005. Experimental measurement of the photonic properties of icosahedral quasicrystals. *Nature*, **436**(7053), 993–996.

[317] Markoš, P. 1999. Probability distribution of the conductance at the mobility edge. *Phys. Rev. Lett.*, **83**, 588–591.

[318] Markoš, P. and Soukoulis, C. M. 2005. Intensity distribution of scalar waves propagating in random media. *Phys. Rev. B*, **71**(5), 054201.

[319] Markushev, V. M., Zolin, V. F., and Briskina, Ch. M. 1986. Powder laser. *Zh. Prikl. Spektrosk*, **45**, 847–850.

[320] Maruo, S., Nakamura, O., and Kawata, S. 1997. Three-dimensional microfabrication with two-photon-absorbed photopolymerization. *Opt. Lett.*, **22**(2), 132.

[321] Maxwell Garnett, J. C. 1904. Colours in metal glasses and in metal films. *Philos. Trans. Roy. Soc. A*, **203**, 385–420.

[322] Mazurenko, D. A., Kerst, R., Dijkhuis, J. I., *et al.* 2003. Ultrafast optical switching in three-dimensional photonic crystals. *Phys. Rev. Lett.*, **91**, 213903.

[323] McCabe, D. J., Tajalli, A., Austin, D. R., *et al.* 2011. Spatio-temporal focusing of an ultrafast pulse through a multiply scattering medium. *Nature Commun.*, **2**, 447.

[324] Mehta, M. L. 2004. *Random Matrices, Third Edition*. Academic Press, New York.

[325] Meisel, D. C., Diem, M., Deubel, M., *et al.* 2006. Shrinkage pre-compensation of holographic three-dimensional photonic crystal templates. *Adv. Mater.*, **18**(22), 2964.

[326] Mello, P. A., Akkermans, E., and Shapiro, B. 1988. Macroscopic approach to correlations in the electronic transmission and reflection from disordered conductors. *Phys. Rev. Lett.*, **61**, 459–462.

[327] Mello, P. A., Pereyra, P., and Kumar, N. 1988. Macroscopic approach to multichannel disordered conductors. *Ann. Phys.*, New York, **181**, 290–317.

[328] Melloni, A., Morichetti, F., and Martinelli, M. 2003. Optical slow wave structures. *Opt. Photonics News*, **14**, 44–48.

[329] Melloni, A. and Morichetti, F. 2009. The long march of slow photonics. *Nature Photon.*, **3**(3), 119.

[330] Mermin, N. D. and Wagner, H. 1966. Absence of ferromagnetism or antiferromagnetism in one- or two-dimensional isotropic Heisenberg models. *Phys. Rev. Lett.*, **17**, 1133–1136.

[331] Miller, D. A. B. 2000. Rationale and challenges for optical interconnects to electronic chips. *Proc. IEEE*, **88**(6), 728–749.

[332] Miller, D. A. B. 2009. Device requirements for optical interconnects to silicon chips. *Proc. IEEE*, **97**(7), 1166–1185.

[333] Milner, V. and Genack, A. Z. 2005. Photon localization laser: Low-threshold lasing in a random amplifying layered medium via wave localization. *Phys. Rev. Lett.*, **94**, 073901.

[334] Milonni, P. W. 1994. *The Quantum Vacuum: An Introduction to Quantum Electrodynamics*. Academic Press, Boston.

[335] Mirlin, A. D. 2000. Statistics of energy levels and eigenfunctions in disordered systems. *Phys. Rep.*, **326**, 259–382.

[336] Mnaymneh, K. and Gauthier, R. C. 2007. Mode localization and band-gap formation in defect-free photonic quasicrystals. *Opt. Express*, **15**(8), 5089.

[337] Mookherjea, S. and Oh, A. 2007. Effect of disorder on slow light velocity in optical slow-wave structures. *Opt. Lett.*, **32**, 289–291.

[338] Mookherjea, S., Park, J. S., Yang, S. H., and Bandaru, P. R. 2008. Localization in silicon nanophotonic slow-light waveguides. *Nature Photon.*, **2**(2), 90–93.

[339] Mookherjea, S. and Schneider, M. A. 2011. Avoiding bandwidth collapse in long chains of coupled optical microresonators. *Opt. Lett.*, **36**(23), 4557–4559.

[340] Mookherjea, S. and Yariv, A. 2002. Coupled resonator optical waveguides. *IEEE J. Sel. Top. Quantum Electron.*, **8**, 448–456.

[341] Moretti, L. and Mocella, V. 2007. Two-dimensional photonic aperiodic crystals based on Thue–Morse sequence. *Opt. Express*, **15**(23), 15314–15323.

[342] Moretti, L., Rea, I. Rotiroti, L., *et al.* 2006. Photonic band gaps analysis of Thue–Morse multilayers made of porous silicon. *Opt. Express*, **14**(13), 6264–6272.

[343] Morichetti, F., Ferrari, C., Canciamilla, A., and Melloni, A. 2012. The first decade of coupled resonator optical waveguides: Bringing slow light to applications. *Laser Photon. Rev.*, **6**(1), 74–96.

[344] Morichetti, F., Canciamilla, A., and Melloni, A. 2010. Statistics of backscattering in optical waveguides. *Opt. Lett.*, **35**(11), 1777–1779.

[345] Mosk, A. P., Lagendijk, A., Lerosey, G., and Fink, M. 2012. Controlling waves in space and time for imaging and focusing in complex media. *Nat. Photon.*, **6**, 283–292.

[346] Moss, T. S. 1959. *Optical Properties of Semiconductors*. Butterworth, London.

[347] Mott, N. F. 1970. Conduction in non-crystalline systems IV. Anderson localization in a disordered lattice. *Philos. Mag.*, **22**, 7–29.

[348] Mujumdar, S., Ricci, M., Torre, R., and Wiersma, D. S. 2004. Amplified extended modes in random lasers. *Phys. Rev. Lett.*, **93**(5), 53903.

[349] Muttalib, K. A. and Wölfle, P. 1999. One-sided log-normal distribution of conductances for a disordered quantum wire. *Phys. Rev. Lett.*, **83**, 3013–3016.

[350] Muzykantskii, B. A. and Khmelnitskii, D. E. 1995. Nearly localized states in weakly disordered conductors. *Phys. Rev. B*, **51**, 5480–5483.

[351] Nazarov, Y. V. 1994. Limits of universality in disordered conductors. *Phys. Rev. Lett.*, **73**, 134–137.

[352] Nielsen, M. A. and Chuang, I. L. 1959. *Quantum Computation and Quantum Information*. Cambridge University Press, Cambridge.

[353] Nieuwenhuizen, Th. M. and van Rossum, M. C. 1995. Intensity distribution of waves transmitted through a multiple scattering medium. *Phys. Rev. Lett.*, **74**, 2674–2677.

[354] Nikolaev, I. S., Vos, W. L., and Koenderink, A. F. 2009. Accurate calculation of the local density of optical states in inverse-opal photonic crystals. *J. Opt. Soc. Am. B*, **26**, 987–997.

[355] Noda, S., Fujita, M., and Asano, T. 2007. Spontaneous-emission control by photonic crystals and nanocavities. *Nature Photon.*, **1**, 449–458.

[356] Noda, S., Tomoda, K., Yamamoto, N., and Chutinan, A. 2000. Full three-dimensional photonic bandgap crystals at near-infrared wavelengths. *Science*, **289**, 604–606.

[357] Noginov, M. A., Egarievwe, S. U., Noginova, N., Caulfield, H. J., and Wang, J. C. 1999. Interferometric studies of coherence in a powder laser. *Opt. Mater.*, **12**(1), 127–134.

[358] Noh, H., Yang, J. K., Boriskina, S. V., *et al.* 2011. Lasing in Thue–Morse structures with optimized aperiodicity. *Appl. Phys. Lett.*, **98**(20), 201109.

[359] Nori, F. and Rodriguez, J. P. 1986. Acoustic and electronic properties of one-dimensional quasicrystals. *Phys. Rev. B*, **34**, 2207–2211.

[360] Notomi, M., Suzuki, H., Tamamura, T., and Edagawa, K. 2004. Lasing action due to the two-dimensional quasiperiodicity of photonic quasicrystals with a Penrose lattice. *Phys. Rev. Lett.*, **92**(12), 123906.

[361] Notomi, M., Kuramochi, E., and Tanabe, T. 2008. Large-scale arrays of ultrahigh-q coupled nanocavities. *Nature Photon.*, **2**(12), 741–747.

[362] Novotny, L. and Hecht, B. 2006. *Principles of Nano-Optics*. Cambridge University Press, Cambridge.

[363] Nozaki, K. and Baba, T. 2004. Quasiperiodic photonic crystal microcavity lasers. *Appl. Phys. Lett.*, **84**(24), 4875–4877.

[364] Nozaki, K. and Baba, T. 2006. Lasing characteristics of 12-fold symmetric quasi-periodic photonic crystal slab nanolasers. *Jpn. J. Appl. Phys.*, **45**(8A), 6087–6090.

[365] O'Brien, J. L., Furusawa, A., and Vuckovic, J. 2009. Photonic quantum technologies. *Nature Photon.*, **3**, 687–695.

[366] O'Faolain, L., White, T. P., O'Brien, D., *et al.* 2007. Dependence of extrinsic loss on group velocity in photonic crystal waveguides. *Opt. Express*, **15**(20), 13129–13138.

[367] Ogawa, S., Imada, M., Yoshimoto, S., Okano, M., and Noda, S. 2004. Control of light emission by 3D photonic crystals. *Science*, **305**, 227–229.

[368] Ogawa, S., Ishizaki, K., Furukawa, T., and Noda, S. 2008. Spontaneous emission control by 17 layers of three-dimensional photonic crystals. *Electronics Lett.*, **44**, 377.

[369] Oton, C. J., Dal Negro, L., Gaburro, Z., *et al.* 2003. Light propagation in one-dimensional porous silicon complex systems. *Phys. Stat. Sol. (a)*, **197**, 298–302.

[370] Pappu, R. B., Taylor, J., and Gershenfeld, N. 2002. Physical one-way functions. *Science*, **297**, 2026.

[371] Park, H.-G., Kim, S.-H., Kwon, S.-H., *et al.* 2004. Electrically driven single-cell photonic crystal laser. *Science*, **305**(5689), 1444–1447.

[372] Patra, M. 2002. Theory for photon statistics of random lasers. *Phys. Rev. A*, **65**(4), 043809.

[373] Patterson, M., Hughes, S., Combrie, S., *et al.* 2009. Disorder-induced coherent scattering in slow-light photonic crystal waveguides. *Phys. Rev. Lett.*, **102**(25), 253903.

[374] Pavesi, L., Panzarini, G., and Andreani, L. C. 1998. All-porous silicon-coupled microcavities: Experiment versus theory. *Phys. Rev. B*, **58**, 15794–15800.

[375] Payne, B., Yamilov, A., and Skipetrov, S. E. 2010. Anderson localization as position-dependent diffusion in disordered waveguides. *Phys. Rev. B*, **82**, 024205.

[376] Pedrotti, F. L., Pedrotti, L. M., and Pedrotti, L. S. 2006. *Introduction to Optics*. Benjamin-Cummings Pub Co.

[377] Pellandini, P., Stanley, R. P., Houdré, R., *et al.* 1997. Dual-wavelength laser emission from a coupled semiconductor microcavity. *Appl. Phys. Lett.*, **71**, 864–866.

[378] Pendry, J. B. 1987. Quasi-extended electron states in strongly disordered systems. *J. Phys. C*, **20**, 733.

[379] Pendry, J. B. 1991. Catching moonbeams. *Nature*, **351**, 438–439.

[380] Pendry, J. B., Mackinnon, A., and Pretre, A. B. 1990. Maximal fluctuations – a new phenomenon in disordered systems. *Physica A*, **168**, 400–407.

[381] Pérez-Álvarez, R., García-Moliner, F., and Velasco, V. R. 2001. Some elementary questions in the theory of quasiperiodic heterostructures. *J. Phys. Condens. Mat.*, **13**(15), 3689.

[382] Peter, E., Senellart, P., Martrou, D., *et al.* 2005. Exciton-photon strong-coupling regime for a single quantum dot embedded in a microcavity. *Phys. Rev. Lett.*, **95**, 067401.

[383] Petrov, A., Krause, M., and Eich, M. 2009. Backscattering and disorder limits in slow light photonic crystal waveguides. *Opt. Express*, **17**(10), 8676–8684.

[384] Piéchon, F. 1996. Anomalous diffusion properties of wave packets on quasiperiodic chains. *Phys. Rev. Lett.*, **76**, 4372–4375.

[385] Plerou, V. and Wang, Z. 1998. Conductances, conductance fluctuations, and level statistics on the surface of multilayer quantum Hall states. *Phys. Rev. B*, **58**, 1967–1979.

[386] Poddubny, A. N. and Ivchenko, E. L. 2010. Photonic quasicrystalline and aperiodic structures. *Physica E*, **42**(7), 1871–1895.

[387] Polson, R. C. and Vardeny, Z. V. 2004. Random lasing in human tissues. *Appl. Phys. Lett.*, **85**(7), 1289–1291.

[388] Polson, R. C. and Vardeny, Z. V. 2005. Organic random lasers in the weak-scattering regime. *Phys. Rev. B*, **71**(4), 045205.

[389] Polson, R. C. and Vardeny, Z. V. 2010. Cancerous tissue mapping from random lasing emission spectra. *J. Opt.*, **12**(2), 024010.

[390] Pompe, G., Rappen, T., Wehner, M., Knop, F., and Wegener, M. 1995. Transient response of a short-cavity semiconductor laser. *Phys. Stat. Solidi B*, **188**(1), 175–180.

[391] Poon, J. K. S., Scheuer, J., Mookherjea, S., *et al.* 2004. Matrix analysis of microring coupled-resonator optical waveguides. *Opt. Express*, **12**(1), 90–103.

[392] Poon, J. K. S., Zhu, L., DeRose, G., and Yariv, A. 2006. Transmission and group delay of microring coupled-resonator optical waveguides. *Opt. Lett.*, **31**, 456–458.

[393] Popoff, S. M., Lerosey, G., Carminati, R., *et al.* 2010. Measuring the transmission matrix in optics: An approach to the study and control of light propagation in disordered media. *Phys. Rev. Lett.*, **104**, 100601.

[394] Povey, I. M., Whitehead, D., Thomas, K., *et al.* 2006. Photonic crystal thin films of GaAs prepared by atomic layer deposition. *Appl. Phys. Lett.*, **89**, 104103.

[395] Povinelli, M. L., Johnson, S. G., Lidorikis, E., Joannopoulos, J. D., and Soljacic, M. 2004. Effect of a photonic band gap on scattering from waveguide disorder. *Appl. Phys. Lett.*, **84**(18), 3639–3641.

[396] Purcell, E. M. 1946. Spontaneous emission probabilities at radio frequencies. *Phys. Rev.*, **69**, 681.

[397] Qi, M., Lidorikis, E., Rakich, P. T., *et al.* 2004. A three-dimensional optical photonic crystal with designed point defects. *Nature*, **429**, 538–542.

[398] Qiu, F., Peng, R. W., Huang, X. Q., *et al.* 2003. Resonant transmission and frequency trifurcation of light waves in Thue–Morse dielectric multilayers. *Europhys. Lett.*, **63**(5), 853–859.

[399] Qiu, F., Peng, R. W., Huang, X. Q., *et al.* 2007. Omnidirectional reflection of electromagnetic waves on Thue–Morse dielectric multilayers. *Europhys. Lett.*, **68**(5), 658–663.

[400] Raedt, H. D., Lagendijk, A., and de Vries, P. 1989. Transverse localization of light. *Phys. Rev. Lett.*, **62**, 47–50.

[401] Ramanan, V., Nelson, E., Brzezinski, A., Braun, P. V., and Wiltzius, P. 2008. Three dimensional silicon-air photonic crystals with controlled defects using interference lithography. *Appl. Phys. Lett.*, **92**, 173304.

[402] Rechtsman, M. C., Jeong, H.-C., Chaikin, P. M., Torquato, S., and Steinhardt, P. J. 2008. Optimized structures for photonic quasicrystals. *Phys. Rev. Lett.*, **101**(7), 73902.

[403] Redding, B., Choma, M. A., and Cao, H. 2011. Spatial coherence of random laser emission. *Opt. Lett.*, **36**(17), 3404–3406.

[404] Redding, B., Choma, M. A., and Cao, H. 2012. Speckle-free laser imaging using random laser illumination. *Nature Photon.*, **6**(6), 355–359.

[405] Reithmaier, J. P., Sęk, G., Löffler, A., *et al.* 2004. Strong coupling in a single quantum dot-semiconductor microcavity system. *Nature*, **432**, 197–200.

[406] Reyntjens, S. and Puers, R. 2001. A review of focused ion beam applications in microsystem technology. *J. Micromech. Microeng.*, **11**(4), 287.

[407] Rinne, S. A., García-Santamaría, F., and Braun, P. V. 2007. Embedded cavities and waveguides in three-dimensional silicon photonic crystals. *Nature Photon.*, **2**, 52–56.

[408] Robertson, W. M., Arjavalingam, G., Meade, R. D., *et al.* 1992. Measurement of photonic band structure in a two-dimensional periodic dielectric array. *Phys. Rev. Lett.*, **68**, 2023–2026.

[409] Rodriguez, A. W., McCauley, A. P., Avniel, Y., and Johnson, S. G. 2008. Computation and visualization of photonic quasicrystal spectra via Bloch's theorem. *Phys. Rev. B*, **77**(10), 104201.

[410] Roichman, Y. and Grier, D. G. 2005. Holographic assembly of quasicrystalline photonic heterostructures. *Opt. Express*, **13**(14), 5434.

[411] Ruijgrok, P. V., Wüest, R., Rebane, A. A., Renn, A., and Sandoghdar, V. 2010. Spontaneous emission of a nanoscopic emitter in a strongly scattering disordered medium. *Opt. Express*, **18**, 6360.

[412] Sakoda, K. 1995. Symmetry, degeneracy, and uncoupled modes in two-dimensional photonic lattices. *Phys. Rev. B*, **52**, 7982–7986.

[413] Sanchez-Gil, J. A., Freilikher, V., Yurkevich, I., and Maradudin, A. A. 1998. Coexistence of ballistic transport, diffusion, and localization in surface disordered waveguides. *Phys. Rev. Lett.*, **80**(5), 948.

[414] Sapienza, L., Thyrrestrup, H., Stobbe, S., *et al.* 2010. Cavity quantum electrodynamics with Anderson-localized modes. *Science*, **327**, 1352–1355.

[415] Sapienza, R., Bondareff, P., Pierrat, R., *et al.* 2011. Long-tail statistics of the Purcell factor in disordered media driven by near-field interactions. *Phys. Rev. Lett.*, **106**, 163902.

[416] Sapienza, R., Costantino, P., Wiersma, D. S., *et al.* 2003. Optical analogue of electronic Bloch oscillations. *Phys. Rev. Lett.*, **91**, 263902.

[417] Sapienza, R., García, P. D., Bertolotti, J., *et al.* 2007. Observation of resonant behavior in the energy velocity of diffused light. *Phys. Rev. Lett.*, **99**(23), 233902.

[418] Sarma, R., Yamilov, A., Neupane, P., Shapiro, B., and Cao, H. 2014. Probing long-range intensity correlations inside disordered photonic nanostructures. arxiv.org/abs/1405.6339.

[419] Scheffold, F. and Maret, G. 1998. Universal conductance fluctuations of light. *Phys. Rev. Lett.*, **81**, 5800–5803.

[420] Schilling, J., White, J., Scherer, A., *et al.* 2005. Three-dimensional macroporous silicon photonic crystal with large photonic band gap. *Applied Physics Letters*, **86**, 011101.

[421] Schuurmans, F. J. P., de Lang, D. T. N., Wegdam, G. H., Sprik, R., and Lagendijk, A. 1998. Local-fields effects on spontaneous emission in a dense supercritical gas. *Phys. Rev. Lett.*, **80**, 5077–5080.

[422] Schwartz, T., Bartal, G., Fishman, S., and Segev, M. 2007. Transport and Anderson localization in disordered two-dimensional photonic lattices. *Nature*, **446**, 52–55.

[423] Sebbah, P., Hu, B., Genack, A. Z., Pnini, R., and Shapiro, B. 2002. Spatial field correlation: The building block of mesoscopic fluctuations. *Phys. Rev. Lett.*, **88**, 123901.

[424] Sebbah, P., Hu, B., Klosner, J., and Genack, A. Z. 2006. Extended quasimodes within nominally localized random waveguides. *Phys. Rev. Lett.*, **96**, 183902.

[425] Sebbah, P. and Vanneste, C. 2002. Random laser in the localized regime. *Phys. Rev. B*, **66**(14), 144202.

[426] Shapira, O. and Fischer, B. 2005. Localization of light in a random grating array in a single mode fiber. *J. Opt. Soc. of Am. B*, **22**, 2542–2552.

[427] Shapiro, B. 1999. New type of intensity correlation in random media. *Phys. Rev. Lett.*, **83**, 4733–4735.

[428] Shechtman, D., Blech, I., Gratias, D., and Cahn, J. W. 1984. Metallic phase with long-range orientational order and no translational symmetry. *Phys. Rev. Lett.*, **53**, 1951–1953.

[429] Sheng, P. 2005. *Introduction to Wave Scattering, Localization and Mesoscopic Phenomena*. Springer, Berlin.

[430] Sheng, P. 1995. *Introduction to Wave Scattering, Localization and Mesoscopic Phenomena*. Academic Press, New York.

[431] Shi, Z., Davy, M., Wang, J., and Genack, A. Z. 2013. Focusing through random media in space and time: a transmission matrix approach. *Opt. Lett.* **38**, 2714.

[432] Shi, Z. and Genack, A. Z. 2012. Transmission eigenvalues and the bare conductance in the crossover to Anderson localization. *Phys. Rev. Lett.*, **108**, 043901.

[433] Shi, Z. and Genack, A. Z. 2014. Modal makeup of transmission eigenchannels. arxiv.org/abs/1406.3673.

[434] Shi, Z., Wang, J., and Genack, A. Z. 2014. Microwave conductance in random waveguides in the cross-over to Anderson localization and single-parameter scaling. *Proceedings of the National Academy of Sciences* (PNAS) **111**, 2926.

[435] Shipman, P. D. and Newell, A. C. 2004. Phyllotactic patterns on plants. *Phys. Lev. Lett.*, **92**, 168102.

[436] Shir, D., Liao, H., Jeon, S., *et al.* 2008. Three-dimensional nanostructures formed by single step, two-photon exposures through elastomeric penrose quasicrystal phase masks. *Nano Letters*, **8**(8), 2236.

[437] Shir, D. J., Nelson, E. C., Chanda, D., *et al.* 2010. Dual exposure, two-photon, conformal phase mask lithography for three dimensional silicon inverse woodpile photonic crystals. *J. Vac. Sci. Technol. B*, **28**, 783–788.

[438] Siegman, A. E. 1986. *Lasers*. University Science Books, Sausalito, USA.

[439] Sigalas, M. M., Soukoulis, C. M., Chan, C. T., Biswas, R., and Ho, K. M. 1999. Effect of disorder on photonic band gaps. *Phys. Rev. B*, **59**, 12767–12770.

[440] Sigler, L. E. 2002. *Fibonacci's Liber Abaci*. Springer-Verlag, New York.

[441] Skipetrov, S. E. and van Tiggelen, B. A. 2003. *Wave Scattering in Complex Media, from Theory to Applications*. NATO series II, vol. 107. Kluwer, Dordrecht.

[442] Skipetrov, S. E. and van Tiggelen, B. A. 2004. Dynamics of weakly localized waves. *Phys. Rev. Lett.*, **92**, 113901.

[443] Skipetrov, S. E. and van Tiggelen, B. A. 2006. Dynamics of Anderson localization in open 3D media. *Phys. Rev. Lett.*, **96**, 043902.

[444] Slevin, K. and Ohtsuki, T. 1997. The Anderson transition: Time reversal symmetry and universality. *Phys. Rev. Lett.*, **78**, 4083–4086.

[445] Smith, P. W. 1970. Mode-locking of lasers. *Proc. IEEE*, **58**(9), 1342–1357.

[446] Smolka, S., Thyrrestrup, H., Sapienza, L., *et al.* 2011. Probing the statistical properties of Anderson localization with quantum emitters. *New J. Phys.*, **13**, 063044.

[447] Sokoloff, J. B. 1987. Anomalous electrical conduction in quasicrystals and Fibonacci lattices. *Phys. Rev. Lett.*, **58**, 2267–2270.

[448] Soukoulis, C. M. (ed.). 1996. *Photonic Band Gap Materials. Proceedings of the NATO Advanced Study Institute on Photonic Band Gap Materials*. Kluwer, Dordrecht.

[449] Soukoulis, C. M. (ed.). 2001. *Photonic Crystals and Light Localization in the 21st Century*. Kluwer, Dordrecht.

[450] Soukoulis, C. M. and Economou, E. N. 1982. Localization in one-dimensional lattices in the presence of incommensurate potentials. *Phys. Rev. Lett.*, **48**, 1043–1046.

[451] Soukoulis, C. M., Wang, X., Li, Q., and Sigalas, M. M. 1999. What is the right form of the probability distribution of the conductance at the mobility edge? *Phys. Rev. Lett.*, **82**, 668.

[452] Sözüer, H. S., Haus, J. W., and Inguva, R. 1992. Photonic bands: Convergence problems with the plane-wave method. *Phys. Rev. B*, **45**, 13962–13972.

[453] Sperling, T., Buehrer, W., Aegerter, C. M., and Maret, G. 2012. Direct determination of the transition to localization of light in three dimensions. *Nature Photon.*, **7**(1), 48–52.

[454] Sprik, R., van Tiggelen, B. A., and Lagendijk, A. 1996. Optical emission in periodic dielectrics. *Europhys. Lett.*, **35**, 265–270.

[455] Stanley, R. P., Houdré, R., Oesterle, U., Ilegems, M., and Weisbuch, C. 1994. Coupled semiconductor microcavities. *Appl. Phys. Lett.*, **65**, 2093–2095.

[456] Stano, P. and Jacquod, P. 2013. Suppression of interactions in multimode random lasers in the Anderson localized regime. *Nature Photon.*, **7**, 66–71.

[457] Starykh, O. A., Jacquod, P. R. J., Narimanov, E. E., and Stone, A. D. 2000. Signature of dynamical localization in the resonance width distribution of wave-chaotic dielectric cavities. *Phys. Rev. E*, **62**, 2078–2084.

[458] Staude, I., McGuinness, C., Frölich, A., *et al.* 2012. Waveguides in three-dimensional photonic bandgap materials for particle-accelerator on a chip architectures. *Opt. Express*, **20**, 5607–5612.

[459] Staude, I., Thiel, M., Essig, S., *et al.* 2010. Fabrication and characterization of silicon woodpile photonic crystals with a complete bandgap at telecom wavelengths. *Opt. Lett.*, **35**, 1094–1096.

[460] Staude, I., von Freymann, G., Essig, S., Busch, K., and Wegener, M. 2011. Waveguides in three-dimensional photonic-bandgap materials by direct laser writing and silicon double inversion. *Opt. Lett.*, **36**, 67.

[461] Steinbach, F., Ossipov, A., Kottos, T., and Geisel, T. 2000. Statistics of resonances and of delay times in quasiperiodic Schrödinger equations. *Phys. Rev. Lett.*, **85**, 4426–4429.

[462] Stephen, M. J. and Cwilich, G. 1987. Intensity correlation functions and fluctuations in light scattered from a random medium. *Phys. Rev. Lett.*, **59**, 285–287.

[463] Steurer, W. and Sutter-Widmer, D. 2007. Photonic and phononic quasicrystals. *J. Phys. D: Appl. Phys.*, **40**(13), R229.

[464] Stone, A. D., Mello, P. A., Muttalib, K., and Pichard, J. L. 1991. Random matrix theory and maximum entropy models for disordered conductors. Pages 369–448 in: Altshuler, B. L., Lee, P. A., and Webb, R. A. (eds), *Mesoscopic Phenomena in Solids*. Elsevier, Amsterdam.

[465] Storzer, M., Gross, P., Aegerter, C. M., and Maret, G. 2006. Observation of the critical regime near Anderson localization of light. *Phys. Rev. Lett.*, **96**, 063904.

[466] Stoytchev, M. and Genack, A. Z. 1997. Measurement of the probability distribution of total transmission in random waveguides. *Phys. Rev. Lett.*, **79**, 309–312.

[467] Stoytchev, M. and Genack, A. Z. 1999. Observations of non-Rayleigh statistics in the approach to photon localization. *Opt. Lett.*, **24**(4), 262–264.

[468] Takahashi, S., Okano, M., Imada, M., and Noda, S. 2006. Three-dimensional photonic crystals based on double-angled etching and wafer-fusion techniques. *Appl. Phys. Lett.*, **89**, 123106.

[469] Takahashi, S., Suzuki, K., Okano, M., *et al.* 2009. Direct creation of three-dimensional photonic crystals by a top-down approach. *Nat. Mater.*, **8**, 721–725.

[470] Tandaechanurat, A., Ishida, S., Aoki, K., *et al.* 2009. Demonstration of high-Q (>8600) three-dimensional photonic crystal nanocavity embedding quantum dots. *Appl. Phys. Lett.*, **94**, 171115.

[471] Tandaechanurat, A., Ishida, S., Guimard, D., *et al.* 2010. Lasing oscillation in a three-dimensional photonic crystal nanocavity with a complete bandgap. *Nature Photon.*, **5**, 91–94.

[472] Tang, L. and Yoshie, T. 2011. Light localization in woodpile photonic crystal built via two-directional etching. *IEEE J. Quant. Elec.*, **47**, 1028–1035.

[473] Tétreault, N., Míguez, H., and Ozin, G. A. 2004. Silicon inverse opal – a platform for photonic bandgap research. *Adv. Mater.*, **16**, 1471–1476.

[474] Tétreault, N., von Freymann, G., Deubel, M., *et al.* 2006. New route to three-dimensional photonic bandgap materials: Silicon double inversion of polymer templates. *Adv. Mater.*, **18**, 457–460.

[475] Texier, C. and Comtet, A. 1999. Universality of the Wigner time delay distribution for one-dimensional random potentials. *Phys. Rev. Lett.*, **82**(21), 4220–4223.

[476] Thouless, D. J. 1974. Electrons in disordered systems and the theory of localization. *Phys. Rep.*, **13**, 93–142.

[477] Thouless, D. J. 1977. Maximum metallic resistance in thin wires. *Phys. Rev. Lett.*, **39**, 1167–1169.

[478] Thue, A. 1909. Über Annäherungswerte algebraischer Zahlen. *Journal für die reine und angewandte Mathematik*, **135**, 284–305.

[479] Thyrrestrup, H., Hartsuiker, A., Gérard, J.-M., and Vos, W. L. 2013. Non-exponential spontaneous emission dynamics for emitters in a time-dependent optical cavity. *http://Arxiv.org*, 1301.7612.

[480] Tian, C. S., Cheung, S. K., and Zhang, Z. Q. 2010. Local diffusion theory for localized waves in open media. *Phys. Rev. Lett.*, **105**, 263905.

[481] Tjerkstra, R. W., Woldering, L. A., van den Broek, J. M., *et al.* 2011. Method to pattern etch masks in two inclined planes for three-dimensional nano- and microfabrication. *J. Vac. Sci. Technol. B*, **29**, 061604.

[482] Topolancik, J., Ilic, B., and Vollmer, F. 2007. Experimental observation of strong photon localization in disordered photonic crystal waveguides. *Phys. Rev. Lett.*, **99**, 253901.

[483] Trevino, J., Cao, H., and Dal Negro, L. 2011. Circularly symmetric light scattering from nanoplasmonic spirals. *Nano Lett.*, **11**(5), 2008–2016.

[484] Trevino, J., Liew, S. F., Noh, H., Cao, H., and Dal Negro, L. 2012. Geometrical structure, multifractal spectra and localized optical modes of aperiodic Vogel spirals. *Opt. Express*, **20**(3), 3015–3033.

[485] Tseng, A. A., Chen, K., Chen, C. D., and Ma, K. J. 2003. Electron beam lithography in nanoscale fabrication: Recent development. *IEEE. T. Electron. Pa. M.*, **26**(2), 141–149.

[486] Türeci, H. E., Ge, L., Rotter, S., and Stone, A. D. 2008. Strong interactions in multimode random lasers. *Science*, **320**, 643–646.

[487] van Albada, M. P., de Boer, J. F., and Lagendijk, A. 1990. Observation of long-range intensity correlation in the transport of coherent light through a random medium. *Phys. Rev. Lett.*, **64**, 2787–2790.

[488] van Albada, M. P. and Lagendijk, A. 1985. Observation of weak localization of light in a random medium. *Phys. Rev. Lett.*, **55**, 2692–2695.

[489] van Albada, M. P., van Tiggelen, B. A., Lagendijk, A., and Tip, A. 1991. Speed of propagation of classical waves in strongly scattering media. *Phys. Rev. Lett.*, **66**, 3132–3135.

[490] van Coevorden, D. V., Sprik, R., Tip, A., and Lagendijk, A. 1997. Photonic band structure of atomic lattices. *Phys. Rev. Lett.*, **77**, 2412–2415.

[491] van de Hulst, H. C. 1957. *Light Scattering by Small Particles*. Dover, New York.

[492] van den Broek, J. M., Woldering, L. A., Tjerkstra, R. W., *et al.* 2012. Inverse-woodpile photonic band gap crystals with a cubic diamond-like structure made from single-crystalline silicon. *Adv. Func. Mater.*, **22**, 25–31.

[493] van der Beek, T., Barthelemy, P., Johnson, P. M., Wiersma, D. S., and Lagendijk, A. 2012. Light transport through disordered layers of dense gallium arsenide submicron particles. *Phys. Rev. B.*, **85**, 115401.

[494] van der Molen, K. L., Mosk, A. P., and Lagendijk, A. 2006. Intrinsic intensity fluctuations in random lasers. *Phys. Rev. A*, **74**(Nov), 053808.

[495] van der Molen, K. L., Mosk, A. P., and Lagendijk, A. 2007. Quantitative analysis of several random lasers. *Opt. Commun.*, **278**(1), 110–113.

[496] van der Molen, K. L., Tjerkstra, R. W., Mosk, A. P., and Lagendijk, A. 2007. Spatial extent of random laser modes. *Phys. Rev. Lett.*, **98**(14), 143901.

[497] van Driel, A. F., Nikolaev, I. S., Vergeer, P., *et al.* 2007. Statistical analysis of time-resolved emission from ensembles of semiconductor quantum dots: Interpretation of exponential decay models. *Phys. Rev. B*, **75**, 035329.

[498] van Langen, S. A., Brouwer, P. W., and Beenakker, C. W. J. 1996. Nonperturbative calculation of the probability distribution of plane-wave transmission through a disordered waveguide. *Phys. Rev. E*, **53**(2), R1344–R1347.

[499] van Putten, G. E., Akbulut, D., Bertolotti, J., *et al.* 2011. Scattering lens resolves sub-100 nm structures with visible light. *Phys. Rev. Lett.*, **106**, 193905.

[500] van Putten, G. E. and Mosk, A. P. 2010. The information age in optics: Measuring the transmission matrix. *Physics*, **3**, 22.

[501] van Rossum, M. C. W. and Nieuwenhuizen, T. M. 1999. Multiple scattering of classical waves: Microscopy, mesoscopy, and diffusion. *Rev. Mod. Phys.*, **71**, 313–371.

[502] van Tiggelen, B. A., Sebbah, P., Stoytchev, M., and Genack, A. Z. 1999. Delay-time statistics for diffuse waves. *Phys. Rev. E*, **59**(6), 7166.

[503] Vanneste, C. and Sebbah, P. 2005. Localized modes in random arrays of cylinders. *Phys. Rev. E*, **71**(2), 026612.

[504] Vanneste, C., Sebbah, P., and Cao, H. 2007. Lasing with resonant feedback in weakly scattering random systems. *Phys. Rev. Lett.*, **98**, 143902.

[505] Vasconcelos, M. S. and Albuquerque, E. L. 1999. Transmission fingerprints in quasiperiodic dielectric multilayers. *Phys. Rev. B*, **59**(17), 11128–11131.

[506] Vats, N., John, S., and Busch, K. 2002. Theory of fluorescence in photonic crystals. *Phys. Rev. A*, **65**, 043808.

[507] Vellekoop, I. M. and Aegerter, C. M. 2010. Scattered light fluorescence microscopy: Imaging through turbid layers. *Opt. Lett.*, **35**, 1245–1247.

[508] Vellekoop, I. M., Lagendijk, A., and Mosk, A. P. 2010. Exploiting disorder for perfect focusing. *Nature Photon.*, **4**, 320–322.

[509] Vellekoop, I. M. and Mosk, A. P. 2007. Focusing coherent light through opaque strongly scattering media. *Opt. Lett.*, **32**, 2309–2311.

[510] Vellekoop, I. M., van Putten, E. P., Lagendijk, A., and Mosk, A. P. 2008. Demixing light paths inside disordered metamaterials. *Opt. Express*, **16**, 67–80.

[511] Vlasov, Y. A., Bo, X.-Z., Sturm, J. C., and Norris, D. J. 2001. On-chip natural assembly of silicon photonic bandgap crystals. *Nature*, **414**, 289–293.

[512] Vogel, H. 1979. A better way to construct the sunflower head. *Math. Biosci.*, **44**, 179–189.

[513] von Freymann, G., Ledermann, A., Thiel, M., *et al.* 2010. Three-dimensional nanostructures for photonics. *Adv. Funct. Mater.*, **20**, 1038.

[514] Vos, W. L., Koenderink, A. F., and Nikolaev, I. S. 2009. Orientation-dependent spontaneous emission rates of a two-level quantum emitter in any nanophotonic environment. *Phys. Rev. A*, **80**, 053802.

[515] Vos, W. L., Sprik, R., van Blaaderen, A., *et al.* 1996. Strong effects of photonic band structures on the diffraction of colloidal crystals. *Phys. Rev. B*, **53**, 16231–16235.

[516] Vos, W. L. and van Driel, H. M. 2000. Higher order Bragg diffraction by strongly photonic fcc crystals: Onset of a photonic bandgap. *Phys. Lett. A*, **272**, 101–106.

[517] Vos, W. L., van Driel, H. M., Megens, M., Koenderink, A. F., and Imhof, A. 2001. Experimental probes of the optical properties of photonic crystals. Pages 181–198 in: *Proceedings of the NATO ASI "Photonic Crystals and Light Localization in the 21st century."* Kluwer, Dordrecht.

[518] Wang, Ch. and Barrio, R. A. 1988. Theory of the Raman response in Fibonacci superlattices. *Phys. Rev. Lett.*, **61**, 191–194.

[519] Wang, J., Chabanov, A. A., Lu, D. Y., Zhang, Z. Q., and Genack, A. Z. 2010. Dynamics of fluctuations of localized waves. *Phys. Rev. B*, **81**, 241101(R).

[520] Wang, J. and Genack, A. Z. 2011. Transport through modes in random media. *Nature*, **471**, 345–348.

[521] Wang, K. 2006. Light wave states in two-dimensional quasiperiodic media. *Phys. Rev. B*, **73**(23), 235122.

[522] Wang, X. H., Gu, B. Y., Wang, R. Z., and Xu, H. Q. 2003. Decay kinetic properties of atoms in photonic crystals with absolute gaps. *Phys. Rev. Lett.*, **91**, 113904.

[523] Watson, G. H., Fleury, P. A., and McCall, S. L. 1987. Searching for photon localization in the time domain. *Phys. Rev. Lett.*, **58**, 945–948.

[524] Weaver, R. 1993. Anomalous diffusivity and localization of classical waves in disordered media: The effect of dissipation. *Phys. Rev. B*, **47**, 1077–1080.

[525] Webb, R. A., Washburn, S., Umbach, C. P., and Laibowitz, R. B. 1985. Observations of h/e Aharonov–Bohm oscillations in normal-metal rings. *Phys. Rev. Lett.*, **54**, 2696–2699.

[526] Wei, H., Underwood, D. F., Han, S. E., Blank, D. A., and Norris, D. J. 2009. The role of stress in the time-dependent optical response of silicon photonic band gap crystals. *Appl. Phys. Lett.*, **95**, 051910.

[527] Whittaker, D. M. and Culshaw, I. S. 1999. Scattering-matrix treatment of patterned multilayer photonic structures. *Phys. Rev. B*, **60**(4), 2610.

[528] Wiersma, D. 2000. Laser physics: The smallest random laser. *Nature*, **406**(6792), 132–135.

[529] Wiersma, D. S. 2008. The physics and applications of random lasers. *Nature Phys.*, **4**, 359–367.

[530] Wiersma, D. S. 2013. Disordered photonics. *Nature Photon.*, **7**, 188–196.

[531] Wiersma, D. S., Bartolini, P., Lagendijk, A., and Righini, R. 1997. Localization of light in a disordered medium. *Nature*, **390**, 671–673.

[532] Wiersma, D. S. and Cavalieri, S. 2001. Light emission: A temperature-tunable random laser. *Nature*, **414**(6865), 708–709.

[533] Wiersma, D. S. and Lagendijk, A. 1996. Light diffusion with gain and random lasers. *Phys. Rev. E*, **54**, 4256–4265.

[534] Wiersma, D. S., van Albada, M. P., and Lagendijk, A. 1995. Random laser? *Nature*, **373**, 203–204.

[535] Wiersma, D. S., van Albada, M. P., van Tiggelen, B. A., and Lagendijk, A. 1995. Experimental evidence for recurrent multiple scattering events of light in disordered media. *Phys. Rev. Lett.*, **74**, 4193–4196.

[536] Wigner, E. P. 1951. On the statistical distribution of the widths and spacing of nuclear resonance levels. Page 790 in: *Proc. Cambridge Phil. Soc.*

[537] Wijnhoven, J. E. G. J., Bechger, L., and Vos, W. L. 2001. Fabrication and characterization of large macroporous photonic crystals in titania. *Chem. Mater.*, **13**, 4486–4499.

[538] Wijnhoven, J. E. G. J. and Vos, W. L. 1998. Preparation of photonic crystals made of air spheres in titania. *Science*, **281**, 802–804.

[539] Wilson, K. G. 1971. Renormalization group and critical phenomena. I. Renormalization group and the Kadanoff scaling picture. *Phys. Rev. B*, **4**, 3174–3183.

[540] Woldering, L. A., Mosk, A. P., Tjerkstra, R. W., and Vos, W. L. 2009. The influence of fabrication deviations on the photonic band gap of three-dimensional inverse woodpile nanostructures. *J. Appl. Phys.*, **105**, 093108.

[541] Woldering, L. A., Tjerkstra, R. W., Jansen, H. V., Setija, I. D., and Vos, W. L. 2008. Periodic arrays of deep nanopores made in silicon with reactive ion etching and deep UV lithography. *Nanotechnology*, **19**, 145304.

[542] Woldeyohannes, M. and John, S. 1999. Coherent control of spontaneous emission near a photonic band edge: A qubit for quantum computation. *Phys. Rev. A*, **60**, 5046–5068.

[543] Wolf, P. E. and Maret, G. 1985. Weak localization and coherent backscattering of photons in disordered media. *Phys. Rev. Lett.*, **55**, 2696–2699.

[544] Wu, X., Fang, W., Yamilov, A., *et al.* 2006. Random lasing in weakly scattering systems. *Phys. Rev. A*, **74**(5), 053812.

[545] Xia, F., Sekaric, L., and Vlasov, Y. A. 2007. Ultracompact optical buffers on a silicon chip. *Nature Photon.*, **1**(1), 65–71.

[546] Xia, F., Rooks, M., Sekaric, L., and Vlasov, Y. 2007b. Ultra-compact high order ring resonator filters using submicron silicon photonic wires for on-chip optical interconnects. *Opt. Express*, **15**(19), 11934–11941.

[547] Xu, J., Ma, R., Wang, X., and Tam, W. Y. 2007. Icosahedral quasicrystals for visible wavelengths by optical interference holography. *Opt. Express*, **15**(7), 4287.

[548] Xu, Y., Lee, R. K., and Yariv, A. 2000. Propagation and second-harmonic generation of electromagnetic waves in a coupled-resonator optical waveguide. *J. Opt. Soc. Am. B*, **17**(3), 387–400.

[549] Yablonovitch, E. 1987. Inhibited spontaneous emission in solid-state physics and electronics. *Phys. Rev. Lett.*, **58**, 2059–2062.

[550] Yablonovitch, E., Gmitter, T. J., and Leung, K. M. 1991. Photonic band structure: The face-centered-cubic case employing nonspherical atoms. *Phys. Rev. Lett.*, **67**, 2295–2298.

[551] Yablonovitch, E., Gmitter, T. J., Meade, R. D., *et al.* 1991. Donor and acceptor modes in photonic band structure. *Phys. Rev. Lett.*, **67**, 3380–3383.

[552] Yamilov, A., Wu, X., Cao, H., and Burin, A. L. 2005. Absorption-induced confinement of lasing modes in diffusive random media. *Opt. Lett.*, **30**, 2430–2432.

[553] Yang, J. K., Boriskina, S. V., Noh, H., *et al.* 2010. Demonstration of laser action in a pseudorandom medium. *Appl. Phys. Lett.*, **97**(22), 223101.

[554] Yang, S. and Astratov, V. N. 2009. Spectroscopy of coherently coupled whispering-gallery modes in size-matched bispheres assembled on a substrate. *Opt. Lett.*, **34**(13), 2057–2059.

[555] Yariv, A., Xu, Y., Lee, R. K., and Scherer, A. 1999. Coupled-resonator optical waveguide: A proposal and analysis. *Opt. Lett.*, **24**(11), 711–713.

[556] Yariv, A. and Yeh, P. 1983. *Optical Waves in Crystals: Propagation and Control of Laser Radiation*. Wiley, New York.

[557] Ye, D.-X., Yang, Z.-P., Chang, A. S., *et al.* 2007. Experimental realization of a well-controlled 3D silicon spiral photonic crystal. *J. Phys. D*, **40**, 2624–2628.

[558] Yin, J., Huang, X., Liu, S., and Hu, S. 2007. Photonic bandgap properties of 8-fold symmetric photonic quasicrystals. *Opt. Commun.*, **269**(2), 385.

[559] Yoshie, T., Scherer, A., Hendrickson, J., *et al.* 2004. Vacuum Rabi splitting with a single quantum dot in a photonic crystal nanocavity. *Nature*, **432**, 201–203.

[560] Zhang, S., Lockerman, Y., and Genack, A. Z. 2010. Mesoscopic speckle. *Phys. Rev. E*, **82**, 051114.

[561] Zhang, S., Park, J., Milner, V., and Genack, A. Z. 2008. Delocalization transition in dimensional crossover in random layered media. *Phys. Rev. Lett.*, **101**, 183901.

[562] Zhang, X., Zhang, Z.-Q., and Chan, C. T. 2001. Absolute photonic band gaps in 12-fold symmetric photonic quasicrystals. *Phys. Rev. B*, **63**(8), 081105.

[563] Zhang, X. D., Zhang Z. Q., and Chan, C. T. 2001. Absolute photonic band gaps in 12-fold symmetric photonic quasicrystals. *Phys. Rev. B*, **63**, 081105.

[564] Zhang, Z. Q., Chabanov, A. A., Cheung, S. K., Wong, C. H., and Genack, A. Z. 2009. Dynamics of localized waves. *Phys. Rev. B*, **79**, 144203.

[565] Zhu, J. X., Pine, D. J., and Weitz, D. 1991. Internal reflection of diffusive light in random media. *Phys. Rev. A*, **44**, 3948–3959.

[566] Zhukovsky, S. V., Chigrin, D. N., and Kroha, J. 2006. Low-loss resonant modes in deterministically aperiodic nanopillar waveguides. *J. Opt. Soc. Am. B*, **23**(10), 2265–2272.

[567] Zito, G., Piccirillo, B., Santamato, E., *et al.* 2008. Two-dimensional photonic quasicrystals by single beam computer-generated holography. *Opt. Express*, **16**(8), 5164.

[568] Zoorob, M. E., Charlton, M. D. B., Parker, G. J., Baumberg, J. J., and Netti, M. C. 2000. Complete photonic bandgaps in 12-fold symmetric quasicrystals. *Nature*, **404**(6779), 740–743.

[569] Zurek, W. H. 1991. Decoherence and the transition from quantum to classical. *Phys. Today*, **44**, 36–44.

Index